Polymer Nanocomposites by Emulsion and Suspension
Polymerization

RSC Nanoscience & Nanotechnology

Series Editors:
Professor Paul O'Brien, *University of Manchester, UK*
Professor Sir Harry Kroto FRS, *University of Sussex, UK*
Professor Harold Craighead, *Cornell University, USA*

Titles in the Series:

How to obtain future titles on publication:
A standing order plan is available for this series. A standing order will bring delivery of
each new volume immediately on publication.

For further information please contact:
Book Sales Department, Royal Society of Chemistry, Thomas Graham House,
Science Park, Milton Road, Cambridge CB4 0WF, UK
Telephone: +44 (0)1223 420066, Fax: +44 (0)1223 420247,
Email: books@rsc.org
Visit our website at http://www.rsc.org/Shop/Books/

A gift of Professor Paul O'Brien
2010.

Polymer Nanocomposites by Emulsion and Suspension Polymerization

Edited by

Vikas Mittal
BASF SE, Polymer Research, Ludwigshafen, Germany

RSCPublishing

RSC Nanoscience & Nanotechnology No. 16

ISBN: 978-1-84755-225-9
ISSN: 1757-7136
A catalogue record for this book is available from the British Library

Published by The Royal Society of Chemistry,
Thomas Graham House, Science Park, Milton Road,
Cambridge CB4 0WF, UK

Registered Charity Number 207890

For further information see our web site at www.rsc.org

Preface

Functional hybrids involving the combination of organic and inorganic components are often achieved in order to enhance the properties of the constituent materials and to generate new characteristics by the synergistic combination of the constituents. Polymer–clay nanocomposites are one such class of functional hybrids, the advent of which has led to tremendous research activity in this direction. Significant improvements in the composite properties have been reported at very low filler volume fractions, thus allowing the composites to retain both transparency and low weight. Thus, a large number of polymer matrices have been used to achieve nanocomposites and a wide range of synthetic methods have been developed to achieve nanoscale dispersion of filler in the matrix phase. Of the various methods developed in recent years for obtaining nanocomposites, the emulsion and suspension routes of synthesis provide benefits not available in other methods. The use of a dispersion medium (water) always allows the homogeneous viscosity control of the polymerization reaction and the heat can also be optimally transmitted from the polymerizing medium. Controlled living polymerization methods can also be effectively employed in emulsions in order to obtain well-defined polymer chains. Water-soluble monomers can also be polymerized using inverse polymerization modes. The aim of this book is to present the advances in recent years in synthetic methods of nanocomposite generation in emulsion and suspension. Another aim is to present the wide range of polymer systems used to generate such nanocomposites. The use of various filler types is also highlighted.

Chapter 1 provides a brief overview on the synthesis of polymer nanocomposites in emulsion and suspension. Other synthetic methodologies such as *in situ* polymerization and melt intercalation are also described in comparison with emulsion and suspension polymerization. Chapter 2 describes the use of layered double hydroxides for the synthesis of nanocomposites in both emulsion and suspension. Properties and potential applications of such composites are considered. Chapter 3 describes nanocomposite synthesis in the inverse

RSC Nanoscience & Nanotechnology No. 16
Polymer Nanocomposites by Emulsion and Suspension Polymerization
Edited by Vikas Mittal
© Royal Society of Chemistry 2011
Published by the Royal Society of Chemistry, www.rsc.org

emulsion mode in comparison with the direct emulsion mode. Chapter 4 describes PMMA clay nanocomposites obtained by emulsifier-free emulsion polymerization. Structural evaluation and physical and mechanical properties of the composites are described. Chapter 5 explains the generation of acrylic clay nanocomposites for use as pressure-sensitive adhesives. Environmentally friendly biodegradable polymer clay nanocomposites are reported in Chapter 6. Suspension polymerization of inverse emulsions is the focus of Chapter 7. The synthesis of polymer nanocomposites by gamma irradiation is reported in Chapter 8. Apart from clay as filler, composites with metal nanoparticles and carbon nanotubes are also discussed. The use of magnesium hydroxide as filler for the synthesis of polymer nanocomposites in emulsions is reported in Chapter 9. Chapter 10 describes the use of miniemulsion polymerization for the synthesis of polymer nanocomposites. Kinetics of nanocomposite synthesis, properties and applications of the resulting latexes are included. The combination of silica and clay is demonstrated in Chapter 11 for the synthesis of PAN nanocomposites in emulsions. The use of controlled living polymerization techniques such as RAFT is described for the synthesis of polymer nano-composites in miniemulsions in Chapter 12. State-of-the-art and recent advances in emulsifier-free latexes stabilized with clay platelets are discussed in Chapter 13.

It gives me immense pleasure to thank the Royal Society of Chemistry for publishing the book and for their kind support during the project. I dedicate this book to my mother for being a constant source of inspiration. Heartfelt thanks are due to my wife Preeti for her continuous help in co-editing the book and for her ideas for improving the manuscript.

Vikas Mittal
Ludwigshafen

Contents

RSC Nanoscience & Nanotechnology No. 16
Polymer Nanocomposites by Emulsion and Suspension Polymerization
Edited by Vikas Mittal
© Royal Society of Chemistry 2011
Published by the Royal Society of Chemistry, www.rsc.org

CHAPTER 1

Polymer Nanocomposites in Emulsion and Suspension: an Overview*

VIKAS MITTAL

BASF SE, Polymer Research, 67069, Ludwigshafen, Germany

1.1 Polymer Nanocomposites

Functional hybrids of inorganic-organic materials are developed on a regular basis in order to achieve synergistic combinations of the constituents' properties to achieve better performance.[1,2] Clay (layered silicate)-based polymer nanocomposites are one such class of functional hybrids, where the compatibility between the clay platelets and polymer matrices leads to the nanoscale dispersion of nanometer-thick clay platelets in the polymer matrices. This leads to the generation of very large amount of interfacial contact between the inorganic and organic phases, thus helping to achieve enhancement of the properties at a much lower volume fraction of filler than with the conventional composites.[3] This allows the generation of lightweight composites which are also transparent, thus expanding the application potential of these materials. Substantial improvements in properties such as strength, modulus, thermal stability, flame retardancy and decreased gas permeability at very low filler contents as compared with the conventional composites[4–12] have been reported. Fillers with other morphologies, such as all three dimensions of the particles on the nanometer scale (spherical),[13,14] two dimensions on the nanometer scale and the third in the range of micrometers (carbon nanotubes or whiskers),[15,16]

---

*This review work was carried out at Institute of Chemical and Bio-engineering, Department of Chemistry and Applied Biosciences, ETH Zurich, Switzerland

---

RSC Nanoscience & Nanotechnology No. 16
Polymer Nanocomposites by Emulsion and Suspension Polymerization
Edited by Vikas Mittal
© Royal Society of Chemistry 2011
Published by the Royal Society of Chemistry, www.rsc.org

etc., have also been extensively reported for nanocomposite synthesis. However, layered silicate (aluminosilicate) platelet materials with two finite dimensions in the range of micrometers, and the third dimension on the nanometer scale,[17] have been the materials of choice in a large number of studies.

Owing to its low cost, the incorporation of layered silicate materials into polymer materials for enhancement of properties has been known for over half a century. In 1950, reinforcement of elastomers was achieved by the incorporation of organically modified silicates by Carter *et al.*[18] An aqueous solution of poly(vinyl alcohol) was also reported to be reinforced with clays by Greenland.[19] The real scientific thrust to this technology was achieved in the early 1990s when Toyota researchers developed polyamide-based nano-composites,[20,21] by dispersing the layered silicate clay platelets in the polymer matrix at the nanometer scale, and thus as a result achieving tremendous improvements in properties. The filler platelets were organically modified before incorporation in the matrix and the polymer was generated in situ in the presence of clay for nanocomposite synthesis. However, this technique was quickly followed by the melt intercalation method reported by Giannelis and co-workers.[22,23] This technique allowed one to work directly with high molecular weight polymers, which were melted and mixed with filler at high temperature under the action of shear. As high temperature is required for the filler mixing in this method, it is important to avoid thermal degradation of the polymer and the organic modifiers bound to filler by choosing the optimum compounding temperature and time.

Montmorillonite, belonging to the family of layered silicates (aluminosilicates), has been commonly used for compounding with polymers. Montmorillonite is an expandable dioctahedral smectite belonging to the family of the 2:1 phyllosilicates[24,25] with the general formula $M_x(Al_{4-x}Mg_x)Si_8O_{20}(OH)_4$. One layer or platelet in this material is formed by a central octahedral (Al) sheet sandwiched between two tetrahedral (Si) sheets. Isomorphic substitutions of aluminum by magnesium in the octahedral sheet generate negative charges, which are compensated for by alkaline earth or hydrated alkali metal cations, as shown in Figure 1.1.[26,27] These aluminosilicate layers are roughly 1 nm thick and, owing to electrostatic forces of attraction, form thick stacks. These stacks thus have interlayers or galleries between the two layers, where the alkaline earth or hydrated alkali metal cations reside. The amount of substitutions in the silicate crystals leads to a corresponding amount of negative charges on the surface and edges of the aluminosilicate layers. This is represented by a term called layer charge density and the montmorillonites have a mean layer charge density of 0.25–0.5 equiv. mol^{-1}.

Although the layered silicate fillers are inexpensive and represent high-quality fillers for incorporation in the polymer matrices, the compatibility of the layered silicates with all the polymer matrices cannot be guaranteed. The silicate layers are polar in nature and these high-energy hydrophilic surfaces are incompatible with many polymers, whose low-energy surfaces are hydrophobic. Hence organic modification of the silicate fillers is generally required in

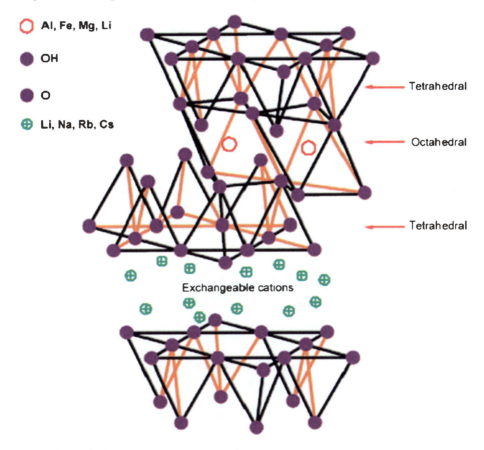

Al, Fe, Mg, Li

OH

O

Li, Na, Rb, Cs

Tetrahedral

Octahedral

Tetrahedral

Exchangeable cations

Figure 1.1 Crystal structure of 2:1 layered silicates. Reproduced from references 26 and 27 with permission from Elsevier.

order to decrease the surface energy and to enhance the compatibility with the polymer matrices. Exchange of surface alkaline earth- or hydrated alkali metal cations with the organic ammonium ions has commonly been used to organically modify the surface of layered silicate platelets.[28–31] Alkylammonium ions such as octadecyltrimethylammonium and dioctadecyldimethylammonium ions have been conventionally used for the organic modification of silicates. The surface modification also leads to an increase in the interlayer spacing in the clay platelets, which also helps in the easy intercalation of polymer chains in them. Owing to the low charge density of the montmorillonite platelets, the electrostatic forces holding them are weak. Therefore, the stacks can be easily broken in water to disperse them as single platelets. This then allows the exchange of surface inorganic cations with the required organic ammonium cations. Other layered silicate minerals such as mica have much higher layer charge density and, as a result, the platelets are very strongly bound to each other. Therefore, it is not possible to disperse the mica platelets completely in water, hence the cation exchange in this case is not optimal.

During the cation exchange on the surface of the layered silicate platelets, the cationic head group of the alkylammonium moiety preferentially resides at the layer surface, leaving the organic tail radiating away from the surface. It has been reported that two parameters define the equilibrium layer spacing in a given temperature range: the cation-exchange capacity of the layered silicate (a function of layer charge), driving the packing of the chains, and the chain length of organic tails.[20] Molecular dynamics simulations have also been used to gain further insights into the structures of the layered silicate in addition to the positioning and behavior of ammonium ions after cation exchange and as a function of temperature, as shown in Figure 1.2 for octadecyltrimethylammonium- and dioctadecyldimethylammonium-modified mica platelets.[32]

1.2 Synthesis Methodologies for Polymer Nanocomposites

The following techniques are commonly used for the synthesis of polymer nanocomposites.[26,33]

1.2.1 Template Synthesis

In this technique, organic and inorganic components are not mixed directly, but the inorganic material is synthesized in the presence of the polymer matrix. This method is different from commonly used nanocomposite synthesis techniques where either the organic component is generally synthesized in the presence of the silicate filler or both organic and inorganic parts are simply compounded together. The template synthesis method represents an attractive technique as the synthesis of silicate layers in the presence of polymer can eliminate the filler dispersion problems associated with other methods; however, there are also some drawbacks.[26] High temperature is often required for the synthesis of nanocomposite materials and not all polymer materials can withstand this. Also, filler aggregation is still noted even though it is synthesized *in situ*. Double-layer hydroxide-based nanocomposites have been synthesized by using this route *in situ* in an aqueous solution containing the polymer and the silicate building blocks.[34,35]

1.2.2 *In Situ* Intercalative Polymerization

In this mode of polymerization, the monomer is used to swell the layered silicate mineral. As the monomer has low molecular weight, it is more mobile in nature and can diffuse easily into the filler interlayers. During polymerization, the clay interlayers becomes trapped by the forming polymer chains and an

Figure 1.2 Snapshots of octadecyltrimethylammonium (C_{18})-modified mica and dioctadecyldimethylammonium ($2C_{18}$) modified mica in MD after 400 ps. (a) C_{18}-mica, 20 °C; (b) C_{18}-mica, 100 °C; (c) $2C_{18}$-mica, 20 °C; (d) $2C_{18}$-mica, 100 °C. A major conformational change (corresponding to two phase transitions) in C_{18}-layered mica can be seen, whereas in $2C_{18}$-mica, only some more *gauche* conformations are present at 100 °C and no order–disorder transition occurs due to close packing. Reproduced from reference 32 with permission from the American Chemical Society.

exfoliated or intercalated morphology is achieved. It is important to control carefully the reaction conditions such as polymerization time, temperature and swelling time to achieve optimum extent of polymerization in and out of the layers.

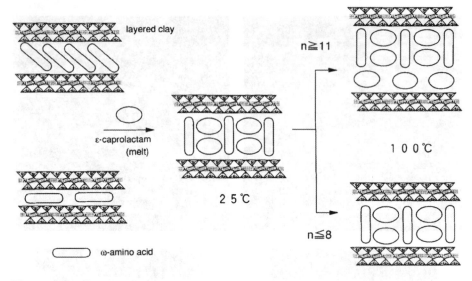

Figure 1.3 Organically modified montmorillonite intercalation with caprolactam. Reproduced from reference 36 with permission from Elsevier.

   Both unmodified and organically modified montmorillonites have been used for *in situ* polymerization. In the earlier studies by Toyota researchers, montmorillonites modified with amino acids of different chain lengths were used for swelling by caprolactam. A schematic representation of the process is shown in Figure 1.3.[36] The polymer was formed by ring-opening polymerization of caprolactam. *In situ* polymerization has also been reported for the synthesis of polyalkene nanocomposites.[37] To achieve polyalkene nanocomposites, a Ziegler–Natta or any other coordination catalyst is anchored to the surface of the layered silicates. This can then be directly used for the polymerization of alkenes such as ethylene and propylene from the surface of the silicate. Here the immobilization of the catalyst does not require any cation exchange on the surface of the silicate; rather, the immobilization is carried out by the electrostatic interactions of the catalytic materials with the MAO initially anchored to the filler surface. Other thermoset polymer nanocomposites have also been reported by using this technique.

1.2.3   Melt Intercalation

Melt intercalation has been a method of choice for the commercial generation of polymer nanocomposites as the technique does not require extensive synthesis and work-up processes. As mentioned earlier, the polymer is melted at high temperature and the melt is then mixed with the filler under shear. The temperature is chosen so as to achieve optimum viscosity in the polymer melt, which is able to withstand shear from the compounder, and also allows good

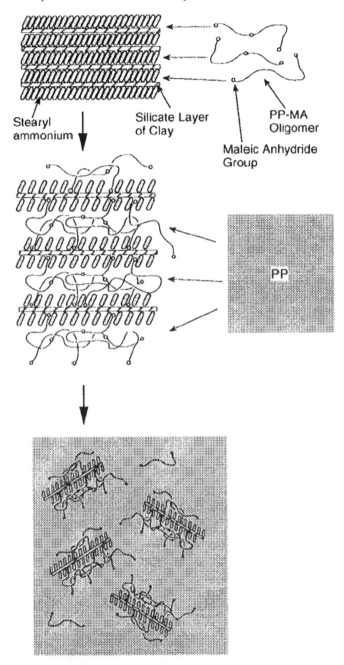

Stearyl ammonium

Silicate Layer of Clay

PP-MA Oligomer

Maleic Anhydride Group

PP

Figure 1.4 Intercalation of polymer in the silicates in the presence of PP-g-MA using the melt intercalation approach. Reproduced from reference 44 with permission from the American Chemical Society.

mixing with the filler. Also, the compounding temperature should be high enough to allow efficient mixing of the organic–inorganic phase, but should not be so high that it starts to degrade the polymer and organic modification on the filler. The compounding time should also be controlled so as to avoid excessive exposure to high temperature. Another advantage of this method is that the polymer can intercalate between the interlayers even if it is thermodynamically not feasible, because under the action of shear, the filler stacks are broken and the filler platelets are kinetically trapped in the polymer melt. Thus, owing to the combination of benefits such as being an environmentally friendly method with no requirements for solvents or complex reactions, simplicity and economic viability, the melt intercalation method has been widely used for the synthesis of polymer nanocomposites with a large number of polymer materials.[38–42]

For polar polymer matrices such as polyamides, the better surface energy match between the organic and inorganic components leads to intercalation of polymer chains inside the clay layers. However, hydrophobic polymers such as polyalkenes suffer from polarity mismatch with the inorganic components. Partial polarization of the polymer matrix is generally achieved to improve the compatibility between the organic and inorganic phases by the use of compatibilizers such as polypropylene grafted with maleic anhydride (PP-g-MA). Figure 1.4 demonstrates the polymer intercalation in the interlayers in the presence of a compatibilizer. Most of the reported polypropylene nanocomposites incorporating organically modified montmorillonite (OMMT) prepared by the melt compounding approach use a low molecular weight PP-g-MA compatibilizer.[43–49]

1.2.4 Intercalation of Polymer or Prepolymer from Solution

To achieve nanocomposites by this method, a common solvent in which the organically modified silicate can be dispersed and also in which the polymer is soluble is chosen. The solvent is subsequently evaporated or the system is precipitated, which leads to sandwiching of the polymer chains between the clay sheets. A schematic of the process is shown in Figure 1.5.[26] The polymer chains lose entropy owing to diffusion inside the silicate interlayers; however, such a process is still thermodynamically viable, owing to the gain in the entropy by the solvent molecules due to desorption from the silicate interlayers.[50,51]

Although this technique is mostly used for the intercalation of water-soluble polymers such as poly(vinyl alcohol), poly(ethylene oxide), poly(acrylic acid) and polyvinlypyrrolidone,[52–55] nanocomposites have also been synthesized in organic solvents.[56] Emulsion and suspension polymerization of monomers in the presence of layered silicates also falls into this category and has also been reported to generate intercalated and exfoliated nanocomposites using monomers such as methyl methacrylate and styrene, as explained in the next section.

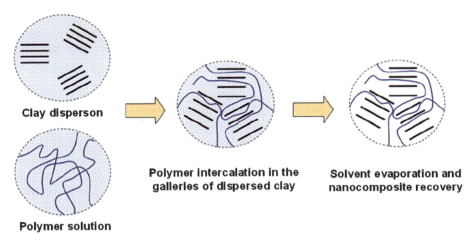

Clay disperson

Polymer solution

Polymer intercalation in the galleries of dispersed clay

Solvent evaporation and nanocomposite recovery

Figure 1.5 Polymer adsorption from solution. Reproduced from reference 26 with permission from Elsevier.

1.3 Polymer Nanocomposites in Emulsion and Suspension

To achieve polymer nanocomposites in emulsion and suspension, the monomer is suspended in water along with various amounts of silicates in the presence of emulsifier in emulsion polymerization and suspension stabilizers in suspension polymerization. The monomer is then polymerized either in emulsion micelles or in suspension monomer droplets, leading to the generation of the nanocomposites with part of the silicate embedded inside the polymer particles and part adsorbed on the surface of the particles. In emulsion polymerization, the initiators used are generally water soluble, in contrast to suspension polymerization, where water-insoluble initiators are used. The behavior or role of the emulsifier in emulsion polymerization is different from that in suspension polymerization. The emulsifier in the emulsion polymerization, in addition to stabilizing the monomer droplets, also forms micelles once the concentration of the stabilizer exceeds the critical micelle concentration. The internal environment of a micelle is totally hydrophobic and is very favorable for monomer polymerization. The size of the polymer particles in the case of emulsion polymerization is dependent on the amount of surfactant, whereas the size of monomer droplets and hence polymer particles in suspension polymerization can be controlled by varying the ratio of monomer to the dispersion medium, *i.e.* water, the speed of agitation to generate droplets and the amount of stabilizing agents. The main advantage of generating polymer nanocomposites in emulsion or suspension is the that as the dispersion medium, *i.e.* water, surrounds every forming polymer particle, hence there is no problem with viscosity enhancement and heat can also be dissipated without problem, thus allowing suitable conditions for filler exfoliation without causing any extensive thermal damage to the polymer or the organic modification on the filler surface. Surfactant-free emulsion polymerization can also be achieved to generate nanocomposites; the mechanism of particle nucleation in this case involves homogeneous nucleation compared with micellar nucleation in the case of emulsified emulsion polymerization.

In the mid-1990s, Lee and Jang[57] reported a poly(methyl methacrylate)–clay composite with intercalated morphology obtained by emulsion polymerization of methyl methacrylate in the presence of sodium montmorillonite. No specific modification of the monomer or filler was carried out. The resulting nanocomposite latex particles were exposed to boiling toluene for 5 days to remove unbound polymer. From the IR spectra of the extracted composites, the presence of the characteristic group frequencies of both poly(methyl methacrylate) and montmorillonite was confirmed. The absorption bands at 3630, 1050 and $600-400\,cm^{-1}$ could be associated with –OH stretching of the lattice water, Si–O stretching and Al–O stretching and Si–O bending, respectively. The IR bands at $3000-2900\,cm^{-1}$ ($-CH_3$ stretching), $1735\,cm^{-1}$ (C=O stretching) and $1300-1100\,cm^{-1}$ (C–O stretching) were reported to be due to the characteristic frequencies of poly(methyl methacrylate). The basal plane spacing of the filler was also observed to increase in the extricated composite materials and this increase corresponded to the increasing polymer content in the system, indicating that the composites were intercalated with polymer. In general, an increase of 0.54 nm was observed in the basal plane spacing, which was comparable to the 0.68 nm obtained from the intercalation of alkylammonium montmorillonite with polystyrene, but was lower than the 0.83 nm obtained from the functionalization of montmorillonite by poly(methyl methacrylate) containing side-chain ammonium cation. The mechanical and thermal properties of the composites were also enhanced on incorporation of the clay. The onset of the thermal decomposition in composites was observed to shift to higher temperature compared with pure polymer. Also, no clear differential scanning calorimetry endotherm was observed for the intercalated polymer as compared with pure poly(methyl methacrylate), indicating that the segmental motion of the polymer chains in the filler interlayers was confined. The composite containing 10 wt% clay was also observed to exhibit superior mechanical properties. A tensile strength of 62 MPa and tensile modulus of 4.64 GPa were observed for the composite, compared with 53.9 MPa and 3.06 GPa for the pure poly(methyl methacrylate) polymer, respectively.

Lee and Jang subsequently reported the synthesis of epoxy–clay nanocomposites by emulsion polymerization.[58] In this study also, no coupling or modifying agent was used for the polymer or montmorillonite. To generate epoxy polymer, equimolar quantities of bisphenol A and an epoxy prepolymer ($n = 0.2$) were polymerized in an emulsion medium in the presence of sodium montmorillonite. It was observed through X-ray diffractograms of the acetone extracted composite material that an increase in basal plane spacing from 0.96 to 1.64 nm occurred, indicting the intercalation of epoxy polymer in the filler interlayers. The nanocomposites were also observed to have enhanced thermal stabilities. It was further reported that an overwhelming fraction of the

Figure 1.6   SEM images of (a) polystyrene, (b) polystyrene–unmodified MMT, (c) polystyrene–PhSiMMT, (d) polystyrene–VSiMMT, (e) polystyrene–C8SiMMT and (f) polystyrene–C18SiMMT latexes. Reproduced from reference 59 with permission from Elsevier.

nanocomposite contained intercalated clay. The authors also opined that the strong fixation of the polymer to the inorganic surfaces occurred due to the cooperative formation of the ion–dipole force acting between polar functional groups of the polymer chain and the interlayer ions.

Ianchis *et al.* recently reported the surfactant-free emulsion polymerization of styrene in the presence of unmodified montmorillonite and also montmorillonite modified with silane-based surface modifications.[59] These modifications were trimethylethoxysilane (Me3ES), vinyldimethylethoxysilane (VMe2ES), phenyldimethylethoxysilane (PhMe2ES), octyldimethylmethoxysilane (C8Me2MS) and octadecyldimethylmethoxysilane (C18Me2MS). Figure 1.6 shows the SEM images for the pure polystyrene particles and also composite particles containing modified and unmodified montmorillonite. It was observed that for the particles synthesized in the presence of unmodified montmorillonite, the interaction of the clay with the polymer was weak and the particles maintained their spherical shape, similar to the particles without filler. The filler platelets were localized at the particle–water interface and significant aggregation was observed. When more hydrophobic montmorillonites (*i.e.* modified montmorillonites) were used for the composite synthesis, it was observed in the micrographs that the modified montmorillonite particles were mainly localized into the polymer particles. The shape of the resulting particles was also very different from the spherical morphologies. The FTIR and XRD analyses of solid materials further proved the presence of montmorillonite particles within the polymer matrix. The zeta potential of the latexes containing modified montmorillonites also became more negative as compared with the latex without filler or latex containing unmodified montmorillonite. This indicated that the modified clay platelets acted as stabilizers for the polystyrene, similar to Pickering emulsions. There was no change in the zeta potential of the latex containing unmodified montmorillonite and latex without filler, indicating that the unmodified filler did not influence the charges on the surface of the latex particles.

Clarke *et al.*[60] reported the suspension polymerization of methyl methacrylate to achieve polymer–clay nanocomposites. The modified filler with trade name Cloisite 15a (C15a) organically modified with dimethyl dehydrogenated tallow quaternary ammonium salt was used. A suspending agent or stabilizer, which was a 1% aqueous solution of poly(2-methyl-2-propenoic acid) sodium salt, and 2,2′-azobis(2-methylpropionitrile) (AIBN) initiator were employed in the study. A comonomer of ethyl propenoate and a chain-transfer agent (CTA) of 1-dodecanethiol were also used in the suspension polymerization. During the polymerization, the pH of the aqueous phase was maintained at 10 by buffer solution. It was observed that increasing the amount of filler used for suspension polymerization caused a corresponding increase in the molecular weight of the polymer, but subsequently it reached plateau. A plateau value of approximately $800\,000\,\mathrm{g\,mol}^{-1}$ with a filler content >0.5 wt% was observed. The presence or absence of a chain transfer agent during the polymerization produced similar molecular weights, which was due to the inactiveness of the chain transfer agent CTA in the presence of clay. The chain transfer agent reacted with the Lewis acid moieties situated on the rim of the clay. The TEM images indicated that a good degree of dispersion of filler was obtained during the suspension polymerization process, although there were still regions where no clay was present. The basal plane spacing in the composites increased by 0.55 nm on average compared with the modified montmorillonite, indicating the intercalation of the clay galleries with the polymer.

Yuan *et al.*[61] applied the emulsion polymerization approach to prepare nanocomposites of polyurethane with clay. Industrial grade montmorillonite with a cation-exchange capacity of 90 mequiv. per 100 g was used. Two different surface modifications, with dimethyldistearylbenzylammonium chloride (DMDSBA) and dimethyldistearylammonium chloride (DMDSA), were compared for their influence on the performance and properties of the resulting nanocomposites. Hydrated methyl diphenyl isocyanate (H-MDI) and polyols (MW 400 containing carbonyl groups) were the building blocks for the polyurethane synthesis. Initially, the reaction between the clay, polyols and isocyanate was carried out in acetone to form a pale, transparent solution, and this solution after cooling and neutralization of the carbonyl groups with triethylamine was transferred into water with vigorous stirring. It was observed by TEM that DMDSBA-modified MMT was well dispersed in polyurethane and the filler particles had an average size of 100 nm. On the other hand, DMDSA-modified MMT in the polyurethane matrix showed larger sizes of about 500 nm owing to aggregation. Thus, even through the processing conditions for both the modified clays were similar, DMDSA-modified MMT always showed a larger size in the polymer matrix than that of DMDSBA-modified filler, indicating that the DMDSBA-modified MMT was easily dispersible in polyurethane. The possible reason for this behavior may be the better compatibility of DMDSBA modification with the polymer. The wide-angle X-ray diffraction results also confirmed the microscopy observations. The tensile stress was observed to increase with increasing amount of modified montmorillonite in the polyurethane composites, indicating reinforcement effects of the filler on the polymer. The result was also confirmed by the increase in the glass transition temperature of the polymer in the presence of filler. The other tensile properties such as tensile strength and elongation at break, however, were observed to decrease on the addition of montmorillonite to the polymer matrix. The gas barrier properties of the polyurethane nanocomposites were also measured and it was observed that the DMDSBA modification performed better than the DMDSA modification. The oxygen permeability coefficient was observed to decrease fourfold on addition of 8% montmorillonite modified with DMDSBA. This was also related to the better dispersion of the DMDSBA-modified montmorillonite in the polyurethane matrix, as observed with TEM and wide-angle X-ray diffraction methods.

Khatana *et al.*[62] reported the used of clay as suspension stabilizer for the suspension polymerization of methyl methacrylate to generate nanocomposites. Poly(vinyl alcohol) (PVA) was used as suspension stabilizer and its performance was compared with clay in stabilizing the suspension. Uniform beads were obtained when methyl methacrylate was polymerized without clay and in the presence of PVA and the dispersion was observed to be milky owing to the partial solubility of the methyl methacrylate monomer in water and the formation of very fine poly(methyl methacrylate) particles. Similarly, the dispersion of methyl methacrylate with C20A-modified clay was also milky; however, in the methyl methacrylate–sodium montmorillonite system, with and without PVA, the suspensions were much clearer owing to the strong interaction between the unmodified filler surface and methyl methacrylate. It was observed that the type of clay used in the suspension polymerization influenced the shape of the resulting particles. Fine particles were formed for the case when sodium montmorillonite was used, both in the presence and absence of PVA. However, when modified montmorillonite was used in the presence of PVA, both beads and clusters were formed, and only clusters were formed when no PVA was used. In the case of the sodium montmorillonite–methyl methacrylate polymerization system in the presence of PVA, the resulting uniform particles indicated that the combination of PVA and Na$^+$ clay platelets in water prior to the methyl methacrylate polymerization was able to stabilize effectively the methyl methacrylate droplets suspended in water. The resulting uniform particles even in the absence of PVA for the sodium montmorillonite–methyl methacrylate polymerization system confirmed that the clay platelets could be used effectively as a suspension stabilizer and no additional PVA was required. The presence of silicon and aluminum peaks in the energy-dispersive spectra of the particles polymerized with and without PVA indicated the formation of poly(methyl methacrylate) nanocomposites with sodium montmorillonite. The glass transition temperature of poly(methyl methacrylate) was observed to increase in the composites from 103 °C for the pure polymer. In the case of poly(methyl methacrylate) composites with modified clay, the glass transition temperature was observed to increase to 108 °C. On the other hand, the enhancement of the glass transition temperature in the case of sodium montmorillonite-filled composites was more significant. The glass transition temperature was observed to increase to 124 and 121 °C in the nanocomposites synthesized with and without PVA, respectively.

Sedlakova et al.[63] reported the synthesis of polystyrene–clay and poly(butyl methacrylate)–clay nanocomposites using seeded emulsion polymerization. Ammonium persulfate (APS) was used as initiator and sodium dodecyl sulfate (SDS) was used as surfactant. Two different montmorillonites were used for the study: one with a cation-exchange capacity (CEC) of 145 mequiv. per 100 g and the other with a CEC of 92.6 mequiv. per 100 g. The montmorillonites were ion exchanged with [2-(acryloyloxy)ethyl]trimethylammonium chloride to render them organophilic. Different clay loadings and different solid contents were generated during the emulsion polymerization. The particle size of the polystyrene latex was observed to increase from 68 nm for neat particles (without clay) to larger sizes when clay was added. The particle size was observed to have a slight dependence on the solid content in the latex. The presence of clay had a strong influence on the mean particle size as increasing the concentration of the clay led to an increase in the mean particle size. At higher clay concentrations and solid contents, occurrence of aggregation was also observed.

Similar behaviors were observed for poly(butyl methacrylate) latexes. The morphology of the polymer nanocomposites formed was observed to be exfoliated. The nanocomposites were also prepared with unmodified montmorillonites and the dispersion of the filler was observed to be poorer as compared with the nanocomposites prepared with the modified clays.

Diaconu et al.[64] reported that the miniemulsion polymerization process achieved water-borne acrylic clay nanocomposite latexes with a high solid content. A cationic macromonomer, 2-methacryloylethylhexadecyldimethyl-ammonium bromide, as shown in Figure 1.7, and a cationic macroinitiator, cationic acrylic–styrene oligomer end-capped with a nitroxide, were used to modify the montmorillonite organically in order to enhance the compatibility between the clay platelets and the acrylic polymer matrix. Both the functional modifications could be successfully exchanged on the surface of the montmorillonite platelets, thus allowing potential polymer grafting either to or from the surface. The use of organically modified montmorillonites resulted in the synthesis of stable and coagulum-free 30% solids content water-borne acrylic polymer–clay nanocomposites. It was observed with TEM and wide-angle X-ray diffraction that partial exfoliation of the filler could be achieved, which resulted in enhanced mechanical, thermal and barrier properties compared with

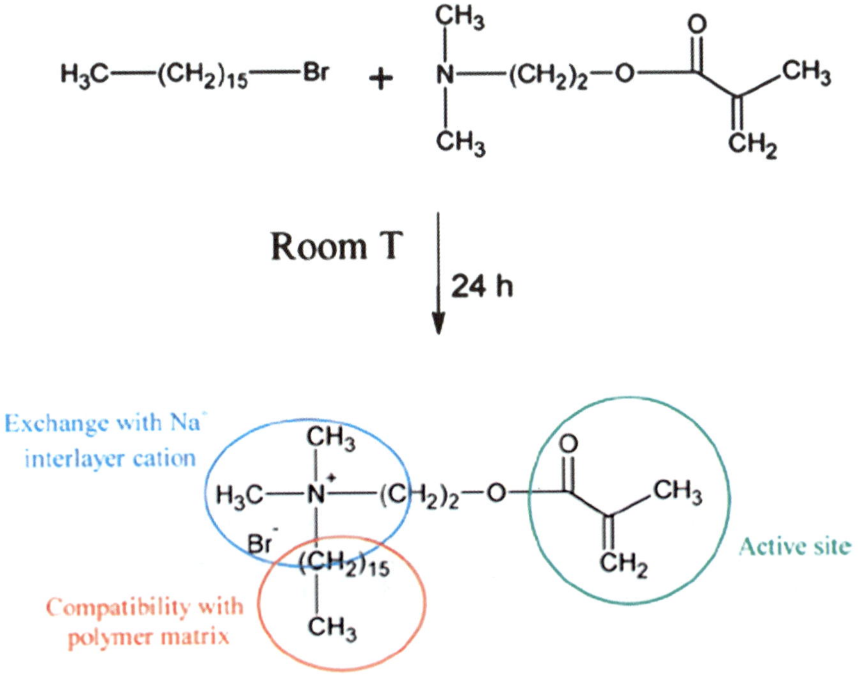

Figure 1.7 Cationic macromonomer, 2-methacryloylethylhexadecyldimethylammonium bromide, used to exchange on the surface of the filler platelets. Reproduced from reference 64 with permission from the American Chemical Society.

the pure polymer. However, the authors could not confirm the tethering of the acrylic polymer chains to the clay platelets.

Kajtna and Sebenik[65] described the synthesis of pressure-sensitive adhesives based on polyacrylates by suspension polymerization. The monomers used were 2-ethylhexyl acrylate (2-EHA) and ethyl acrylate (EA), and dibenzoyl peroxide (DBP) was used as initiator. Surface-active agents [modified ester of sulfocarboxylic acid (SCA) and ethoxylated oleyl alcohol (EOA)], chain transfer agent (*n*-dodecanethiol) (CTA) and suspension stabilizer [poly(vinyl alcohol) (PVA)] were also used. Various organically modified montmorillonites were used as fillers. It was observed that the kinetics of suspension polymerization were independent of the presence of the montmorillonite in the system, as shown in Figure 1.8.

The adhesive properties of the pressure-sensitive adhesive formulations were observed to be strongly influenced by the presence of clay. On using the modified montmorillonites, the peel strength and tack were reported to decrease with increasing clay content, whereas the shear strength of the composites increased. On using the unmodified clay, only a moderate effect on these properties was observed. An increase in the storage modulus was observed on

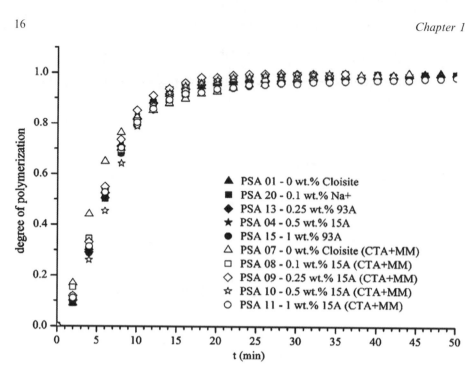

Figure 1.8 Time conversion curves for the suspension polymerization trials using unmodified montmorillonite and montmorillonites modified with different surface modifiers. Reproduced from reference 65 with permission from Elsevier.

addition of clay, irrespective of the type of clay used in the suspension polymerization.

The effect of clay loading on the morphology and properties of poly(styrene-*co*-butyl acrylate)–clay nanocomposites prepared with emulsion polymerization was reported by Greesh *et al.*[66] 2-Acrylamido-2-methyl-1-propanesulfonic acid (AMPS) was used as a clay modifier. Both batch and semi-batch modes of emulsion polymerization were reported for the composite synthesis. It was reported on the basis of small-angle X-ray scattering results that the morphology of the generated nanocomposites was strongly dependent on the filler content. For the batch emulsion polymerization route, the nanocomposites containing 1 and 3% of the filler were observed to have intercalated morphology. However, the diffraction peak indicating the presence of ordered clay structures became broad at higher clay contents, indicating the generation of partially exfoliated morphologies. At a 10% clay loading in the composite, no X-ray diffraction peak was observed in the diffractograms. These findings were also confirmed by TEM. Thus, the clay-to-polymer ratio was observed to influence directly the generated morphology and hence the properties of nanocomposites. The generation of exfoliated filler morphologies in the composites at higher loadings is unusual and it was opined that owing to the presence of a large number of clay platelets at higher filler loadings in close proximity to each other, the movement of platelets generates energy by friction. This energy was reported to lead to greater mobility in the clay platelets, thus causing their random orientation. Better thermal stability was also observed in the generated nanocomposite materials. For the nanocomposites generated with semi-batch addition of monomers during the polymerization process, similar morphologies of exfoliated structures at higher filler loadings were also observed. This indicated that the synthesis process did not affect the morphology and properties of the nanocomposites. The composites were also prepared at a 15% filler loading by semi-batch emulsion polymerization. However, at this filler loading, the morphology obtained was again intercalated, in contrast to exfoliated morphologies at higher filler loadings around 10%.

Effenberger *et al.*[67] reported a redox initiation system for the synthesis of polyacrylate–montmorillonite nanocomposites by emulsion polymerization. The montmorillonite was surface modified with three different amino acids, alanine, leucine and phenylalanine, in the presence of hydrochloric acid and the edges of the platelets were subsequently modified with methyltriethoxysilane coupling agent. *In situ* redox emulsion polymerization of the monomers glycidyl methacrylate (GMA) and methyl methacrylate (MMA) was carried out. It was observed that weight loss in the nanocomposite samples was observed at higher temperatures as compared with the pure polymer, indicating better thermal stability in the presence of clay. It was also reported that increasing the content of filler also enhanced the thermal stability of these materials. It was observed by microscopic analysis that the filler was more homogeneously dispersed in PMMA than in PGMA matrix.

Ruggerone *et al.*[68] prepared highly filled polystyrene nanocomposites by emulsion polymerization. Figure 1.9 shows the TEM images of the latex particles containing varying extents of laponite clay. It was reported that though occasionally laponite platelets were also observed in the aqueous phase, their strong tendency to adhere to the surface of the polymer particles was demonstrated. Uniform dispersion of laponite platelets was observed in the polystyrene matrix and partial exfoliation was observed.

Lin *et al.*[69] studied the mechanism of formation of poly(methyl methacrylate)–clay nanocomposite particles by emulsion polymerization. It was observed that by using soap-free emulsion polymerization conditions, it was possible to synthesize exfoliated nanocomposite latex. It was found that neither methyl methacrylate monomer nor potassium persulfate (KPS) initiator were absorbed into the montmorillonite interlayers. It was subsequently established that the polymerizing ionic radicals in the aqueous phase at the beginning of the polymerization reaction diffuse into the filler interlayers and organize into disk-like micelles in the interlayer regions owing to the confined space in the interlayers. Filler exfoliation by diffusing ionic radicals is achieved as more and more disc-like micelles are formed. Thus, the exfoliation of the filler is over before the micellization stage is completed. Subsequently, the disk-like micelles become polymerization loci for monomers. It was also reported that the number of disk-like micelles was substantially higher than

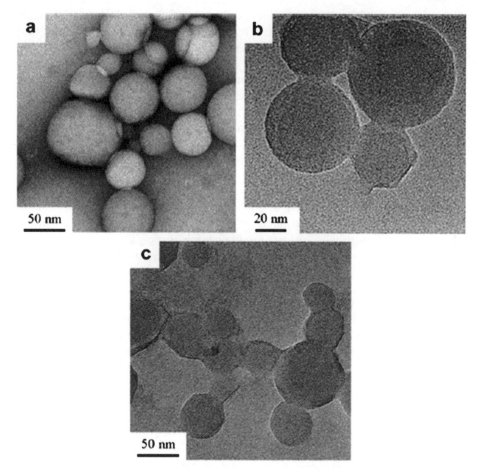

Figure 1.9 TEM images of the latexes with (a) no clay, (b) 10% clay and (c) 20% clay.
Reproduced from reference 68 with permission from Elsevier.

that of the spherical micelles formed in the conventional soap-free emulsion polymerization; therefore, the conversion rate of methyl methacrylate to form poly(methyl methacrylate)–MMT nanocomposite latex was faster. The particle size and polymer molecular weight in the poly(methyl methacrylate)–clay nanocomposite particles was similar to those of neat poly(methyl methacrylate) particles, but the shape of the nanocomposite particles was irregular owing to the exfoliated clay platelets.

The use of a reversible addition–fragmentation chain transfer (RAFT)-controlled polymerization method to synthesize polystyrene–clay nanocomposites in miniemulsion was reported by Samakande et al.[70] For this purpose, the RAFT-grafted montmorillonite clays N,N-dimethyl-N-(4-{[(phenylcarbonothioyl)thio]-methyl}benzyl)ethanammonium-MMT (PCDBAB-MMT) and N-[4-({[(dodecyl-thio)carbonothioyl]thio}methyl)benzyl]-N,N-dimethylethanammonium-MMT (DCTBAB-MMT), were used in different amounts.

Sodium dodecyl sulfate (SDS) surfactant, hexadecane co-surfactant and AIBN initiator were used for the miniemulsion polymerization process. It was observed that the molecular weight of the polymer decreased on increasing the amount of clay in the polymerization system owing to an increase in RAFT agent with increasing amount of clay which had been modified with RAFT agent. Polydispersity index values in the range 1.4–1.5 were observed as the clay loading was increased for both PCDBAB- and DCTBAB-modified montmorillonites. Increasing the clay loading also increased the particle size of the resulting nanocomposite particles, indicating that more space was required to accommodate the swelling clay within the polymer particle. It was further observed from the microscopic analysis that the particle size distribution was fairly narrow, thus confirming that either little or no secondary nucleation occurred during the polymerization process in the presence of modified montmorillonites. The filler platelets had partially exfoliated to intercalated morphology and the microstructure was also affected by the amount of clay loading. The thermomechanical properties of the composites were observed to be better than those of pure polystyrene polymer. Enhanced storage moduli in the glassy state were observed for all the nanocomposite materials owing to the high aspect ratio of the filler platelets and the interaction between the polymer chains and clay layers.

Ali *et al.*[71] synthesized anisotropic polymer–inorganic nanocomposite particles by using RAFT-based starved feed emulsion polymerization. Gibbsite inorganic platelets were encapsulated by the polymerization of butyl acrylate and methyl methacrylate controlled by surface-adsorbed RAFT copolymers. The amphipathic random RAFT copolymers in the aqueous phase acted as stabilizers and could adsorb on the oppositely charged gibbsite surface by electrostatic interactions. Once the polymerization was initiated, the RAFT copolymer chains adsorbed on the surface of gibbsite extended by undergoing rapid transfer and incremental growth on the particle surface, which subsequently resulted in a uniform coating over the entire surface. During the course of polymerization, the RAFT copolymer chains dissolved in the aqueous phase were also reported to adsorb on the growing surface by polymerization around the gibbsite platelets, thus providing further stabilization. The RAFT copolymers were also reported to have the capability to self-assemble in the aqueous phase, thus forming centers for secondary particle nucleation. These centers can generate new particles, which were reported to also adsorb subsequently on the surface of the larger particles. Figure 1.10 shows the encapsulated gibbsite platelets using various RAFT copolymers. It was also possible to control the layer thickness by controlling the amount of monomer fed to the system.

Greesh *et al.*[72] described an interesting study on the effect of clay surface modification on the microstructure and properties of polymer–clay nanocomposites synthesized by using free radical emulsion polymerization. Copolymers of styrene and butyl acrylate were synthesized in the presence of montmorillonites modified with various surface modifiers in order to relate their chemical structure, degree of interaction within the clay gallery surface

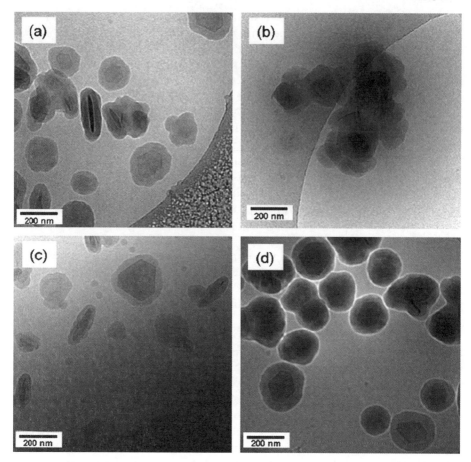

Figure 1.10   Encapsulated gibbsite particles obtained by emulsion polymerization of
methyl methacrylate and butyl acrylate using different RAFT copoly-
mers: (a) BA_5-*co*-AA_{10}, (b) $BA_{2.5}$-*co*-AA_{10}, (c) $BA_{7.5}$-*co*-AA_{10} and (d)
BA_5-*co*-AA_5 (BA = butyl acrylate and AA = acrylic acid). Reproduced
from reference 71 with permission from the American Chemical Society.

and ability to copolymerize with monomers to the morphology and properties
of the nanocomposite particles. Modifiers used in the study were sodium 1-
allyloxy-2-hydroxypropylsulfonate (COPS), 2-acrylamido-2-methyl-1-propa-
nesulfonic acid (AMPS), *N*-isopropylacrylamide (NIPA) and sodium 11-
methacryloyloxyundecan-1-yl sulfate (MET). It was observed that the morphology of the generated nanocomposites was strongly dependent on the type
of modification present on the surface of the filler platelets. The authors
reported that the modification must fulfill a few requirements for the exfoliated
structures to be obtained, *e.g.* ability to interact significantly with the clay
surface, compatibility with the monomer system and good reactivity with the
monomer system. On the other hand, intercalated or partially exfoliated

nanocomposite morphologies were obtained when one or two of these requirements were fulfilled. Similar observations have also been reported by Osman *et al.*[73] for epoxy–clay nanocomposites generated by solution polymerization. It was observed that nanocomposites containing 10% clay modified with COPS, NIPA and MET had intercalated to partially exfoliated structures. On the other hand, AMPS-modified clay was observed to have a fully exfoliated structure. Also, the nanocomposites had much better thermal stability, storage moduli and glass transition temperatures than pure copolymer material. Interestingly, the intercalated nanocomposites were reported to have better thermal stability than fully exfoliated ones and intermediate behavior was observed for partially exfoliated nanocomposites. On the other hand, the exfoliated morphologies generated maximum enhancement in storage moduli and glass transition temperature.

Diaconu *et al.*[74] also reported the use of pristine sodium montmorillonite for the synthesis of very high solids poly(methyl methacrylate-*co*-butyl acrylate) latexes by using seeded emulsion polymerization. Figure 1.11 shows the particle size distributions for the various latexes synthesized without or with clay following different synthetic methodologies. For the pure polymer latex (EP1), the distribution was unimodal and narrow. However, for the nanocomposite latexes, the distributions were broad and tended to have shoulders towards the

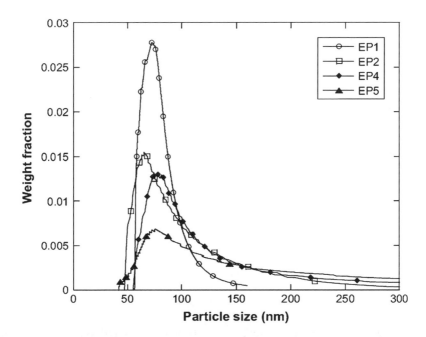

Figure 1.11   Particle size distributions of the latexes synthesized without clay (EP1) and with clay using different synthetic methodologies. Reproduced from reference 74 with permission from Elsevier.

higher particle sizes owing to the presence of clay. For comparison with the *in situ*-produced nanocomposites, the nanocomposite latexes were also produced by physical mixing of clay with preformed polymer particles. It was observed that the *in situ*-produced nanocomposite latexes were superior to physical blends in mechanical, thermal and permeability properties. It was further observed by TEM and X-ray scattering that the dispersion of the clay in the blend was also not optimal. By using semi-batch seeded emulsion polymerization techniques, nanocomposite latexes with 45 wt% solids content with intercalated filler morphologies and enhanced mechanical properties could be produced.

Bouanani *et al.*[75] also reported the use of miniemulsion polymerization for the encapsulation of acid-treated montmorillonite within a fluorinated cyclosiloxane [1,3,5-tris(trifluoropropylmethyl)cyclotrisiloxane]. Figure 1.12 shows the X-ray diffractograms of the pristine montmorillonite, acid-treated montmorillonite and nanocomposite before and after polymerization. The diffraction peak for the pristine montmorillonite was observed at 13.5 Å, which was shifted to 18.01 Å° after acid treatment. The authors opined that the observed larger basal plane spacing may be a result of interlamellar water layers due to the acid treatment and thus reflects the change in interlayer cations and their

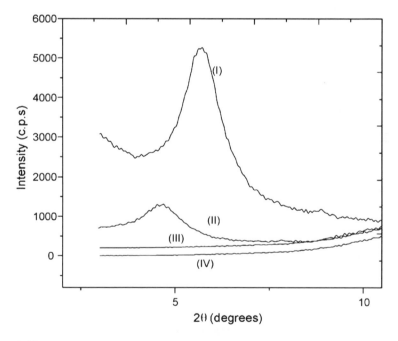

Figure 1.12 X-ray diffraction spectra of (I) pure montmorillonite, (II) acidified montmorillonite, (III) nanocomposite before polymerization and (IV) nanocomposite after polymerization. Reproduced from reference 75 with permission from Elsevier.

associated hydration states. Fully exfoliated morphology of the clay platelets was observed in the nanocomposites before and after polymerization as no diffraction peak was observed for these materials.

Nanocomposite latexes synthesized using spherical inorganic particles as compared with platy clay particles have also been reported using emulsion polymerization. In one such study, Schmid *et al.*[76] synthesized poly-(styrene-*co*-butyl acrylate)–silica nanocomposite particles using a cationic azo initiator and a glycerol-modified silica sol. The synthesis methodology used led to the generation of core–shell morphology where the silica particles were adsorbed on the surface of the particles. By using various ratios of monomers, the glass transition temperature of the resulting copolymer in the nano-composite latex could be changed, which subsequently affected their minimum film-forming temperature. It was observed that formulations comprising at least 50 wt% styrene were colloidally stable, whereas acrylate-rich formulations invariably led to particle aggregation (Figure 1.13). The colloidally stable nanocomposite particles were also observed to have narrow particle size distributions.

Duan *et al.*[77] also reported the synthesis of poly(*N*-isopropylacrylamide)–silica composite microspheres by using inverse Pickering suspension polymerization with various sizes of silica particles as stabilizers. Figure 1.14 shows examples of such microgels stabilized by silica particles with mean diameters of 53, 301, 500 and 962 nm. To generate these nanocomposite structures, droplets of an aqueous solution of *N*-isopropylacrylamide were first dispersed in toluene and then stabilized by silica particles. The monomer was subsequently polymerized to obtain polymer–silica composite microspheres. It was also observed that the thermo-responsive behavior of the polymer was not affected in the presence of silica, as a lower critical solution temperature of 32 °C for the poly(*N*-isopropylacrylamide) was also observed in the polymer–silica microspheres.

Yei *et al.*[78] described the synthesis of polystyrene–clay nanocomposites using emulsion polymerization and specifically modified montmorillonite. Two different surface modifications were used to organophilize the montmorillonite, with cetylpyridinium chloride (CPC) and the CPC–α-cyclodextrin (CD) inclusion complex. Glass transition temperatures of 102 and 106 °C were observed for CPC–clay and inclusion complex–clay nanocomposites, compared with 100 °C for pure polystyrene.

The observation of better dispersion of the filler in inclusion complex-intercalated clay nanocomposites also indicates that the modification was able to retard chain movement more effectively than that of the CPC–clay-modified nanocomposite. The CPC–α-CD-intercalated clay nanocomposite was the most thermally stable compared with pure polymer and also CPC–clay-modified nanocomposite. The nanocomposite prepared from the CPC–α-CD-inter-calated clay exhibited a 5% weight loss temperature 33 °C higher than that of the pure polystyrene as compared with 18 °C for the CPC-intercalated clay nanocomposite. Figure 1.15 shows the TEM low- and high-magnification images of the CPC–α-CD-intercalated clay nanocomposite.

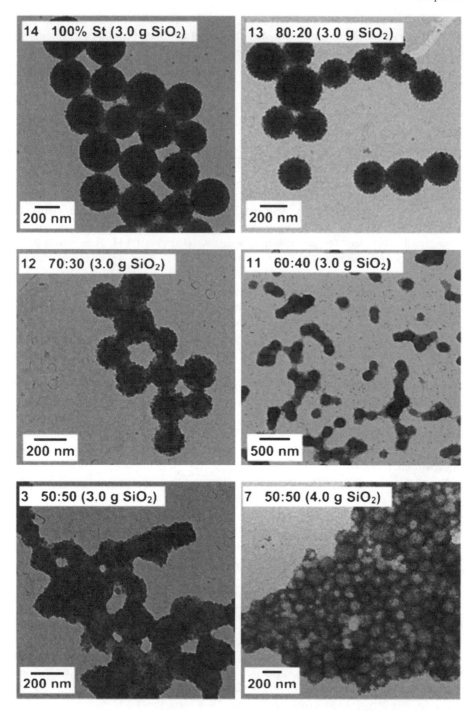

Figure 1.13   Transmission electron micrographs of nanocomposite latex particles prepared by copolymerization of various mass ratios of styrene to *n*-butyl acrylate in the presence of a glycerol-modified silica sol. Reproduced from reference 76 with permission from the American Chemical Society.

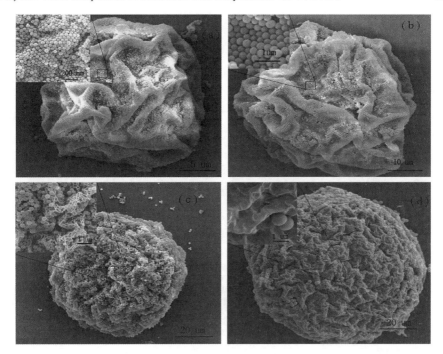

Figure 1.14 Microgel-silica microspheres fabricated using different sizes of silica and different microgel cross-linking density: (a) 53, (b) 301, (c) 500 and (d) 962 nm silica. Reproduced from reference 77 with permission from the American Chemical Society.

Li *et al.*[79] reported the use of montmorillonite modified with zwitterion aminoundecanoic acid (AUA) to synthesize polystyrene–clay nanocomposites by emulsion polymerization. Protonation of the zwitterionic acid molecules led to the exchange of the cations in the interlayers of montmorillonite and the carboxyl groups were subsequently ionized in the alkaline aqueous medium to permit exfoliation of the clay. The resulting polystyrene–montmorillonite nanocomposites were observed to have an exfoliated microstructure, confirmed by X-ray diffraction and TEM. Figure 1.16 depicts the exfoliated microstructure of the nanocomposites (curve c in the X-ray diffractograms and the TEM micrograph). As a reference, the nanocomposites were also generated by using cetyltrimethylammonium bromide as filler modifier and curve b in Figure 1.16 represents the X-ray diffractogram of this composite, thus indicating only an intercalated morphology. The exfoliated nanocomposites, on the other hand, had much higher enhancements in modulus, glass transition temperature and thermal stability as compared with pure polystyrene and the intercalated composites.

Yang *et al.*[80] reported the emulsion and suspension polymerization of styrene in the presence of montmorillonite modified with various modifying agents such as cetyltrimethylammonium chloride, cetylpyridinium bromide,

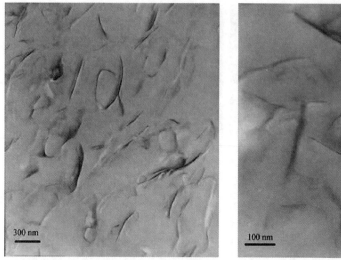

Figure 1.15 TEM images of the CPC–α-CD-intercalated clay nanocomposite at two magnifications. Reproduced from reference 78 with permission from Elsevier.

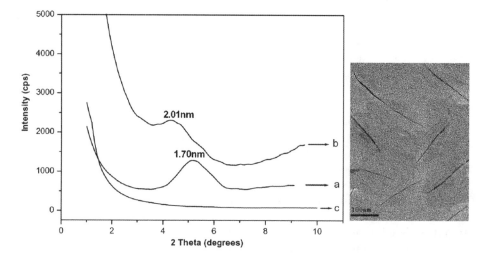

Figure 1.16 Left: X-ray diffractograms of composites with (a) sodium montmorillonite, (b) cetyltrimethylammonium bromide and (c) zwitterion. Right: TEM image of the polystyrene nanocomposites containing montmorillonite modified with zwitterion. Reproduced from reference 79 with permission from Elsevier.

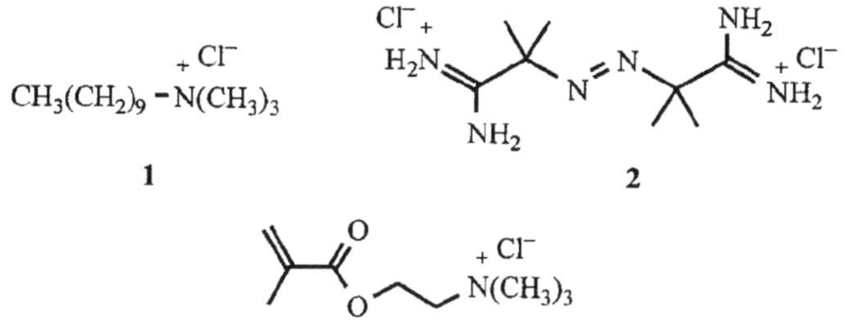

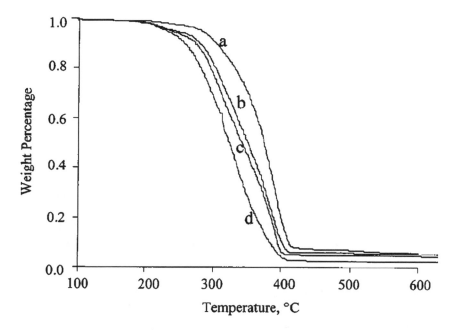

Figure 1.17 Structures of the organic modifiers **1–3** used to ion exchange on the surface of montmorillonite platelets. Reproduced from reference 82 with permission from the American Chemical Society.

Figure 1.18 TGA thermograms of (a) nanocomposite with 5 wt% montmorillonite–**2**, (b) nanocomposite with 5 wt% montmorillonite–**3**, (c) PMMA macrocomposite and (d) nanocomposite with 5 wt% montmorillonite–**1**. Reproduced from reference 82 with permission from the American Chemical Society.

dicocoalkyldimethylammonium chloride and di(hydrogenated tallow-alkyl)dimethylammonium chloride. The montmorillonite modified with di(hydrogenated tallow-alkyl)dimethylammonium chloride showed the maximum increase in basal plane spacing. The properties of the composites were enhanced as compared with the pure polymer. Essawy *et al.*[81] also reported similarly the emulsion and suspension polymerization of methyl methacrylate to generate poly(methyl methacrylate)–clay nanocomposites. Cetyltrimethylammonium bromide (CTAB) and cetylpyridinium chloride (CPC) were used as surface modifiers for the clay platelets. The exchange of CPC molecules in the inter-layer expanded the basal plane spacing from 12.19 to 21.47 Å. and in the case of CTAB modification, the basal plane spacing expanded to 19.35 Å. Huang and Brittain[82] further reported the synthesis of poly(methyl methacrylate) nano-composites by emulsion and suspension polymerization by using mont-morillonites modified with the modifications shown in Figure 1.17. The generated composites generally had much better properties than the pure polymer. Figure 1.18 shows an example of the TGA thermograms of the poly(methyl methacrylate) and poly(methyl methacrylate)–clay nanocompo-sites synthesized by suspension polymerization.

Acknowledgment

This review work was carried out at Institute of Chemical and Bio-engineering, Department of Chemistry and Applied Biosciences, ETH Zurich, Switzerland.

References

1. H. S. Katz and J. V. Milewski, *Handbook of Fillers for Plastics*, Van Nostrand, New York, 1987.
2. R. Rothon, *Particulate-filled Polymer Composites*, Longman, Harlow, 1995.
3. Y. Brechet, J. Y. Cavaille, E. Chabert, L. Chazeau, R. Dendievel, L. Flandin and C. Gauthier, *Adv. Eng. Mater.*, 2001, **3**, 571.
4. T. Lan, P. D. Kaviratna and T. J. Pinnavaia, *Chem. Mater.*, 1994, **6**, 573.
5. I.-J. Chin, T. Thurn-Albrecht, H.-C. Kim, T. P. Russell and J. Wang, *Polymer*, 2001, **42**, 5947.
6. S. K. Lim, J. W. Kim, I.-J. Chin, Y. K. Kwon and H. J. Choi, *Chem. Mater.*, 2002, **14**, 1989.
7. Z. Wang and T. J. Pinnavaia, *Chem. Mater.*, 1998, **10**, 3769.
8. P. B. Messersmith and E. P. Giannelis, *J. Polym. Sci., Part A: Polym. Chem.*, 1995, **33**, 1047.
9. K. Yano, A. Usuki and A. Okada, *J. Polym. Sci., Part A: Polym. Chem.*, 1997, **35**, 2289.
10. H. Shi, T. Lan and T. J. Pinnavaia, *Chem. Mater.*, 1996, **8**, 1584.
11. E. P. Giannelis, *Adv. Mater.*, 1996, **8**, 29.
12. P. C. LeBaron, Z. Wang and T. J. Pinnavaia, *Appl. Clay Sci.*, 1999, **15**, 11.
13. J. E. Mark, *Polym. Eng. Sci.*, 1996, **36**, 2905.
14. E. Reynaud, C. Gauthier and J. Perez, *Rev. Metall./Cah. Inf. Technol.*, 1999, **96**, 169.
15. P. Calvert, in *Carbon Nanotubes*, ed. T. W. Ebbesen, CRC Press, Boca Raton, FL, 1997, pp. 277–292.

16. V. Favier, G. R. Canova, S. C. Shrivastava and J. Y. Cavaille, *Polym. Eng. Sci.*, 1997, **37**, 1732.
17. V. Mittal, Organic Modifications of Clay and Polymer Surfaces for Specialty Applications, PhD Thesis, ETH Zurich, 2006.
18. L. W. Carter, J. G. Hendricks and D. S. Bolley, *US Pat.*, 2 531 396, 1950.
19. D. J. Greenland, *J. Colloid Sci.*, 1963, **18**, 647.
20. K. Yano, A. Usuki, A. Okada, T. Kurauchi and O. Kamigaito, *J. Polym. Sci., Part A: Polym. Chem.*, 1993, **31**, 2493.
21. Y. Kojima, K. Fukumori, A. Usuki, A. Okada and T. Kurauchi, *J. Mater. Sci. Lett.*, 1993, **12**, 889.
22. R. A. Vaia, H. Ishii and E. P. Giannelis, *Chem. Mater.*, 1993, **5**, 1694.
23. V. Mehrotra and E. P. Giannelis, *Mater. Res. Soc. Symp. Proc.*, 1990, **171**, 39.
24. S. W. Bailey, *Reviews in Mineralogy*, Virginia Polytechnic Institute and State University, Blacksburg, VA, 1984.
25. G. W. Brindley and G. Brown, *Crystal Structures of Clay Minerals and Their X-ray Identification*, Mineralogical Society, London, 1980.
26. S. Pavlidoua and C. D. Papaspyrides, *Prog. Polym. Sci.*, 2008, **33**, 1119.
27. G. Beyer, *Plast. Addit. Compd.*, 2002, **4**, 22.
28. B. K. G. Theng, *The Chemistry of Clay–Organic Reactions*, Wiley, New York, 1974.
29. K. Jasmund and G. Lagaly, *Tonminerale und Tone Struktur*, Steinkopff, Darmstadt, 1993.
30. T. J. Pinnavaia, *Science*, 1983, **220**, 365.
31. R. Krishnamoorti and E. P. Giannelis, *Macromolecules*, 1997, **30**, 4097.
32. H. Heinz, H. J. Castelijns and U. W. Suter, *J. Am. Chem. Soc.*, 2003, **125**, 9500.
33. M. Alexandre and P. Dubois, *Mater. Sci. Eng. R: Rep.*, 2000, **28**, 1.
34. O. C. Wilson Jr, T. Olorunyolemi, A. Jaworski, L. Borum, D. Young, A. Siriwat, E. Dickens, E. Oriakhi and M. Lerner, *Appl. Clay Sci.*, 1999, **15**, 265.
35. C. O. Oriakhi, I. V. Farr and M. M. Lerner, *Clays Clay Miner.*, 1997, **45**, 194.
36. A. Okada and A. Usuki, *Mater. Sci. Eng.*, 1995, **C3**, 109.
37. P. Dubois, M. Alexandre, F. Hindryckx and R. Jerome, *J. Macromol. Sci. Rev. Macromol. Chem. Phys.*, 1998, **C38**, 511.
38. T. McNally, W. R. Murphy, C. Y. Lew, R. J. Turner and G. P. Brennan, *Polymer*, 2003, **44**, 2761.
39. C. H. Davis, L. J. Mathias, J. W. Gilman, D. A. Schiraldi, J. R. Shields, P. Trulove, T. E. Sutto and H. C. Delong, *J. Polym. Sci., Part B: Polym. Phys.*, 2002, **40**, 2661.
40. S. K. Srivastava, M. Pramanik and H. Acharya, *J. Polym. Sci., Part B: Polym. Phys.*, 2006, **44**, 471.
41. S.-Y. A. Shin, L. C. Simon, J. B. P. Soares and G. Scholz, *Polymer*, 2003, **44**, 5317.
42. T. G. Gopakumar, J. A. Lee, M. Kontopoulou and J. S. Parent, *Polymer*, 2002, **43**, 5483.

43. M. Kato, A. Usuki and A. Okada, *J. Appl. Polym. Sci.*, 1997, **66**, 1781.
44. M. Kawasumi, N. Hasegawa, M. Kato, A. Usuki and A. Okada, *Macromolecules*, 1997, **30**, 6333.
45. N. Hasegawa, M. Kawasumi, M. Kato, A. Usuki and A. Okada, *J. Appl. Polym. Sci.*, 1998, **67**, 87.
46. E. Manias, A. Touny, L. Wu, K. Strawhecker, B. Lu and T. C. Chung, *Chem. Mater.*, 2001, **13**, 3516.
47. Q. Zhang, Q. Fu, L. Jiang and Y. Lei, *Polym. Int.*, 2000, **49**, 1561.
48. A. Oya, Y. Kurokawa and H. Yasuda, *J. Mater. Sci.*, 2000, **35**, 1045.
49. W. Xu, G. Liang, W. Wang, S. Tang, P. He and W. P. Pan, *J. Appl. Polym. Sci.*, 2003, **88**, 3225.
50. H. R. Fischer, L. H. Gielgens and T. P. M. Koster, *Acta Polym.*, 1999, **50**, 122.
51. B. K. G. Theng, *Formation and Properties of Clay–Polymer Complexes*, Elsevier, Amsterdam, 1979.
52. E. Ruiz-Hitzky, P. Aranda, B. Casal and J. C. Galvan, *Adv. Mater.*, 1995, **7**, 180.
53. N. Ogata, S. Kawakage and T. Ogihara, *J. Appl. Polym. Sci.*, 1997, **66**, 573.
54. J. Billingham, C. Breen and J. Yarwood, *Vibr. Spectrosc.*, 1997, **14**, 19.
55. R. Levy and C. W. Francis, *J. Colloid Interface Sci.*, 1975, **50**, 442.
56. H. G. Jeon, H.-T. Jung, S. W. Lee and S. D. Hudson, *Polym. Bull.*, 1998, **41**, 107.
57. D. C. Lee and L. W. Jang, *J. Appl. Polym. Sci.*, 1996, **61**, 1117.
58. D. C. Lee and L. W. Jang, *J. Appl. Polym. Sci.*, 1998, **68**, 1997.
59. R. Ianchis, D. Donescu, C. Petcu, M. Ghiurea, D. F. Anghel, G. Stanga and A. Marcu, *Appl. Clay Sci.*, 2009, **45**, 164.
60. N. Clarke, L. R. Hutchings, I. Robinson, J. A. Elder and S. A. Collins, *J. Appl. Polym. Sci.*, 2009, **113**, 1307.
61. G.-L. Yuan, W.-M. Li, S. Yin, F. Zou, K.-C. Long and Z.-F. Yang, *J. Appl. Polym. Sci.*, 2009, **114**, 1964.
62. S. Khatana, A. K. Dhibar, S. Sinha Ray and B. B Khatua, *Macromol. Chem. Phys.*, 2009, **210**, 1104.
63. Z. Sedlakova, J. Plestil, J. Baldrian, M. Slouf and P. Holub, *Polym. Bull.*, 2009, **63**, 365.
64. G. Diaconu, M. Micusik, A. Bonnefond, M. Paulis and J. R. Leiza, *Macromolecules*, 2009, **42**, 3316.
65. J. Kajtna and U. Sebenik, *Int. J. Adhes. Adhes.*, 2009, **29**, 543.
66. N. Greesh, P. C. Hartmann and R. D. Sanderson, *Macromol. Mater. Eng.*, 2009, **294**, 206.
67. F. Effenberger, M. Schweizer and W. S. Mohamed, *J. Appl. Polym. Sci.*, 2009, **112**, 1572.
68. R. Ruggerone, C. J.G. Plummer, N. N. Herrera, E. Bourgeat-Lami and J.-A. E. Manson, *Eur. Polym. J.*, 2009, **45**, 621.
69. K.-J. Lin, C.-A. Dai and K.-F. Lin, *J. Polym. Sci., Part A: Polym. Chem.*, 2009, **47**, 459.

70. A. Samakande, R. D. Sanderson and P. C. Hartmann, *J. Polym. Sci., Part A: Polym. Chem.*, 2008, **46**, 7114.

71. S. I. Ali, J. P. A. Heuts, B. S. Hawkett and A. M. van Herk, *Langmuir*, 2009, **25**, 10523.

72. N. Greesh, P. C. Hartmann, V. Cloete and R. D. Sanderson, *J. Polym. Sci., Part A: Polym. Chem.*, 2008, **46**, 3619.

73. M. A. Osman, V. Mittal, M. Morbidelli and U. W. Suter, *Macromolecules*, 2004, **37**, 7250.

74. G. Diaconu, M. Paulis and J. R. Leiza, *Polymer*, 2008, **49**, 2444.

75. F. Bouanani, D. Bendedouch, P. Hemery and B. Bounaceur, *Colloids Surf. A*, 2008, **317**, 751.

76. A. Schmid, P. Scherl, S. P. Armes, C. A. P. Leite and F. Galembeck, *Macromolecules*, 2009, **42**, 3721.

77. L. Duan, M. Chen, S. Zhou and L. Wu, *Langmuir*, 2009, **25**, 3467.

78. D.-R. Yei, S.-W. Kuo, H.-K. Fu and F.-C. Chang, *Polymer*, 2005, **46**, 741.

79. H. Li, Y. Yu and Y. Yang, *Eur. Polym. J.*, 2005, **41**, 2016.

80. W.-T. Yang, T.-H. Ko, S.-C. Wang, P.-I. Shih, M.-J. Chang and G. J. Jiang, *Polym. Comp.*, 2008, **29**, 409.

81. H. Essawy, A. Badran, A. Youssef and A. E.–F. A. E. Hakim, *Polym. Bull.*, 2004, **53**, 9.

82. X. Huang and W. J. Brittain, *Macromolecules*, 2001, **34**, 3255.

CHAPTER 2

Polymer–Layered Double Hydroxide Nanocomposites by Emulsion and Suspension Polymerization

LONGZHEN QIU[a] AND BAOJUN QU[b]

[a] Key Laboratory of Special Display Technology, Ministry of Education, Academe of Opto-Electronic Technology, Hefei University of Technology, Hefei, Anhui Province, China; [b] Department of Polymer Science and Engineering, University of Science and Technology of China, Hefei, Anhui Province, China

2.1 Introduction

An attractive feature of polymer processing is the ease of combining several materials with different characteristics into polymer composites to achieve properties that are often not accessible with the individual components. A typical polymer composite is a combination of polymers and inorganic fillers, in which the fillers such as talc and mica are used to improve the stiffness and toughness of the materials, to enhance their dimensional stability, to enhance their resistance to fire or simply to reduce costs.[1,2] However, a loading of 30–50 wt% of micron-sized filler is conventionally needed for a significant improvement in properties. This high loading of fillers may reduce the processability and mechanical properties of the polymer compounds and impart drawbacks, such as brittleness or opacity, to the resulting composites and thus limit their applications. The emergence of polymer nanocomposites (PNCs) offers the possibility to overcome these drawbacks.

RSC Nanoscience & Nanotechnology No. 16
Polymer Nanocomposites by Emulsion and Suspension Polymerization
Edited by Vikas Mittal
© Royal Society of Chemistry 2011
Published by the Royal Society of Chemistry, www.rsc.org

PNCs are a relatively new class of hybrid materials composed of an organic polymer matrix with dispersed inorganic fillers, which have at least one dimension in the nanometer range.[3] In contrast to conventional micro-composites, where the combined properties are varied as a function of the filler content, the properties of nanocomposites are not directly predictable from the relative weight proportions. At the nanoscale, a large fraction of the filler atoms can reside at the interface and lead to a strong interfacial interaction with the polymer matrix if the nanofillers are well dispersed. As a result, PNCs often exhibits dramatic improvements in performance at very low filler loadings of about 0.5–5 wt% and even show certain new properties that cannot be derived from macrocomposites or counterparts. Therefore, PNCs have attracted extensive interest from both the industrial and academic areas for the past 20 years.[4]

A variety of nanofillers have been applied to fabricate PNCs. Generally, according to the geometric shape, they can be divided into three categories: zero-dimensional (0D) nanoparticles (such as silica and titania), one-dimensional (1D) nanotubes or nanorods (such as carbon nanotubes) and two-dimensional (2D) nanolayers (such as clay). Among various PNCs, those based on 2D nanolayers have been the most extensively investigated,[5,6] because layered nanofillers have much larger aspect ratios and can enhance stiffness and scratch resistance significantly at a much lower loading. They are also impervious to gases and can yield greatly improved barrier properties and also flame retardancy when they are well dispersed and oriented in the composite. For example, in the pioneering work, Usuki *et al.*[7] originally developed nylon 6–layered silicate nanocomposite by a 'monomer intercalation' method, in which the monomer (ε-caprolactam) was first intercalated into the layers of the organic-modified layered silicate and then polymerized by ring-opening polymerization. They found that the modulus of this PNC increased to 1.5 times that of neat nylon 6, the heat distortion temperature increased to 140 °C from 65 °C and the gas barrier effect was doubled at a low loading (2 wt%) of clay.

Layered double hydroxide (LDH) is another interesting layered crystalline filler for nanocomposites formation. Similarly to the well-known layered silicate, LDHs also consist of ultra-thin layers and interchangeable interlayer ions. However, instead of silicate layers, the LDH layers comprise metal hydroxides, and the interlayer ions in LDHs are anions rather than cations. Therefore, LDHs are considered as the mirror image of cationic clays, *i.e.* anionic clays. Because of their highly tunable properties and unique anion-exchange properties, LDHs are considered to be a new emerging class of the most favorable layered crystals for the preparation of multifunctional polymer–layered filler nanocomposites. In this chapter, the major developments in polymer–LDH nanocomposites are highlighted. First, a snapshot is given on the structure, synthesis and organo-modification of LDHs. Then, the different synthesis routes for polymer–LDH nanocomposites are reviewed. Especially emulsion and suspension polymerization methods are discussed in more detail. Finally, the possible applications of polymer–LDH nanocomposites are discussed.

2.2 Layered Double Hydroxides

2.2.1 Structure of LDHs

LDHs, also known as hydrotalcite-like materials, are a class of host–guest materials consisting of positively charged metal oxide or hydroxide sheets with intercalated anions and water molecules. Generally, their chemical structure is represented by the formula $[M_{1-x}^{2+}M_x^{3+}(OH)_2]^{x+}[A^{m-}]_{x/m}^{x-} \cdot 2H_2O$, where M^{2+} is a divalent metal ion (Mg^{2+}, Zn^{2+}, etc.), M^{3+} is a trivalent metal ion (Al^{3+}, Cr^{3+}, etc.), A^{m-} is an anion with valency m (CO_3^{2-}, Cl^-, NO_3^-, etc.) and the value of x is equal to the molar ratio of M^{2+} to $(M^{2+} + M^{3+})$ and is generally in the range 0.2–0.33.[8]

Since the first discovery of their natural occurrence (*i.e.* as the mineral hydrotalcite) in Sweden in 1842, LDHs have been known for over 150 years.[9] Although the stoichiometry of hydrotalcite, $[Mg_6Al_2(OH)_{16}]CO_3ñ4H_2O$, was first correctly determined by Manasse in 1915, it was not until pioneering single-crystal X-ray diffraction (XRD) studies on mineral samples were carried out by Allmann and Taylor in the 1960s that the main structural features of LDHs were understood.[8]

The typical structure of LDH materials is presented in Figure 2.1. It can be seen that the structure of LDH is based on that of magnesium hydroxide (brucite), in which Mg^{2+} ions are arranged in sheets, each magnesium ion being octahedrally surrounded by six hydroxide groups while each hydroxide spans three magnesium ions. Isomorphous substitution of some fraction of the divalent ions by a trivalent ion of comparable size (*e.g.* Al^{3+}, Fe^{3+}) forms mixed metal layers, *i.e.* $[M_{1-x}^{2+}M_x^{3+}(OH)_2]^{x+}$, with a net positive charge. The electrical neutrality is maintained by the anions located in the interlayer domains containing water molecules, wherein these water molecules are connected to both the metal hydroxide layers and the interlayer anions through extensive hydrogen bondings. The presence of anions and water molecules leads to an enlargement of the basal spacing from 0.48 nm in brucite to about 0.78 nm in Mg–Al LDH. Because the interlayer ions are confined in the interlayer space by relatively weak electrostatic forces, they can be removed

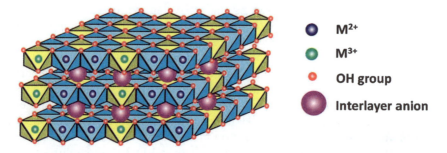

- M^{2+}
- M^{3+}
- OH group
- Interlayer anion

Figure 2.1 Layered crystal structure of layered double hydroxide.

without destroying the layered structures of LDHs. This anion-exchange capacity is a key feature of LDHs, which makes them unique as far as inorganic materials are concerned.

2.2.2 Properties of LDHs

Due to their unique layered structure and highly tunable chemical composition based on different metal species and interlayer anions, LDHs have many interesting properties:

- *Unique anion-exchanging ability:* Ion-exchangeable layered compounds can be classified into two groups depending on the reactivities that they show: cation-exchange or anion-exchange properties. The compounds belonging to the former group include smectite clay minerals, transition metal dichalcogenides and oxides and metal phosphates. However, the compounds in the latter group are relatively rare. LDHs are studied most extensively among them.

- *Good anion-exchange capacities:* The anion-exchange capacity (AEC) of LDHs may vary in a wide range depending on the divalent to trivalent metal ratio present in the hydroxide layer. It ranges between 450 and 200 mequiv. per 100 mg for LDH, whereas smectite-type materials usually present a limited exchange capacity close to 100 mequiv. per 100 mg. In the latter case, it is associated with an average area per charge of 70 $\mathring{A}^2$ per charge, whereas it ranges between 25 and 40 $\mathring{A}^2$ per charge for LDH-type materials.[9]

- *Easy synthesis:* Layered silicates are widespread in Nature, but generally difficult to synthesize, whereas LDHs are mostly synthetic and easily prepared. They can be synthesized with a wide range of M^{3+}–M^{2+} cation combinations and with a wide variety of anions in the interlayer.

- *High bound water content:* The LDHs consist of brucite-like layers. Therefore, like brucite, LDHs undergo endothermic decomposition, thereby releasing the bound water and producing a metal oxide residue. This shows their potential as flame retardants.

- *'Memory effect':* The partial dehydroxylation of LDHs under moderate calcination leads to amorphous mixed oxides usually named layered double oxides (LDOs), which can be regenerated back to the original structure by contacting with solutions containing various anions.

- Non-toxic and biocompatible.

Based on these properties, LDHs are considered as very important layered crystals with potential applications in catalysis,[10,11] controlled drugs release,[12–14] gene therapy,[15–17] improvement of heat stability and flame retardancy of polymer composites,[18,19] controlled release or adsorption of pesticides,[20] preparation of novel hybrid materials for specific applications, such as visible luminescence,[21,22] UV/photo-stabilization,[23,24] and magnetic nanoparticle synthesis[25] and wastewater treatment.[26,27]

2.2.3 Synthesis of LDHs

LDHs can be synthesized by various techniques, such as coprecipiation,[28,29] structure reconstruction,[30,31] sol–gel method,[32] anion-exchange method,[33] and so on.[34,35] Detailed descriptions of the processes are available in the literature and reviews.[36] However, this does not mean that it is easy to prepare pure LDHs; rather, it means that different techniques may be adopted to obtain the desired LDH compositions. Among these different preparation methods, coprecipitation is the most common and useful method to prepared large amounts of LDHs.

2.2.4 Organic Modification of LDHs

LDHs have high charge density, which is generally 3–4-fold higher than that of layered silicates. The high charge density and the high contents of interlayer anions and water molecules result in strong hydrophilic properties and thus prevent swelling and exfoliation of the LDH sheets. Therefore, in order to facilitate the intercalation of hydrophobic polymer chains into the gallery of LDHs and to reach a good level of filler dispersion, the organic modification of LDHs is an inevitable step in the process of polymer nanocomposite preparation. Since the LDH layers are positively charged, LDHs can be modified by a wide variety of anionic surfactants, such as fatty acid salts, sulfonates and phosphates. Alkyl sulfonates have been found to be one of the most efficient classes of anionic surfactants for LDH modification. Several methods can be applied to prepare organo-modified LDHs.

Like layered silicate, organo-modified LDHs can be obtained *via* an *ion-exchange method*, in which the interlayer hydrated anions are interchanged with organic anions. Generally, ion-exchange processes should be conducted in a CO_3^{2-}-free environment because the CO_3^{2-} anions have a strong interaction with LDH layers and cannot be replaced by surfactant anions. Concerning the easy synthesis of LDHs, organo-modification can also be conducted during the synthesis process. When mixed metal salts are simultaneously precipitated to form LDHs using a dilute base solution in an aqueous solution containing the desired anionic surfactant, the organic surfactant is incorporated into the interlayer of the LDH. This is termed the *coprecipitation method*. Another promising means for the preparation of organo-modified LDHs is the *reconstruction method* based on the 'memory effect' of LDHs, in which the incorporation of surfactants is achieved by regenerating LDOs in the presence of organic anionic surfactants.

2.3 Polymer–LDH Nanocomposites

The incorporation of polymers into LDHs to form polymer–LDH nanocomposites has been of academic interest for over 20 years. In pioneering work, Sugahar *et al.*[37] prepared polyacrylonitrile (PAN)–MgAl-LDH *via* an *in situ* polymerization, in which the acrylonitrile monomers were intercalated into

MgAl-LDH modified with dodecyl sulfate and subsequently polymerized. Since then, various polymers have been applied to prepare nanohybrids with LDHs and various methods have been developed to achieve this end.

Depending on the ratio of LDH to polymer in the hybrids and the nature of the polymer components, two classes of nanocomposites may be obtained. The first class concerns LDHs as hosts and polymers as gests, in which a single (and sometimes more than one) extended polymer chain is intercalated between the LDH layers, resulting in a well-ordered multilayer morphology built up with alternating polymeric and inorganic layers. These materials are generally termed organoceramics. A large variety of organoceramics have been prepared by intercalating LDHs with different polymer species including poly(styrene sulfonate) (PSS),[38–45] poly(acrylic acid) (PA),[46–50] poly(vinyl sulfonate) (PVS),[42] poly(vinyl alcohol) (PVA),[51,52] poly(ethylene glycol) (PEG),[53,54] poly(α,β-aspartate),[55] poly(sulfopropyl methacrylate),[56] polyaniline[35,57–59] and biopolymers.[60] All the polymers used in organoceramics are water soluble or derived from water-soluble monomers. Therefore, they have good affinity with hydrophilic LDH layers, which is really advantageous for the intercalation of cumbersome polymer molecules into the constrained interlamellar space of LDH. Furthermore, the easy dispersion of these polymers and monomers in water makes the modification methods of LDHs also suitable for the incorporation of polymers and monomers into LDHs, which makes the synthesis methods for organoceramics relatively abundant (as shown in Figure 2.2). Several reviews have covered the preparation and properties of organoceramics based on LDHs.[38,61,62] Here, we focus our discussions on the second class of hybrid materials.

The second class of hybrid materials is polymeric nanocomposites, usually shortened to nanocomposites, in which the LDH layers are dispersed in a

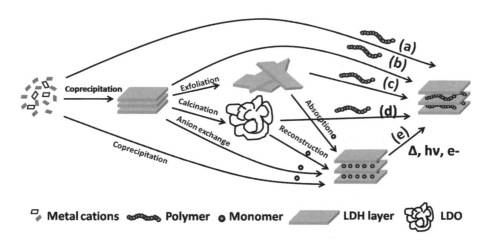

Metal cations   **Polymer**   **Monomer**   **LDH layer**   **LDO**

Figure 2.2   Scheme of the preparation of polymer–LDH organoceramics: (a) coprecipitation; (b) anion exchange; (c) exfoliation–absorption; (d) reconstruction; (e) *in situ* polymerization.

continuous thermoset or thermoplastic polymer matrix. In contrast to the polymers used in organoceramics, the thermoset or thermoplastic polymers are hydrophobic and cannot be dissolved in water. To facilitate the penetration of hydrophobic polymer chains into the interlayer space of LDH, anionic surfactants such as dodecyl sulfate or dodecyl benzene sulfonate are pre-intercalated into the interlayer of LDHs to enlarge the interlayer space and to give the LDH sheet an organophilic property. This can be done by the method described in Section 2.2.3.

2.3.1 Structure of Polymer–LDH Nanocomposites

In general, according to the different dispersion states of the LDH layers in the polymer matrix, two main types of polymer–LDH nanocomposites, *intercalated* and/or *exfoliated nanocomposites*, can be obtained. In an intercalated nanocomposite, the dispersion of the LDH in the polymer matrices is similar to those in traditional microcomposites; however, the interlayer of LDH particles is intercalated by a few molecular layers of polymer. In an exfoliated nanocomposite, the LDH particles are completely exfoliated and the individual LDH layers are uniformly dispersed in a continuous polymer matrix. Usually, the exfoliated nanocomposites attract more attention because they have nanoscale dispersion of high aspect ratio LDH layers in polymer nanocomposites and thus can give some improved properties compared with microdispersed and conventional composites.

X-ray diffraction (XRD) and transmission electron microscopy (TEM) analyses are generally applied for the characterization of polymer–LDH nanocomposites. XRD analysis can provide useful information on the types of the layered structure, *i.e.* intercalated and/or exfoliated structures, because the peaks change with the gallery distance of the LDHs. However, the XRD results cannot provide a complete picture of the state of the nanocomposites. Moreover, for the exfoliated nanocomposites, LDH layers are often randomly dispersed in the polymer matrices and the basal spaces are generally larger than 10 nm, which is beyond the limit that the XRD can analyze. Therefore, to obtain complete information on the materials, actual images, such as TEM images, are required.

2.3.2 Conventional Strategies for Preparing Polymer–LDH Nanocomposites

To prepare polymer–LDH nanocomposites, several strategies have been employed, which are presented in Figure 2.3.

2.3.2.1 Direct Intercalation

The thermoplastic polymers can be introduced into organically modified LDHs as a whole *via* solution intercalation or melt intercalation.

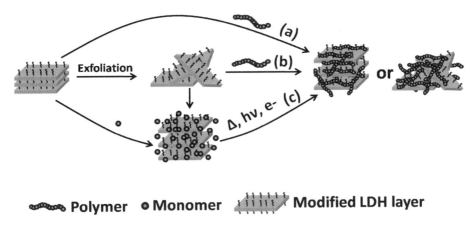

Figure 2.3 Scheme of the preparation of polymer–LDH nanocomposites: (a) direct intercalation; (b) exfoliation–absorption; (e) *in situ* polymerization.

In the solution intercalation process, organically modified LDHs are first dispersed in a solution containing the desired polymer, followed by aging with stirring under a nitrogen atmosphere to accomplish polymer intercalation. This method was first used to disperse ZnAl-LDH into polyethylene (PE).[63–66] By refluxing the mixture of dodecyl sulfate-modified LDH and a xylene solution of PE, exfoliated PE–LDH nanocomposites were obtained. Subsequently, Qiu *et al.*[67] studied the influence of the preparation conditions on the structure of polystyrene (PS)–ZnAl-LDH composites by the same method. It was found that the exfoliation of LDH in PS is improved by decreasing the LDH content and extending the refluxing time. More interestingly, the preparation method has a decisive effect on the structure of the nanocomposite: slow evaporation of the solvent only leads to an intercalated structure, whereas rapid precipitation favors the exfoliated structure. This method has been applied to prepare other nanocomposites based on different polymers such as poly(propylene carbonate),[68] poly(3-hydroxybutyrate),[69–71] poly(vinyl chloride),[72] syndiotactic polystyrene,[73] polyurethane[74] and poly[(3-hydroxybutyrate)-*co*-(3-hydroxy-valerate)].[75]

When a mixture of the polymer and layered filler is annealing, statically or under shear, above the softening point of the polymer, the polymer may crawl into the interlayer space and form either an intercalated or an exfoliated nanocomposite; this is called *melt intercalation*. This method has been widely used to prepare polymer–layered filler nanocomposites because it has many advantages, such as no solvent needed, environmentally friendly and compatible with current industrial processes. Although the melt intercalation method for preparing polymer–LDH nanocomposites was first reported by Nichols and Chou[76] in 1999, it did not attract more interest until the last 5 years. Costa and co-workers[77–81] developed a two-step melt compounding process to prepare low-density polyethylene (LDPE)–MgAl-LDH nanocomposites, where a master bath was prepared by mixing dodecyl benzene sulfonate-modified LDH with maleic anhydride grafted polyethylene followed by dilution in LDPE resin. The XRD and TEM results revealed that only part of the LDH was exfoliated in the polymer matrix. In another study, Costantino *et al.*[82] reported that PE–ZnAl nanocomposite can be obtained when stearate anions are used as modifier. The presence of LDH layers shields PE from thermal oxidation, shifting the temperature range of volatilization towards that of thermal degradation in nitrogen, and leading to a reduction of 55% in the heat release rate during combustion. Du and Qu[83,84] found that the amount of surfactant pre-intercalated into the interlayer of LDH can also affect the structure of nanocomposites prepared by melt intercalation. When the pre-intercalated surfactant, dodecyl sulfate, was about half of the interlay nitrate anions in the LDH, exfoliated linear LDPE–MgAl-LDH nanocomposite was obtained. The same group also prepared nanocomposites based on different polyolefins such as high-density polyethylene (HDPE),[85] polypropylene (PP),[86–88] nylon 6[19] and ethylene vinyl acetate[89–91] by melt intercalation. The photo-oxidative behavior, mechanical properties, thermal properties and flammability of these nanocomposites have been systematically characterized. Recently, Wilkie's group[92–99] studied the compatibility of organically modified MgAl-LDH with PE, PP, PS and poly(methyl methacrylate) (PMMA) by melt blending of the polymers with the LDH. They found that the dispersion of LDH in polymer matrix was dependent on the polarity of the polymer used. A nanocomposite based on the polar polymer PMMA shows the best dispersion of LDH. Other reported polar polymer–LDH nanocomposites prepared by melt intercalation method include nylon,[100] poly(L-lactide),[101] poly(ethylene terephthalate),[102] polycaprolactone[103] and poly(*p*-dioxanone).[104]

2.3.2.2   *Exfoliation–Absorption*

In the exfoliation–absorption process, the host lattice is first exfoliated into single layers using a suitable solvent in which the polymer is soluble. When the solvent is evaporated (or the mixture precipitated), the reassembly of the one-dimensional sheets is prevented by the polymer matrix and thus forms nano-composites. Therefore, the first key factor of the exfoliation–absorption process is to obtain the exfoliated LDH layers in a certain solution system before the polymers or monomers are introduced. However, it is very difficult for the LDHs to delaminate into single sheets due to strong electrostatic interaction between LDH layers and guest anions, which result from the high charge density of LDH layers associated with a high content of intercalated hydrated anions, compared with the smectite clays and some other layered inorganic solids. Investigation on LDH exfoliation began recently and only a few studies have been successful so far.[29,105–111] The modification of LDHs with organic anions was found to be effective for the delamination of the LDHs in some non-aqueous solvents. Adachi-Pagano and co-workers[29,105] first reported that ZnAl-LDH containing dodecyl sulfate can be delaminated after refluxing in butanol at 120 °C. The subsequent reassembly of the exfoliated layers with PSS

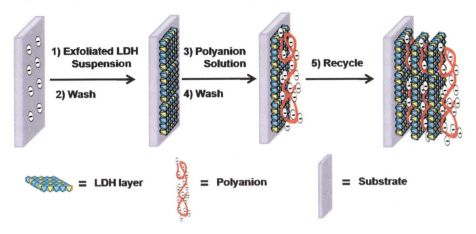

Figure 2.4 Schematic illustration of layer-by-layer self-assembly process for anionic polymer–LDH nanocomposites.

resulted in an organoceramic.[38] Subsequently, Hibino and co-workers[106–108] reported that LDHs modified with amino acids were easily exfoliated in a simple solvent, formamide, at room temperature. Using this method, Hu and co-workers prepared intercalated PVA–LDH nanocomposites[112] and exfoliated PMMA–LDH nanocomposites.[113]

More recently, Sasaki's group[110,111] developed a new approach for the direct delamination of pristine LDH into well-defined nanosheets using formamide. These results will open up new application fields for the LDHs. Using exfoliated LDH nanosheets as building blocks, the layer-by-layer self-assembly of the LDH nanosheets and anionic polymers to produce multi-layered nanocomposite films was demonstrated (Figure 2.4). In this way, homogeneous LDH films such as MgAl-LDH–PSS,[110] CoAl-LDH–PSS[111] and MgAl-LDH–poly(*p*-phenylene)[114] and heterogeneous LDH films[115] such as (CoAl-LDH–PSS–MgAl-LDH–PSS)$_{n/2}$, (MgAl-LDH–PSS/NiAl-LDH–PSS)$_{n/2}$ and (CoAl-LDH–PSS–NiAl-LDH–PSS)$_{n/2}$, where *n* refers to the number of LDH–PSS bilayers, were fabricated.

2.3.2.3 In Situ Polymerization

In this method, the organo-modified LDH is swollen within the liquid monomer solution. When polymerization is initiated, the intercalated monomer can convert to polymer intercalated to the interlayer of LDH and thus forms a nano-composite. Normally, only intercalated nanocomposites were obtained *via* this method if the organo-modified LDH was simply dispersed in acrylate and styrene monomers.[116–119] O'Leary *et al.*[109] reported the first exfoliated polymer–LDH nanocomposite prepared by the *in situ* polymerization method. In their work, MgAl-LDH intercalated with dodecyl sulfate (DS) also underwent exfoliation with high shearing in the acrylate monomer of 2-hydroxyethyl methacrylate (HEMA) at 70 °C, which was then polymerized to polyacrylate with the LDH still in the delaminated form. Qu and co-workers[118,120] reported that *in situ* polymerization initiated by the anionic radical initiator pre-intercalated in the interlayer of LDH is an effective method to prepare highly exfoliated polymer–LDH nanocomposites. In another study, Wang *et al.*[121] prepared exfoliated PMMA–MgAl-LDH nanocomposite by a two-stage bulk polymerization of methyl methacrylate in the presence of 10-undecenoate-intercalated LDH. The flame retardancy of this material was also characterized.[122,123] A similar approach was applied to prepare poly(styrene-*co*-methyl methacrylate)–LDH nanocomposites and poly(ethylene terephthalate).[124]

Moreover, *in situ* polymerization is the only way to prepare thermoset polymer nanocomposites because a thermoset polymer can be neither melted nor dissolved. The first thermoset polymer–LDH nanocomposite was reported by Hsueh and Chen,[125] which was prepared from a mixture of aminobenzoate-modified MgAl-LDH and polyamic acid (polyimide precursor) in *N,N*-dimethylacetamide. In a similar way, epoxy–LDH nanocomposites were also obtained.[126–128]

From the above discussion, it can be seen that the method adopted to prepare nanocomposites is highly dependent on the nature of the polymer. When the polymers or monomers are water soluble, they can be incorporated into the pristine LDH without any organo-modification due to their good affinity with the LDH. Additionally, the aqueous environment is compatible with the condition for the synthesis of LDH materials. Therefore, water-soluble polymer–LDH nanocomposites can be prepared using some special methods such as *in situ* synthesis, ion exchange and reconstruction. In the case of water-insoluble polymers and monomers, their nanocomposites are usually prepared in organosolvent (solution intercalation method, exfoliation–absorption method and *in situ* polymerization method) or molten polymer (melt intercalation method). However, emulsion polymerization and suspension polymerization are methods that allow the incorporation of a water insoluble polymer into an LDH in water. The following sections are devoted to polymer–LDH nanocomposites obtained *via* emulsion polymerization and suspension polymerization.

2.4 Polymer–LDH Nanocomposites Prepared by Emulsion and Suspension Polymerization

2.4.1 Emulsion and Suspension Polymerization

Both emulsion and suspension polymerization are heterogeneous polymerization techniques, in which the polymerization occurs in the monomer-swollen latexes and monomer droplets dispersed in a continuous liquid phase, usually water, respectively. The emulsion polymerization technique involves a water-soluble initiator, a water-insoluble monomer and a micelle-forming surfactant. The monomer is emulsified and stabilized in a continuous aqueous phase by the surfactant. Excess surfactant creates micelles, which are also swollen by a small amount of monomers, in the water. When the water-soluble initiators are heated or irradiated, free radicals are introduced into the water phase where they react with monomer in the micelles. Polymerization then takes place in the monomer-swollen micelles, which are converted to latex particles after initiation. This results in a reaction medium consisting of submicron polymer particles swollen with monomer and dispersed in an aqueous phase. The final product, called latex, consists of a colloidal dispersion of polymer particles in water.

Similarly, the suspension system consists of an initiator, monomer and stabilizer in the liquid medium. However, in suspension polymerization, the initiator is oil soluble and the stabilizer is a non-micelle-forming stabilizing agent such as poly(vinyl alcohol) or polyvinylpyrrolidone. The locus of polymerization, in contrast to emulsion polymerization, is in the monomer droplets. The final particles have a size between 20 μm and 2 mm. Therefore, suspension polymerization differs from emulsion polymerization in the larger size of the particles, the applied stabilizing agent and the kind of initiator employed.

Many advantages are shown by emulsion and suspension polymerization. First, the continuous water phase can act as an excellent heat conductor and allow the heat to be removed from the system. This is an effective way to increase the polymerization rate of many reactions. Second, the viscosity remains close to that of water and is not dependent on molecular weight since the polymer molecules are contained within the particles. Third, emulsion polymerization is unique in the sense that an increase in molar mass can be achieved without reducing the rate of polymerization. Therefore, high molecular weight polymers can be obtained at fast polymerization rates. By contrast, there is a tradeoff between molecular weight and polymerization rate in bulk and solution free-radical polymerization.

2.4.2 Preparation of Polymer–LDH Nanocomposites *via* Emulsion and Suspension Polymerization

Emulsion polymerization and suspension polymerization have been widely studied to promote intercalation of water-insoluble polymers into layered silicates such as montmorillonite, which is well known to delaminate readily in water owing to the weak forces that stack the layers together.[129–137] The emulsion system consisting of an aqueous medium can contribute to the affinity between the hydrophilic inorganic host and hydrophobic polymer guest by the action of the emulsifier. After polymerization, homogeneously dispersed silicate layers are introduced into the polymer matrices. Therefore, the emulsion and suspension polymerization methods are always ascribed to an exfoliation–absorption strategy for the preparation of polymer–silicate nanocomposites.

Due to their similar structure to layered silicate, LDHs are expected to form nanocomposites by the same strategy. However, a high charge density of LDH layers associated with a high content of guest anions results in strong interlayer electrostatic interactions, which makes the delamination of LDHs much more difficult. Therefore, the preparation of polymer–LDH nanocomposites using emulsion and suspension polymerization is more challenging than that of layered silicate.

Chen *et al.*[138] were the first to use this method to produce PMMA–LDH nanocomposites. The MgAl-LDH was synthesized *in situ* by adding base solution to an emulsion consisting of a metal ion aqueous solution, sodium dodecyl sulfate as an emulsifier, methyl methacrylate as a monomer and benzoyl peroxide as a thermal initiator under N_2 at room temperature. The emulsion polymerization was then carried out at 80 °C for 5 h. The latex obtained was filtered, washed, dried and finally compressed under 10 MPa pressure at 150 °C for 10 min to obtain a transparent sheet. Elemental analysis revealed that the content of LDH in the nanocomposite was as high as 33.9 wt%. Although the XRD pattern of the nanocomposite obtained was almost the same as that of MgAl-LDH intercalated with dodecyl sulfate (MgAl–DS), the intercalated structure of the nanocomposite after calcination at 200 °C for 1 h was basically maintained except for a decrease in peak intensity (Figure 2.5a). In contrast, the pillared structure of MgAl–DS was completely destroyed. These results indicate that the PMMA chains have been intercalated into the interlayers of MgAl–DS–LDH. TEM showed that the nanocomposite consists of dispersed hexagonal particles of size 60–120 nm with a thickness of about 25–40 nm, corresponding to 10–15 stacked layers (Figure 2.5b).

A similar method has also been applied to produce PS–LDH nanocomposites. Ding and Qu[139] prepared ZnAl-LDH by simultaneously adding a Zn–Al salt solution and a sodium hydroxide solution to a transparent aqueous solution consisting of a surfactant of *N*-lauroyl glutamate (LG) and *n*-hexadecane. The previously distilled styrene was added to form an emulsion. Then, a water-soluble initiator, *i.e.* potassium persulfate, was added to initiate free radical polymerization at 80 °C for 6 h. The exfoliated structure was obtained since the diffraction peaks of the LDH component in the composite completely disappeared (Figure 2.6a). In contrast to the face-to-face orientation of the disordered silicate layers observed in the exfoliated polymer–layered silicate nanocomposites, most exfoliated LDH sheets combined with surfactants are dispersed in a disordered manner in the PS matrix and mostly parallel to the grid (Figure 2.6b). Some individual exfoliated layers can still be observed, as indicated by the arrow on the image, which shows that the thickness and lateral sizes of the exfoliated LDH layers are about 1 and 70–100 nm, respectively. In order to elucidate the possible mechanism on the formation of exfoliated PS–LDH nanocomposites *via* emulsion polymerization, the LDH before polymerization was dried and then characterized by XRD. Interestingly, it was found that the (00*l*) diffraction peaks had almost disappeared. This result indicates that the LDH layers are swollen or even exfoliated with the help of the LG surfactant and long-chain *n*-hexadecane in the present emulsion system. Subsequently, when the styrene was added and polymerized, the exfoliated LDH layers can be fixed in the PS matrix to form the exfoliated PS–LDH nanocomposites. On the other hand, the *n*-hexadecane also has an important

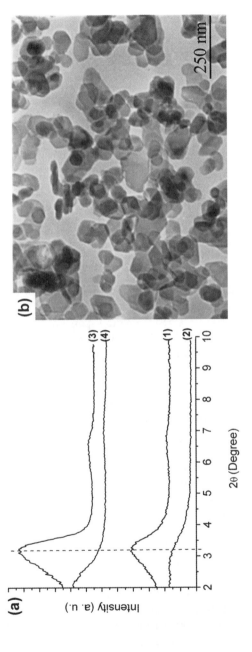

Figure 2.5 (a) XRD patterns of (1) PMMA–MgAl nanocomposite, (2) PMMA–MgAl nanocomposite calcined at 200 °C for 1 h, (3) Mg₃Al–DS and (4) Mg₃Al–DS calcined at 200 °C for 1 h. (b) TEM image of PMMA–MgAl nanocomposite. Reproduced with permission from reference 138.

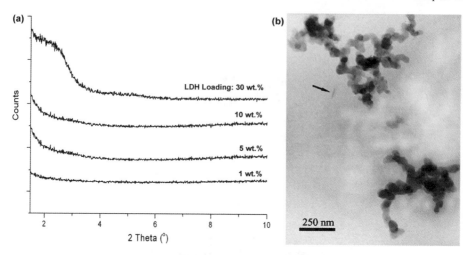

Figure 2.6 (a) Low-angle XRD pattern of PS–LDH with different LDH loadings; (b) TEM images of exfoliated PS–LDH samples with a 1 wt% LDH loading. Reproduced with permission from reference 139.

effect on the exfoliation of LDH layers. The sample prepared without *n*-hexadecane showed an intercalated structure.

Using sodium dodecyl sulfate (SDS) and *n*-pentanol as emulsifier, Zhang *et al.*[140] also prepared PS–MgAl-LDH nanocomposites with an exfoliated structure. They found that, with increase in polymerization temperature and extension of reaction time, the intensity of the (001) peak of LDH decreased and the peak position shifted to low angle (Figure 2.7). When the polymerization temperature is 70 °C and the reaction time is over 6 h, no peaks at $2\theta = 2$–$13°$ are observed in Figure 2.7, which indicates the formation of exfoliated nanocomposites.

From the above discussion, it is concluded that the structure of the nanocomposites obtained from emulsion or suspension polymerization is highly dependent on the reaction conditions. To this end, Ding and Qu[141] investigated the relationship between them. Three main factors were discussed. One is the polymerization method: the emulsion polymerization method was more efficient than the suspension polymerization method in the preparation of the exfoliated PS–LDH nanocomposites, which was reported for the polymer–montmorillonite (MMT) system.[137] Another is the surfactant: by comparing the nanocomposites prepared by using LG and SDS as surfactants, it was found that the LG surfactant favored the exfoliation of LDH layers in the PS–LDH nanocomposites in both suspension and emulsion polymerization. The last is spacers: it was found that a long-chain spacer, *i.e. n*-hexadecane, was more helpful than a short-chain spacer, *i.e. n*-octane, for the exfoliation in the nanocomposites. Schematic diagrams of the formation mechanism of the PS–LDH nanocomposites are shown in Figure 2.8.

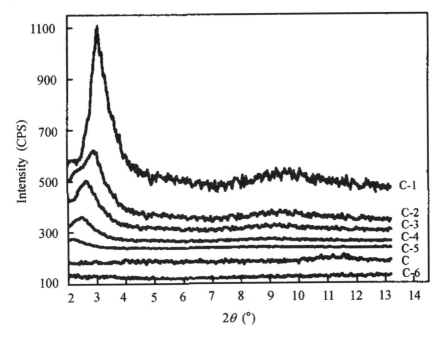

Figure 2.7 XRD patterns of (C-1) PS–LDHs stirred at 25 °C for 30 min, (C-2) PS–LDHs stirred at 25 °C for 6 h, (C-3) PS–LDHs stirred at 40 °C for 1 h, (C-4) PS–LDHs stirred at 40 °C for 6 h, (C-5) PS–LDHs stirred and polymerized at 40 °C for 12 h, (C) PS–LDHs stirred and polymerized at 70 °C for 6 h and (C-6) PS–LDHs stirred and polymerized at 70 °C for 12 h. Reproduced with permission from reference 140.

Except for homopolymers, copolymers and their nanocomposites can also be synthesized using emulsion and suspension polymerization. Recently, the preparation of LDH nanocomposites based on PMMA and its copolymer have been reported.[142] An LDH intercalated with bis(2-ethylhexyl) phosphate (HDEHP) (as shown in Figure 2.9) was prepared by the c-precipitation method and mixed with methyl methacrylate monomer, bis[2-(methacryloyloxy)ethyl] phosphate monomer and benzoyl peroxide initiator to form a homogeneously dispersed system. The mixture was then added dropwise to a water–poly(vinyl alcohol) system and the contents were stirred at 75 °C for 2 h, 85 °C for 2 h and 95 °C for 2 h. XRD characterization showed that the basal spacing of HDEHP–LDH in the homopolymer nanocomposites and copolymer nanocomposites is weak, indicative of disordering. TEM indicated that the global dispersion of HDEHP–LDH in the copolymer is better than that in PMMA (Figure 2.10).

In the common strategy to prepare polymer–LDH nanocomposites, a preintercalation of pristine LDHs by surfactants, such as SDS, is inevitable. Through this step, the organo-modified LDHs are endowed with hydrophobic characteristics ready to be intercalated into polymer chains. However, the introduction of these organic materials with low molecular weight will

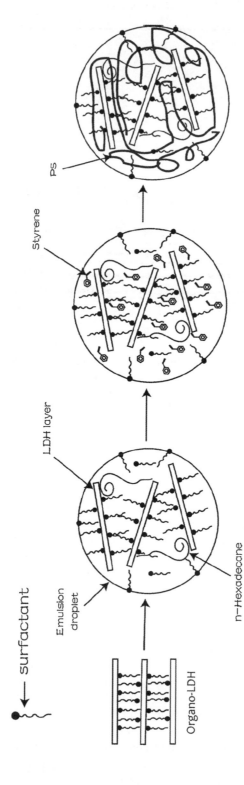

Figure 2.8 Schematic diagrams of the formation procedure of the exfoliated PS–LDH nanocomposites. Reproduced with permission from reference 139.

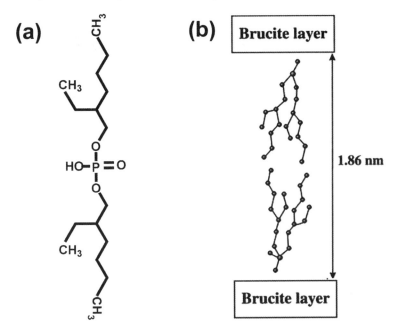

Figure 2.9   (a) Structure of HDEHP and (b) its arrangement in the interlayer of LDHs. Reproduced with permission from reference 142.

adversely affect some properties of the nanocomposites, especially when the LDH content is high. For example, our previous work[65,67,139,141] showed that the thermal properties of the polymer–LDH nanocomposites decreased sharply on increasing the LDH loading from 2–5 wt% to 10–20 wt%. This is because the charred layers are no longer stable enough during the thermal degradation when a relatively large content of SDS is introduced along with the LDH layers. Furthermore, the interaction between the LDH layers and the polymer chains is a weak van der Waals interaction, which makes the structure unstable.

Soap-free emulsion polymerization (SFEP) is an ideal method for the preparation of surfactant-free polymer–LDH nanocomposites. The mechanism of SFEP involves an *in situ* micellization model, where the oligomeric radicals generated by the free radical reaction of an ionic initiator and styrene monomer act as surfactants and assist in forming micelles. Therefore, no surfactants are used during the emulsion polymerization process. The SFEP method has been applied to the preparation of a surfactant-free polystyrene–layered double hydroxide exfoliated nanocomposite.[143] In a typical synthesis, 0.05 g of MgAlNO$_3$ was dispersed in 80 ml of distilled water and sonolyzed for 30 min. The mixture was charged into a 150 ml four-necked reactor equipped with a baffle stirrer, a reflux condenser, a nitrogen inlet and a septum. The temperature of the reactor was raised to 70 °C and then 5 g of styrene and 20 g of aqueous potassium persulfate (KPS) solution (0.4 wt%) were added to the reactor as

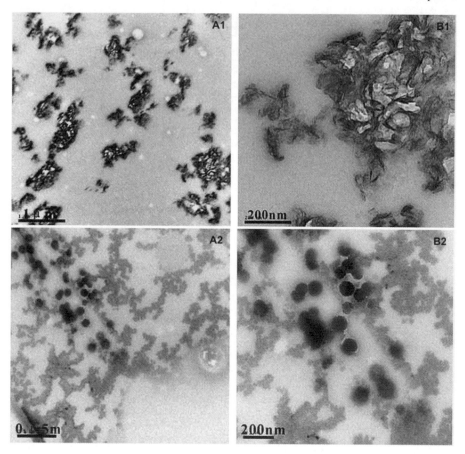

Figure 2.10 TEM images of nanocomposites of PMMA (A1, B1) and copolymer PMMA-*co*-bis[2-(methacryloyloxy)ethyl] phosphate (A2, B2) with HDEHP–LDH prepared from suspension polymerization. Reproduced with permission from reference 142.

initiator. After feeding monomer and initiator, the reactor was kept at the same temperature for 7 h in an oil-bath. The morphological evolution of LDH during the polymerization was studied using TEM analysis (as shown in Figure 2.11). It can be seen from Figure 2.11a that the $MgAlNO_3$–LDH sample before polymerization shows a multiple-layered agglomerate structure 100–200 nm in diameter and 10–30 nm in thickness. After polymerization for 60 min, the interlayer spaces of LDH particles were enlarged, indicating that the PS molecules had been intercalated in the LDH interlayers (Figure 2.11b). The TEM image of the sample after 180 min of polymerization (Figure 2.11c) shows that the LDH layers have lost their tightly stacked structure completely and exist as exfoliated layers in the PS matrix. Figure 2.11d is a TEM image obtained from an ultrathin slice of sample after 420 min of polymerization. It can be seen that the LDH layers are well dispersed in the PS matrix.

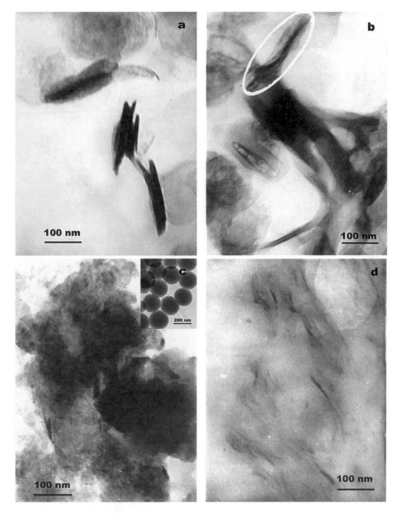

Figure 2.11 TEM images of PS–MgAl-LDH samples with different polymerization times during soap-free emulsion polymerization: (a) 0, (b) 60, (c) 180 and (d) 420 min. Reproduced with permission from reference 143.

Another effective way to overcome the adverse effects of low molecular - weight surfactants is to graft polymer chains on the LDH layers using a reactive modifier instead of a normal surfactant such as SDS. In a recent study, Kovanda *et al.*[144] reported that exfoliated poly(butyl methacrylate)–LDH nanocomposites can be prepared by using reactive LDH fillers in an emulsion polymerization method. The reactive LDH fillers were produced by intercalating monomer anions containing a reactive vinyl group [acrylate, methacrylate (MA), 2-acryl-amido-2-methyl-1-propanesulfonate (AMPS) or 4-vinylbenzoate (VB)], poly-merization initiator [4,4′-azobis(4-cyanopentanoate)] and hydrophobizing agent (dodecyl sulfate) *via* an anion-exchange reaction or a coprecipitation method.

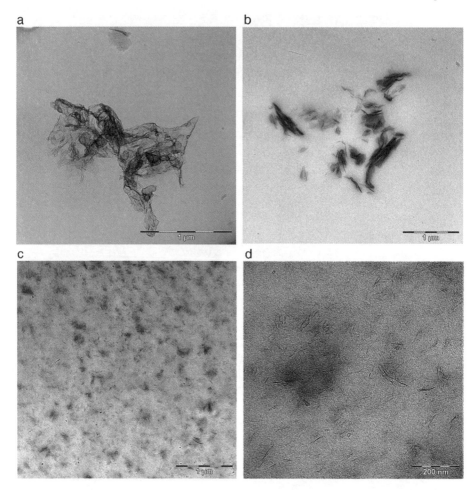

Figure 2.12   TEM images of PBMA membranes with added ZnAl–MA-LDH prepared by (a) emulsion polymerization and (b) solution polymerization in
1-methyl-2-pyrolidone (NMP) or (c) mixture of dimethylformamide
(DMF) and formamide (FA); (d) exfoliated LDH nanoparticles in the
sample prepared by solution polymerization in DMF and FA mixed
solution – image taken at higher magnification. Reproduced with permission from reference 144.

After the intercalated LDHs were applied in *in situ* emulsion polymerization and
solution polymerization, poly(butyl methacrylate)–LDH nanocomposites containing a low concentration (1–5 wt%) of the inorganic filler were obtained.
XRD, TEM and small-angle X-ray scattering (SAXS) studies revealed that the
samples obtained from both emulsion polymerization and solution polymerization had an exfoliated structure, but the best dispersion of LDH nanoparticles was observed from the products prepared by solution polymerization in
a mixture of dimethylformamide and formamide (1:1, v/v) (Figure 2.12).

2.4.3 Properties and Potential Applications of Polymer–LDH Nanocomposites Obtained From Suspension and Emulsion Polymerization

2.4.3.1 Thermal Properties

The thermal stability of a material is characterized mainly by thermogravimetric analysis (TGA), where the sample mass loss due to volatilization of degraded by-products is monitored as a function of temperature. Usually, polymer–LDH nanocomposites have enhanced thermal stability compared with virgin polymers and conventional composites because the well-dispersed LDH layers can act as a superior thermal insulator and mass transport barrier to the volatile products generated during decomposition.

Figure 2.13 depicts the TGA curves of pure PS and PS–LDH samples prepared by emulsion polymerization with LG surfactant and *n*-hexadecane with different LDH contents.[141] All curves for the PS–LDH samples showed similar thermal decomposition behaviors, as mentioned previously. When 50 wt% weight loss was selected as a point of comparison, the thermal

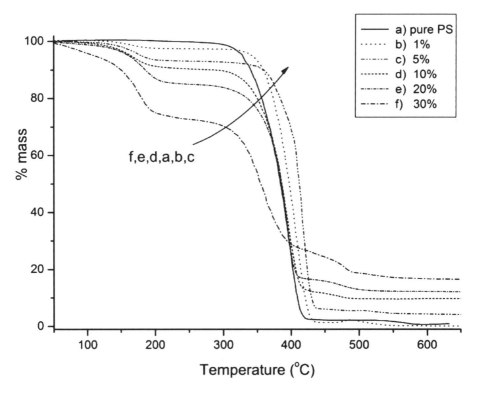

Figure 2.13 TGA curves of pure PS and PS–LDH samples with different LDH contents. Reproduced with permission from reference 141.

decomposition temperatures for pure PS and the PS–LDH nanocomposite samples containing 1, 5, 10, 20 and 30 wt% LDH were determined as 383.8, 396.9, 411.8, 385.8, 385.7 and 354.7 °C, respectively. Apparently, the LDH loading has an important effect on the thermal stability. The thermal decomposition temperature of the PS–LDH samples with 5 and 1 wt% LDH loadings were 28.0 and 13.1 °C higher than that of pure PS, respectively. This enhancement of thermal stability is due to the promotion of the charring process of the polymer matrix and the hindered effect of LDH layers for the diffusion of oxygen and volatile products throughout the composite materials during the thermal decomposition of the composites.[64,66,67]

It should be noted that an obvious weight loss occurs at about 120–300 °C in all TGA curves for PS–LDH nanocomposites. This weight loss has been ascribed to the evaporation of physically absorbed water in the intercalated layers, the loss of hydroxide on LDH layers and the thermal decomposition of LG alkyl chains. In contrast, no weight loss was observed in the temperature range when the surfactant-free PS–LDH nanocomposite prepared by the soap-free emulsion method was heated.[143] This result indicates that the thermal decomposition of LG alkyl chains is the main reason for the weight loss.

2.4.3.2 Fire-retardant Properties

Evaluation of the composite materials in terms of limited oxygen index (LOI) provides first-hand information on the effectiveness of fire-retardant additives. Zhang *et al.*[140] reported on the fire-retardant properties of exfoliated PS–LDH nanocomposites prepared by emulsion polymerization. For comparison, conventional PS–LDH nanocomposites have also been prepared by melt mixing PS with LDH nanoparticles with a diameter of 10–40 nm. Figure 2.14 shows the

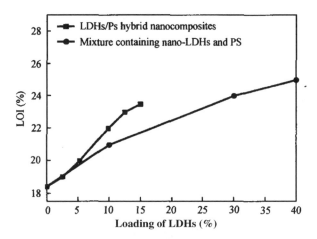

Figure 2.14 Limited oxygen index (LOI) of PS–LDH nanocomposites with different exfoliated LDH nanolayers or LDH nanoparticle loadings. Reproduced with permission from reference 140.

LOI values of both nanocomposites with different LDH loadings. Although the LOI values of both nanocomposites increase with increase in LDH loadings, the rate of increase in exfoliated nanocomposites is much higher than that in conventional nanocomposites. The LOI value of exfoliated nanocomposites containing 14.92 wt% is about 23.5. However, to reach a similar LOI, *i.e.* 24, the LDH loading is as high as 30 wt% in conventional nanocomposites. This result indicates that the exfoliated LDH nanosheets have greater fire-retardant efficiency than the LDH nanoparticles.

Cone calorimetry is another method for evaluating the fire-retardant performance of polymers. During a whole cone calorimetry investigation, a constant external heat flux is maintained to sustain the combustion of the test sample, *i.e.* the test method creates a forced flaming combustion scenario. Therefore, the test results from cone calorimetry are more significant for flammability evaluation than the results obtained from tests where the combustion process is sustained by the heat produced from itself, such as the LOI and UL94 test methods. The parameters that may be evaluated from cone calorimetry include the heat release rate (HRR), the peak of heat release rate (PHRR), the time to ignition (t_{ig}), the volume of smoke (VOS), the total heat released (THR), a measure of the extent to which the entire polymer burns, and the average mass loss rate (AMLR).

The HRR, and in particular the PHRR, have been found to be the most important parameters to evaluate fire retardants. As a typical example, Figure 2.15

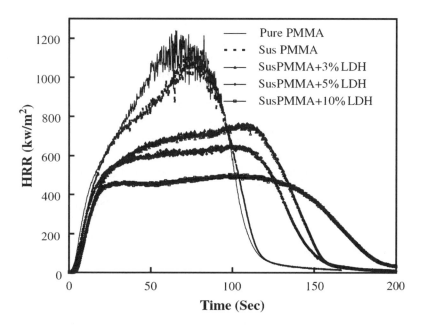

Figure 2.15 Heat release rate curves for PMMA and PMMA–LDH nanocomposite prepared by suspension polymerization with different LDH loadings. Reproduced with permission from reference 142.

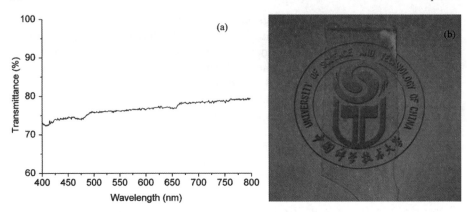

Figure 2.16 (a) Transmittance of PMMA–MgAl-LDH nanocomposite sheet with a thickness of 0.2 mm and (b) photograph of PMMA–MgAl-LDH nanocomposite sheet with a thickness of 0.5 mm on a graphic pattern. Reproduced with permission from reference 138.

shows the HRR plots of PMMA and its nanocomposites with LDH obtained from suspension polymerization.[142] It can be seen that the PHRR decreases significantly and the HRR curve becomes flatter with increase in the content of LDH in the nanocomposite. At a 10 wt% LDH loading, the PHRR of the PMMA nanocomposite is 491 W m^{-2}, which is 56% lower than the 1129 kW m^{-2} of the pristine, commercial PMMA and 55% lower than the 1081 kW m^{-2} of the PMMA obtained by suspension polymerization. Fire retardancy due to the addition of LDH to a polymer may be ascribed to (1) endothermic decomposition of the constituent hydroxides, (2) the presence of water among the combustible gases and (3) the presence of the oxide char formed during the degradation process.

2.4.3.3 Optical Properties

When LDHs are introduced into polymer matrices, the transparency of the polymers is often sacrificed. Only a few transparent polymer–LDH nanocomposite systems can be obtained. Figure 2.16 shows a photograph of a 0.5 mm thick sheet of PMMA–MgAl-LDH nanocomposite with a 33 wt% LDH loading, which indicates the high transparency of the nanocomposite, whereas the PMMA–MgAl-LDH microcomposite is an opaque white sheet even when its thickness is less than 0.1 mm.[138]

2.5 Conclusion

In this chapter, we have dealt with LDHs used as a new kind of inorganic filler and the related polymer–LDH nanocomposites. As nanofillers, the LDH materials combine the features of the conventional metal hydroxide type of fillers, such as magnesium hydroxide and aluminum hydroxide, with the layered silicate type of nanofillers, such as montmorillonite. The former confers excellent flame retardant properties on LDHs because they can undergo endothermic decomposition in the fire and thereby release the bound water and produce metal oxide layered residues to protect the underlayer of polymer from further decomposition. The latter make it possible to achieve a high dispersion status in the polymer matrix through an intercalation process and thus enhance the mechanical and gas barrier properties. In addition, LDHs have many interesting properties, such as anion-exchange ability, relative ease of preparation and a 'memory effect'. Based on these unique properties, a large variety of methods have been developed to incorporate different inorganic or organic anions into LDHs to modify their properties.

The strategies for incorporating polymer chains into the interlayer galleries of LDHs are highly dependent on the hydrophilicity of the polymers. Owing to their good affinity with hydrophilic LDH layers, the water-soluble polymers can be easily intercalated into LDHs to form organoceramics *via* a variety of aqueous intercalation pathways such as coprecipitation, ion exchange, exfoliation–absorption, reconstruction and *in situ* polymerization. On the other hand, for water-insoluble thermoplastic and thermoset polymers, the modification of LDHs with anionic surfactants is an inevitable step in the process of polymer nanocomposites preparation. Usually, the intercalation proceeds in an oil system.

Emulsion and suspension polymerization methods are heterogeneous polymerization techniques in which the polymerization occurs in the monomer-swollen latexes or monomer droplets dispersed in a continuous water phase. Therefore, they provide effective ways to achieve nanocomposites based on water-insoluble polymers and LDHs in aqueous system by separately dispersing hydrophilic LDHs and hydrophobic polymers in water and micelles, respectively, which can be tightly combined by the action of the emulsifier as well. Many advantages are inherent in these methods:

- The production can be made more efficient by conducting the nanocomposite synthesis procedure and the synthesis and modification procedure of LDHs in one system.
- The heat generated by the polymerization reaction can be removed by the continuous water phase.
- The viscosity of the reaction system basically remains constant during the polymerization process.
- An increase in molar mass can be achieved without reducing the rate of polymerization in emulsion polymerization.
- They are environmentally and economically favorable because the reaction proceeds in an aqueous system rather than an oil system.

In conclusion, emulsion and suspension polymerization methods have become promising methods for preparing polymer–LDH nanocomposites based on the above advantages. In the past 5 years, various polymer–LDH nanocomposites have been prepared *via* these methods, with interesting properties such as enhanced thermal stability, fire-retardant properties and optical properties. However, much research remains to be carried out to achieve a more in-depth understanding of the mechanisms of intercalation/exfoliation of LDHs in fabricating polymer–LDH nanocomposites, the influence of LDHs on the emulsion and suspension polymerization and the microstructure–property relationships.

Acknowledgments

The authors wish to acknowledge all the contributions of the references and figures used in this chapter.

References

1. J. C. J. Bart, in *Additives in Polymers: Industrial and Applications*, ed. J. C. J. Bart, John Wiley & Sons, Chichester, 2005, pp. 1–28.
2. B. J. Briscoe, in *The Tribology of Composite Materials: a Preface*, ed. K. Friedrich, Elsevier, Amsterdam, 1993, pp. 3–15.
3. E. P. Giannelis, *Adv. Mater.*, 1996, **8**, 29.
4. R. A. Vaia and J. F. Maguire, *Chem. Mater.*, 2007, **19**, 2736.
5. S. S. Ray and M. Okamoto, *Prog. Polym. Sci.*, 2003, **28**, 1539.
6. M. Alexandre and P. Dubois, *Mater. Sci. Eng., R*, 2000, **28**, 1.
7. A. Usuki, Y. Kojima, M. Kawasumi, A. Okada, Y. Fukushima, T. Kurauchi and O. Kamigaito, *J. Mater. Res.*, 1993, **8**, 1179.
8. D. G. Evans and R. C. T. Slade, *Struct. Bonding*, 2006, **119**, 1.
9. F. Leroux, *J. Nanosci. Nanotechnol.*, 2006, **6**, 303.
10. B. M. Choudary, B. Bharathi, C. V. Reddy, M. L. Kantam and K. V. Raghavan, *Chem. Commun.*, 2001, 1736.
11. B. M. Choudary, S. Madhi, N. S. Chowdari, M. L. Kantam and B. Sreedhar, *J. Am. Chem. Soc.*, 2002, **124**, 14127.
12. S. Bhattacharjee and J. A. Anderson, *Chem. Commun.*, 2004, 554.
13. A. Li, L. Qin, D. Zhu, R. Zhu, J. Sun and S. Wang, *Biomaterials*, 2009, **31**, 748.
14. J. H. Choy, S. Y. Kwak, Y. J. Jeong and J. S. Park, *Angew. Chem. Int. Ed.*, 2000, **39**, 4042.
15. C. Del Hoyo, *Appl. Clay Sci.*, 2007, **36**, 103.
16. L. Desigaux, M. Ben Belkacem, P. Richard, J. Cellier, P. Leone, L. Cario, F. Leroux, C. Taviot-Gueho and B. Pitard, *Nano Lett.*, 2006, **6**, 199.
17. F. Leroux, M. Ben Belkacem, G. Guyot, C. Taviot-Gueho, P. Leone, L. Cario, L. Desigaux and B. Pitard, in *Organic/Inorganic Hybrid Materials – 2004*, ed. C. Sanchez, U. Schubert, R. M. Laine and Y. Chujo, Materials Research Society, Warrendale, PA, 2005, p. 223.
18. L. van der Ven, M. L. M. van Gemert, L. F. Batenburg, J. J. Keern, L. H. Gielgens, T. P. M. Koster and H. R. Fischer, *Appl. Clay Sci.*, 2000, **17**, 25.

19. L. C. Du, B. J. Qu and M. Zhang, *Polym. Degrad. Stabil.*, 2007, **92**, 497.
20. M. Lakraimi, A. Legrouri, A. Barroug, A. de Roy and J. P. Besse, *J. Chim. Phys. Phys. Chim. Biol.*, 1999, **96**, 470.
21. D. Yan, J. Lu, M. Wei, J. Ma, D. G. Evans and X. Duan, *Phys. Chem. Chem. Phys.*, 2009, **11**, 9200.
22. Z. Chang, D. Evans, X. Duan, P. Boutinaud, M. de Roy and C. Forano, *J. Phys. Chem. Solids*, 2006, **67**, 1054.
23. S. C. Guo, D. G. Evans and D. Q. Li, *J. Phys. Chem. Solids*, 2006, **67**, 1002.
24. Y. Tian, G. Wang, F. Li and D. G. Evans, *Mater. Lett.*, 2007, **61**, 1662.
25. A. V. Lukashin, A. A. Vertegel, A. A. Eliseev, M. P. Nikiforov, P. Gornert and Y. D. Tretyakov, *J. Nanopart. Res.*, 2003, **5**, 455.
26. D. Mohan and C. U. Pittman, *J. Hazard. Mater.*, 2007, **142**, 1.
27. L. Lv, J. He, M. Wei, D. G. Evans and Z. L. Zhou, *Water Res.*, 2007, **41**, 1534.
28. J. T. Kloprogge and R. L. Frost, *J. Solid State Chem.*, 1999, **146**, 506.
29. F. Leroux, M. Adachi-Pagano, M. Intissar, S. Chauviere, C. Forano and J. P. Besse, *J. Mater. Chem.*, 2001, **11**, 105.
30. G. Fetter, A. Botello, V. H. Lara and P. Bosch, *J. Porous Mater.*, 2001, **8**, 227.
31. Y. W. You, H. T. Zhao and G. F. Vance, *Colloid Surf. A*, 2002, **205**, 161.
32. T. Lopez, P. Bosch, M. Asomoza, R. Gomez and E. Ramos, *Mater. Lett.*, 1997, **31**, 311.
33. A. I. Khan and D. O'Hare, *J. Mater. Chem.*, 2002, **12**, 3191.
34. C. Delmas and Y. Borthomieu, *J. Solid State Chem.*, 1993, **104**, 345.
35. E. M. Moujahid, M. Dubois, J. P. Besse and F. Leroux, *Chem. Mater.*, 2002, **14**, 3799.
36. J. He, M. Wei, B. Li, Y. Kang, D. G. Evans and X. Duan, *Struct. Bonding*, 2006, **119**, 89.
37. Y. Sugahar, N. Yokoyama, K. Kuroda and C. Kato, *Ceram. Int.*, 1988, **14**, 163.
38. F. Leroux and J. P. Besse, *Chem. Mater.*, 2001, **13**, 3507.
39. E. M. Moujahid, J. P. Besse and F. Leroux, *J. Mater. Chem.*, 2002, **12**, 3324.
40. E. M. Moujahid, J. P. Besse and F. Leroux, *J. Mater. Chem.*, 2003, **13**, 258.
41. L. Vieille, E. Moujahid, C. Taviot-Gueho, J. Cellier, J. P. Besse and F. Leroux, *J. Phys. Chem. Solids*, 2004, **65**, 385.
42. O. C. Wilson, T. Olorunyolemi, A. Jaworski, L. Borum, D. Young, A. Siriwat, E. Dickens, C. Oriakhi and M. Lerner, *Appl. Clay Sci.*, 1999, **15**, 265.
43. E. M. Moujahid, J. Inacio, J. P. Besse and F. Leroux, *Micropor. Mesopor. Mater.*, 2003, **57**, 37.
44. E. M. Moujahid, F. Leroux, M. Dubois and J. P. Besse, *C. R. Chim.*, 2003, **6**, 259.

45. L. Vieille, C. Taviot-Gueho, J. P. Besse and F. Leroux, *Chem. Mater.*, 2003, **15**, 4369.
46. C. Vaysse, L. Guerlou-Demourgues, C. Delmas and E. Duguet, *Macromolecules*, 2004, **37**, 45.
47. S. Rey, J. Merida-Robles, K. S. Han, L. Guerlou-Demourgues, C. Delmas and E. Duguet, *Polym. Int.*, 1999, **48**, 277.
48. C. Vaysse, L. Guerlou-Demourgues, E. Duguet and C. Delmas, *Inorg. Chem.*, 2003, **42**, 4559.
49. M. Tanaka, I. Y. Park, K. Kuroda and C. Kato, *Bull. Chem. Soc. Jpn.*, 1989, **62**, 3442.
50. C. O. Oriakhi, I. V. Farr and M. M. Lerner, *J. Mater. Chem.*, 1996, **6**, 103.
51. P. B. Messersmith and S. I. Stupp, *J. Mater. Res.*, 1992, **7**, 2599.
52. P. B. Messersmith and S. I. Stupp, *Chem. Mater.*, 1995, **7**, 454.
53. Q. Z. Yang, D. J. Sun, C. G. Zhang, X. J. Wang and W. A. Zhao, *Langmuir*, 2003, **19**, 5570.
54. F. Leroux, P. Aranda, J. P. Besse and E. Ruiz-Hitzky, *Eur. J. Inorg. Chem.*, 2003, 1242.
55. N. T. Whilton, P. J. Vickers and S. Mann, *J. Mater. Chem.*, 1997, **7**, 1623.
56. C. Roland-Swanson, J. P. Besse and F. Leroux, *Chem. Mater.*, 2004, **16**, 5512.
57. E. M. Moujahid, M. Dubois, J. P. Besse and F. Leroux, *Chem. Mater.*, 2005, **17**, 373.
58. T. Challier and R. C. T. Slade, *J. Mater. Chem.*, 1994, **4**, 367.
59. E. Monjahid, M. Dubois, J. P. Besse and F. Leroux, in *Organic/Inorganic Hybrid Materials – 2002*, ed. C. Sanchez, R. M. Laine, S. Yang and C. J. Brinker, Materials Research Society, Warrendale, PA, 2002, p. 99.
60. F. Leroux, J. Gachon and J. P. Besse, *J. Solid State Chem.*, 2004, **177**, 245.
61. F. Leroux and C. Taviot-Gueho, *J. Mater. Chem.*, 2005, **15**, 3628.
62. C. Taviot-Gueho and F. Leroux, *Struct. Bonding*, 2006, **119**, 121.
63. W. Chen and B. J. Qu, *Chin. J. Chem.*, 2003, **21**, 998.
64. W. Chen and B. J. Qu, *Chem. Mater.*, 2003, **15**, 3208.
65. W. Chen and B. J. Qu, *J. Mater. Chem.*, 2004, **14**, 1705.
66. L. Z. Qiu, W. Chen and B. J. Qu, *Polymer*, 2006, **47**, 922.
67. L. Z. Qiu, W. Chen and B. J. Qu, *Polym. Degrad. Stabil.*, 2005, **87**, 433.
68. L. C. Du, B. J. Qu, Y. Z. Meng and Q. Zhu, *Compos. Sci. Technol.*, 2006, **66**, 913.
69. S. F. Hsu, T. M. Wu and C. S. Liao, *J. Polym. Sci., Part B: Polym. Phys.*, 2006, **44**, 3337.
70. S. F. Hsu, T. M. Wu and C. S. Liao, *J. Polym. Sci., Part B: Polym. Phys.*, 2007, **45**, 995.
71. T. M. Wu, S. F. Hsu, Y. F. Shih and C. S. Liao, *J. Polym. Sci., Part B: Polym. Phys.*, 2008, **46**, 1207.
72. J. Liu, G. M. Chen and J. P. Yang, *Polymer*, 2008, **49**, 3923.
73. F. A. He, L. M. Zhang, F. Yang, L. S. Chen and Q. Wu, *J. Polym. Res.*, 2006, **13**, 483.

74. M. Kotal, T. Kuila, S. K. Srivastava and A. K. Bhowmick, *J. Appl. Polym. Sci.*, 2009, **114**, 2691.
75. K. L. Dagnon, H. H. Chen, L. H. Lnnocentini-Mei and N. A. D'Souza, *Polym. Int.*, 2009, **58**, 133.
76. K. L. Nichols and C. J. Chou, *US Pat.*, 5 952 093 1999.
77. F. R. Costa, M. Abdel-Goad, U. Wagenknecht and G. Heinrich, *Polymer*, 2005, **46**, 4447.
78. F. R. Costa, U. Wagenknecht, D. Jehnichen, M. A. Goad and G. Heinrich, *Polymer*, 2006, **47**, 1649.
79. F. R. Costa, B. K. Satapathy, U. Wagenknecht, R. Weidisch and G. Heinrich, *Eur. Polym. J.*, 2006, **42**, 2140.
80. F. R. Costa, U. Wagenknecht and G. Heinrich, *Polym. Degrad. Stabil.*, 2007, **92**, 1813.
81. A. Schonhals, H. Goering, F. R. Costa, U. Wagenknecht and G. Heinrich, *Macromolecules*, 2009, **42**, 4165.
82. U. Costantino, A. Gallipoli, M. Nocchetti, G. Camino, F. Bellucci and A. Frache, *Polym. Degrad. Stabil.*, 2005, **90**, 586.
83. L. C. Du and B. J. Qu, *Chin. J. Chem.*, 2006, **24**, 1342.
84. L. C. Du and B. J. Qu, *J. Mater. Chem.*, 2006, **16**, 1549.
85. P. Ding and B. J. Qu, *J. Polym. Sci., Part B: Polym. Phys.*, 2006, **44**, 3165.
86. P. Ding and B. J. Qu, *Polym. Eng. Sci.*, 2006, **46**, 1153.
87. M. Zhang, P. Ding, B. J. Qu and A. G. Guan, *J. Mater. Process. Technol.*, 2008, **208**, 342.
88. M. Zhang, P. Ding and B. J. Qu, *Polym. Compos.*, 2009, **30**, 1000.
89. L. Ye, P. Ding, M. Zhang and B. J. Qu, *J. Appl. Polym. Sci.*, 2008, **107**, 3694.
90. L. Ye and B. J. Qu, *Polym. Degrad. Stabil.*, 2008, **93**, 918.
91. M. Zhang, P. Ding, L. C. Du and B. J. Qu, *Mater. Chem. Phys.*, 2008, **109**, 206.
92. C. Nyambo, D. Chen, S. P. Su and C. A. Wilkie, *Polym. Degrad. Stabil.*, 2009, **94**, 496.
93. C. Nyambo, E. Kandare, D. Y. Wang and C. A. Wilkie, in *27th Polymer Degradation Discussion Group Conference Held in Honor of Norman Billingham*, Birmingham, 2007, p. 1656.
94. C. Nyambo, E. Kandare and C. A. Wilkie, *Polym. Degrad. Stabil.*, 2009, **94**, 513.
95. C. Nyambo, P. Songtipya, E. Manias, M. M. Jimenez-Gasco and C. A. Wilkie, *J. Mater. Chem.*, 2008, **18**, 4827.
96. C. Nyambo, D. Wang and C. A. Wilkie, *Polym. Adv. Technol.*, 2009, **20**, 332.
97. C. Nyambo and C. A. Wilkie, *Polym. Degrad. Stabil.*, 2009, **94**, 506.
98. C. Manzi-Nshuti, D. Y. Wang, J. M. Hossenlopp and C. A. Wilkie, *J. Mater. Chem.*, 2008, **18**, 3091.
99. C. Manzi-Nshuti, J. M. Hossenlopp and C. A. Wilkie, *Polym. Degrad. Stabil.*, 2009, **94**, 782.

100. M. Zammarano, S. Bellayer, J. W. Gilman, M. Franceschi, F. L. Beyer, R. H. Harris and S. Meriani, *Polymer*, 2006, **47**, 652.
101. P. J. Pan, B. Zhu, T. Dong and Y. Inoue, *J. Polym. Sci., Part B: Polym. Phys.*, 2008, **46**, 2222.
102. W. D. Lee, S. S. Im, H. M. Lim and K. J. Kim, *Polymer*, 2006, **47**, 1364.
103. A. Sorrentino, G. Gorrasi, M. Tortora, V. Vittoria, U. Costantino, F. Marmottini and F. Padella, *Polymer*, 2005, **46**, 1601.
104. M. Zubitur, M. A. Gomez and M. Cortazar, *Polym. Degrad. Stabil.*, 2009, **94**, 804.
105. M. Adachi-Pagano, C. Forano and J. P. Besse, *Chem. Commun.*, 2000, 91.
106. T. Hibino and W. Jones, *J. Mater. Chem.*, 2001, **11**, 1321.
107. T. Hibino, *Chem. Mater.*, 2004, **16**, 5482.
108. T. Hibino and M. Kobayashi, *J. Mater. Chem.*, 2005, **15**, 653.
109. S. O'Leary, D. O'Hare and G. Seeley, *Chem. Commun.*, 2002, 1506.
110. L. Li, R. Z. Ma, Y. Ebina, N. Iyi and T. Sasaki, *Chem. Mater.*, 2005, **17**, 4386.
111. Z. P. Liu, R. Z. Ma, M. Osada, N. Iyi, Y. Ebina, K. Takada and T. Sasaki, *J. Am. Chem. Soc.*, 2006, **128**, 4872.
112. B. G. Li, Y. Hu, R. Zhang, Z. Y. Chen and W. C. Fan, *Mater. Res. Bull.*, 2003, **38**, 1567.
113. B. G. Li, Y. Hu, J. Liu, Z. Y. Chen and W. C. Fan, *Colloid. Polym. Sci.*, 2003, **281**, 998.
114. D. P. Yan, J. Lu, M. Wei, J. B. Han, J. Ma, F. Li, D. G. Evans and X. Duan, *Angew. Chem. Int. Ed.*, 2009, **48**, 3073.
115. J. Bin Han, J. Lu, M. Wei, Z. L. Wang and X. Duan, *Chem. Commun.*, 2008, **50**, 5188.
116. W. Chen and B. J. Qu, *Polym. Degrad. Stabil.*, 2005, **90**, 162.
117. W. Chen, B. J. Qu and L. Feng, *Chem. J. Chin. Univ.*, 2003, **24**, 1920 (in Chinese).
118. L. Z. Qiu, W. Chen and B. J. Qu, *Colloid. Polym. Sci.*, 2005, **283**, 1241.
119. P. J. Fu, G. M. Chen, J. Liu and J. P. Yang, *Mater. Lett.*, 2009, 1725.
120. P. Ding, M. Zhang, J. Gai and B. J. Qu, *J. Mater. Chem.*, 2007, **17**, 1117.
121. G. A. Wang, C. C. Wang and C. Y. Chen, *Polymer*, 2005, **46**, 5065.
122. C. Manzi-Nshuti, J. M. Hossenlopp and C. A. Wilkie, *Polym. Degrad. Stabil.*, 2008, **93**, 1855.
123. C. Manzi-Nshuti, D. Y. Wang, J. M. Hossenlopp and C. A. Wilkie, *Polym. Degrad. Stabil.*, 2009, **94**, 705.
124. Z. Matusinovic, M. Rogosic and J. Sipusic, *Polym. Degrad. Stabil.*, 2009, **94**, 95.
125. H. B. Hsueh and C. Y. Chen, *Polymer*, 2003, **44**, 1151.
126. H. B. Hsueh and C. Y. Chen, *Polymer*, 2003, **44**, 5275.
127. Y. N. Chan, T. Y. Juang, Y. L. Liao, S. A. Dai and J. J. Lin, *Polymer*, 2008, **49**, 4796.
128. M. Zammarano, M. Franceschi, S. Bellayer, J. W. Gilman and S. Meriani, *Polymer*, 2005, **46**, 9314.
129. D. C. Lee and L. W. Jang, *J. Appl. Polym. Sci.*, 1996, **61**, 1117.

130. M. H. Noh and D. C. Lee, *J. Appl. Polym. Sci.*, 1999, **74**, 2811.
131. M. H. Noh, L. W. Jang and D. C. Lee, *J. Appl. Polym. Sci.*, 1999, **74**, 179.
132. M. W. Noh and D. C. Lee, *Polym. Bull.*, 1999, **42**, 619.
133. X. Y. Huang and W. J. Brittain, *Macromolecules*, 2001, **34**, 3255.
134. Y. S. Choi, M. H. Choi, K. H. Wang, S. O. Kim, Y. K. Kim and I. J. Chung, *Macromolecules*, 2001, **34**, 8978.
135. X. Tong, H. C. Zhao, T. Tang, Z. L. Feng and B. T. Huang, *J. Polym. Sci., Part A: Polym. Chem.*, 2002, **40**, 1706.
136. M. Z. Xu, Y. S. Choi, Y. K. Kim, K. H. Wang and I. J. Chung, *Polymer*, 2003, **44**, 6387.
137. D. Y. Wang, J. Zhu, Q. Yao and C. A. Wilkie, *Chem. Mater.*, 2002, **14**, 3837.
138. W. Chen, L. Feng and B. J. Qu, *Solid State Commun.*, 2004, **130**, 259.
139. P. Ding and B. J. Qu, *J. Colloid Interface Sci.*, 2005, **291**, 13.
140. Z. J. Zhang, X. J. Mei, C. H. Xu, L. G. Feng and F. L. Qiu, *Acta Polym. Sin.*, 2005, 589.
141. P. Ding and B. J. Qu, *J. Appl. Polym. Sci.*, 2006, **101**, 3758.
142. L. J. Wang, S. P. Su, D. Chen and C. A. Wilkie, *Polym. Degrad. Stabil.*, 2009, **94**, 1110.
143. L. Z. Qiu and B. J. Qu, *J. Colloid Interface Sci.*, 2006, **301**, 347.
144. F. Kovanda, E. Jindova, K. Lang, P. Kubat and Z. Sedlakova, *Appl. Clay Sci.*, 2010, **48**, 260.

CHAPTER 3

Polymer–Clay Nanocomposite Particles by Direct and Inverse Emulsion Polymerization

WEIHUA (MARSHALL) MING,[a] DIRK-JAN VOORN[b] AND ALEX M. VAN HERK[b]

[a] Nanostructured Polymers Research Center, Materials Science Program, University of New Hampshire, Durham, NH 03824, USA; [b] Laboratory of Polymer Chemistry, Eindhoven University of Technology, P.O. Box 513, 5600 MB Eindhoven, The Netherlands

3.1 Introduction

Nanocomposites can be defined as combinations of different materials where at least one of the components is distributed on the nanoscale. Nanocomposite particles are particles with at least one component being structured on the nanoscale. Several types of inorganic materials have been combined with polymer materials to produce nanocomposite particles, for example, titanium dioxide, clay platelets such as laponite (LRD) and montmorillonite (MMT), calcium carbonate and iron oxides.[1,2] In general, beneficial properties of the composite are expected, mainly in the area of mechanical properties, barrier properties and dispersion properties (better dispersion of the inorganic particles in the polymer matrix). Many research groups are working on incorporation of clay platelets either intercalated or (partially) exfoliated in polymer materials. Application areas of the anticipated materials are often in coatings, adhesives, films for barrier properties and plastics.

In this chapter, we focus on polymer–clay nanocomposite particles prepared by emulsion polymerization and related techniques. One of the crucial aspects

RSC Nanoscience & Nanotechnology No. 16
Polymer Nanocomposites by Emulsion and Suspension Polymerization
Edited by Vikas Mittal
© Royal Society of Chemistry 2011
Published by the Royal Society of Chemistry, www.rsc.org

of the resulting materials is the characterization of the composite material in terms of dispersion and extent of exfoliation of the clay platelets.

Although encapsulation of inorganic nanoparticles goes back to the 1980s,[3-5] the work on clay encapsulation prepared by emulsion polymerization has a younger history, starting around 2003.[6] Many research groups in academia and industry since then have begun research in this area. At scientific polymer conferences, sometimes the majority of (poster) presentations have been devoted to some issues of clays in combination with polymers. Although a lot of high-level scientific work has been produced, also less well-defined work has been published where reactions were performed in the presence of clay and several properties were studied without proper characterization of the resulting material. In general, adding a clay to a polymer will change the thermal stability one way or another. Even a simple physical blend of clay and latex will do so. Also, mechanical properties without doubt will be affected; usually an increase in storage modulus is observed, although sometimes the opposite effect is also seen.[7] As the clay might interact with radicals and might also affect the rate of formation of polymer particles and the monomer conversion rate, the addition of clay will usually be accompanied by a change in reaction rate, particle number and molecular weight in an emulsion polymerization. Depending on the extent of the interaction of the clay surface with the surrounding polymer, a change in glass transition temperature will occur, as described in general by Rittigstein and Torkelson.[8] Without a proper assessment of the function of the clay and the final position of the clay with respect to the latex particles, no good science can be done, unfortunately. Characterization of the final materials is usually done through thermogravimetric analysis (TGA), X-ray diffraction (XRD), dynamic mechanical thermal analysis (DMTA), conversion measurements and transmission electron microscopy (TEM). Only the TEM technique gives direct proof for the location of the clay platelets when particles are examined before film formation.[9] Often researchers are not able to visualize the clay platelets within the nanocomposite particles due to the small thickness of exfoliated clay platelets and their basal plane orientation with respect to the electron beam. For that reason, sometimes thicker synthetic clay platelets such as gibbsite have been used.[10]

In our work, we have tried to obtain exfoliated clay platelets encapsulated by polymer material. The challenges in clay encapsulation are shown schematically in Figure 3.1.

In striving towards encapsulation of exfoliated clay platelets inside latex particles, several conflicting prerequisites have to be met. In a normal emulsion polymerization, the initial polymerization reaction takes place inside micelles (particle nucleation). Most natural clay platelets spontaneously exfoliate in water. The formation of polymer near or on these highly hydrophilic clay surfaces is not very likely, however. For that reason, modification of the clay is undertaken. Depending on the surface charge of the clay, charged reactive molecules that can copolymerize with other monomers can be adsorbed on the clay surface; furthermore, surfactant-like molecules and coupling agents (such as silanes and titanates) can be applied, using reactive hydroxyl groups on the

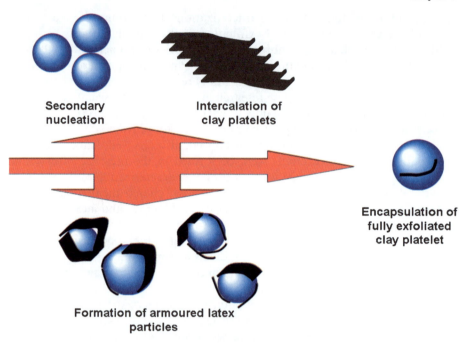

Figure 3.1   Schematic representation of the challenges in encapsulation of clay
platelets through emulsion polymerization and related techniques.

edge of the clay.[9] In many cases, the favorable interaction of the clay with water
is (partially) lost, leading to intercalation again. The actual polymerization in
the presence of clay can lead to clay-encapsulated polymer particles but also
to the formation of a second crop of latex particles (secondary nucleation).
Several ways to minimize the chance of secondary nucleation are at hand,
including the use of a more hydrophobic initiator, slow addition of the
monomers, preventing the presence of micelles, producing a more or less
hydrophobic domain on the (inorganic) surface, offering a large surface area to
the produced aqueous phase oligomers and using less water-soluble monomers.
    In general, clay platelets have a tendency to locate themselves at the oil–
water interface and one of the more common outcomes of an emulsion polymerization in the presence of clay is the formation of armored latex particles,[11]
the clay being at the surface of the latex particles. Although these particles
definitely have interesting properties in themselves, the question is whether they
have optimum properties. Physical blending of preformed latex and exfoliated
clay dispersion will lead to similar structures without chemistry being involved.
Solid particles brought into a mixture of oil and water have a tendency to go to
the interface between the water and oil, whereby the total interfacial energy is
reduced.[12] If one wants to optimize the possible beneficial properties of the high
aspect ratio of the clay in combination with directional alignment of the clay in
the final material, the incorporation of the clay in a non-spherical latex particle

might be of interest. Hence another goal of our work was to produce non-spherical latex particles with directional positioning of the clay platelet inside. One relatively new approach is the use of controlled radical polymerization to locate the formation of polymer on the surface of the clay platelets.[10] Alternatively, miniemulsion polymerization can be a solution to some of the conflicting requirements.[13] The orientation of clay platelets in the final film, for example by using flat latex particles with clay inside, will improve barrier properties more than random orientation of the clay platelets, and gives the opportunity to reach desired properties at even lower levels of clay.[14–16]

Yet another interesting application of clay platelets is to make use of their tendency to position themselves at the oil–water interface,[17] following the so-called Pickering stabilization mechanism.[18] The use of clay platelets as a replacement for the conventional surfactant is an interesting possibility, although the directional alignment of the particles in the final film is no longer possible in that case. This strategy can be applied in both direct and inverse emulsion polymerization, as described below.

In this chapter, we focus on three areas of research: the use of nascent clay as the stabilizing agent in direct emulsion polymerization, the use of hydrophobized clay as the stabilizing agent in inverse emulsion polymerization and encapsulation of clay by means of emulsion polymerization processes.

3.2 Polymer–Clay Nanocomposite Particles by Direct Emulsion Polymerization

There has been extensive interest in preparing polymer–clay nanocomposites by using direct emulsion and miniemulsion polymerizations, which are covered in other chapters of this book. The focus of this section is on using nascent clay particles as stabilizing agents for direct emulsion polymerization.

It is well established that solid particles with special features can self-assemble at the liquid–liquid interface to reduce the interfacial energy between the two immiscible liquids. Bon and co-workers[19] obtained polystyrene latex particles of about 145 nm in diameter via a miniemulsion polymerization process by only using LRD platelets as the Pickering-type stabilizing agent, leading to the formation of latex particles armored with LRD particles. It was concluded that the addition of a proper amount of sodium chloride (0.1 mol l^{-1} of NaCl for 0.5 wt% LRD) was important to render the Pickering stabilization of oils in water by LRD. For particles to be used as a Pickering stabilizing agent, they must possess surface characteristics that allow them to be preferentially located at the oil–water interface, which can be enabled by the addition of NaCl in the case of LRD particles.[19,20] Too high a concentration of NaCl would lead to undesired clay flocculation.

The Pickering miniemulsion polymerization was successfully extended to other monomers including butyl methacrylate, 2-ethylhexyl acrylate and lauryl methacrylate.[21] It appeared that these hydrophobic monomers showed more success than partially water-miscible monomers such as methyl acrylate or methyl methacrylate. Furthermore, due to the presence of clay at the particle surface, retardation effects up to intermediate monomer conversions were observed, especially for smaller latex particles.[21]

There have been many other reports on direct emulsion polymerizations involving nascent clays,[22–26] either using clays as stabilizing agents or with the aim of preparing polymer–clay nanocomposites. In addition, organically modified clays have also been used in direct emulsion polymerization.[27] Improved mechanical and thermal properties, reduced vapor permeability and improved flame retardancy for the prepared polymer–clay composites were reported.

3.3 Polymer–Clay Nanocomposite Particles by Inverse Emulsion Polymerization

Due to the hydrophilic nature of nascent clays, one may envisage that hydrophilic clays can be encapsulated by inverse emulsion polymerization. As discussed in Section 3.2, in a direct emulsion containing hydrophilic clays, clays may be predominately located in the continuous aqueous phase and some may reside at the surface of the micelles (forming a Pickering-type emulsion in the presence or absence of surfactants); after polymerization, armored particles with clay covering the particles have been commonly observed (Figure 3.2a).

In contrast, for an inverse emulsion containing hydrophilic clay particles, there is a possibility that the hydrophilic clay particles may reside inside the monomer micelles (Figure 3.2b), and upon polymerization the clay platelets may, hypothetically, have a chance to be encapsulated effectively inside the polymer particles.

We have tested this hypothesis in the inverse emulsion polymerization of acrylamide (AAm). A typical recipe included AAm as the monomer dissolved in water, cyclohexane as the continuous phase, sodium bis(2-ethylhexyl) sulfosuccinate (AOT) or sorbitan monooleate (Span-80) as the surfactant and MMT or LRD as typical nascent clays. Both oil- and water-soluble azo compounds, such as 2,2′-azobisisobutyronitrile (AIBN) and 2,2′-azobis[2-methyl-*N*-(2-hydroxyethyl)propionamide] (VA-086), were used as initiators. Span-80 appeared to lead to more stable inverse emulsions in the presence of clay than AOT. We first noticed a different kinetic behavior during polymerization in the presence of clay platelets. The conversion–time history for the AAm inverse emulsion polymerization in the presence of MMT showed a significant decrease in the rate of polymerization in comparison with those without clay. With increasing concentration of clay particles, lower rates of polymerizations and lower final monomer conversions were found, together with increased retardation of the polymerization. The clay platelets in the inverse emulsions might act as diffusion barriers for monomer and/or initiator.

The typical particle diameters of LRD- and MMT-containing inverse PAAm latexes were about 70 and 240 nm, respectively. We then examined the location of clays in the final latex particles. To our great surprise, cryogenic TEM

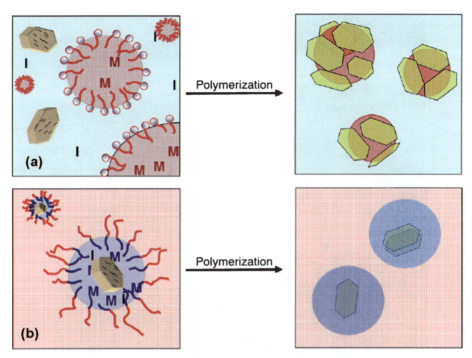

Figure 3.2 Schematic representation of (a) oil-in-water (O/W) emulsion polymerization in the presence of hydrophilic clay platelets and (b) water-in-oil
(W/O) inverse emulsion polymerization of hydrophilic monomer with
hydrophilic clay which, hypothetically, may lead to clay encapsulated
inside the latex particles.

(cryo-TEM) images of PAAm latexes in the presence of MMT demonstrated
clearly that clay platelets were located predominantly at the perimeter of the
latex particles as black, sometimes curved, lines (Figure 3.3). The curved lines
also indicate that the large MMT platelets are fairly flexible and can bend
around the latex particles.

To find out why the hydrophilic platelets were located at the interface between
a hydrophilic polymer and a hydrophobic continuous phase in the inverse
latexes, cryo-TEM was used to examine the locations of the platelets in the
starting inverse emulsions. Conventional TEM requires dried samples and
cannot be used to characterize the starting inverse emulsions containing clay
platelets. We examined a model inverse emulsion comprising water, cyclohexane
and clay particles, with Span-80 as the surfactant. In the absence of clay particles, the cryo-TEM image showed that water droplets of 100–200 nm in
diameter were coated with surfactant molecules. The surfactant-stabilized water
droplets appeared to bundle together and formed 'cauliflower-like' structures.
When either LRD or MMT was added to the inverse emulsion, the 'cauliflower-
like' structure appeared to remain intact, but the clay platelets were located at
the interface between the water droplets and cyclohexane, not inside the water

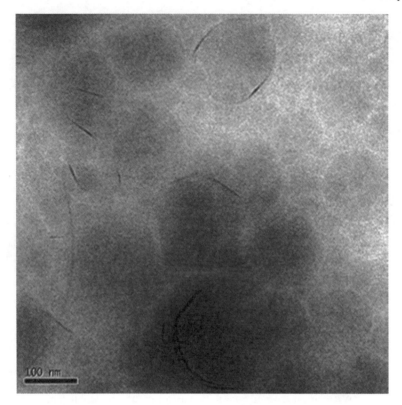

Figure 3.3 Cryo-TEM image of a PAAm latex in the presence of MMT from inverse emulsion polymerization using VA-086 as the initiator. Some MMT platelets may not be visible due to their basal plane orientation relative to the electron beam during the image collection.

compartments. We believe that this may be the primary reason why hydrophilic clays were not encapsulated inside PAAm particles after polymerization.

Considering the fact that clay particles have a tendency to orient themselves at the interface between water and oil phases, we have taken advantage of this by using clay particles as the motif to stabilize inverse emulsions. As discussed earlier, for particles to be used as a Pickering-type stabilizer, they have to possess certain surface characteristics. It appeared that hydrophobized clays were better suited to stabilize inverse emulsions.[17] The long alkyl chain of the quaternary ammonium compounds that are adsorbed on the surface of the hydrophobized clay surface enable the swelling and exfoliation of the platelets in apolar medium.

We used a dimethyl-ditallow-ammonium organically modified montmorillonite (MMT20; commercial name Cloisite 20A) as the stabilizer for an AAm–water inverse emulsion in cyclohexane, without adding any other surfactant.[17] Nascent, unmodified hydrophilic clays were also tried as the sole stabilizer for the inverse emulsions; however, no stable dispersion was obtained. We also attempted to prepare inverse emulsions using dimethyl-ditallow-ammonium chloride (the agent used to hydrophobize the clay) as the surfactant and it turned out that no stable water-in-oil (W/O) emulsions could be obtained, either. Only the hydrophobically modified clay platelets appeared to be able to stabilize W/O emulsions without the addition of surfactants. The typical content of MMT20 was 0.2–0.7 wt% for an inverse emulsion containing about 4 wt% of AAm. The average diameter of the inverse emulsion droplets increased from 500 to 700 nm on decreasing the MMT20 concentration from 0.7 to 0.2 wt%, similar to an emulsion droplet stabilized by a conventional surfactant. When the clay content was too high (1.42 wt%), the inverse emulsion became highly viscous, most likely due to the stacking behavior of the hydrophobic clays in the cyclohexane phase.

After polymerization initiated by either AIBN or VA-086, the MMT20-stabilized PAAm particles ranged from 700 to 980 nm in diameter, significantly larger than the particles made in MMT-containing inversion emulsions stabilized by Span-80.[17] The final latex particles were about 200 nm larger than the initial emulsion droplets, indicating that there may be monomer transportation during the polymerization, whereas there was no significant change in the particle polydispersity index. Again, the particle size depended on the MMT20 content: the higher the clay content, the smaller the particles. During the polymerization, the particle size increased gradually, as revealed by dynamic light scattering (DLS) measurements.

We used electron microscopy to examine the particle morphology. In particular, cryo-TEM can provide true information on the location of clay platelets. As shown in Figure 3.4, particles with fluffy surfaces were observed for MMT20-stabilized inverse latexes. The fluffy structure obviously corresponded to MMT20 platelets located at the particle surface, which were also partially 'stretched' out into the organic medium. Bundles of several stacked platelets were also occasionally found in the organic medium, but most of these platelets were shown to be located around the latex particles. The location of the hydrophobized clay was further confirmed by SEM, where a rugged surface was observed for the particles.[17] The difference between the SEM and cryo-TEM observations was probably due to the SEM sample preparation: the 'fluffy' clay platelets became collapsed on the surface of the latex during drying. In addition, not much difference between the latexes prepared with the water-soluble initiator VA-086 and with the oil-soluble initiator AIBN was observed, in terms of the particle size and morphology.

3.4 Clay Encapsulation by Emulsion Polymerization

3.4.1 General Approaches in Encapsulation of Inorganic Particles (in General) by Emulsion Polymerization

We first describe some general approaches to encapsulate inorganic particles other than clay by emulsion polymerization. Increased interest in this approach

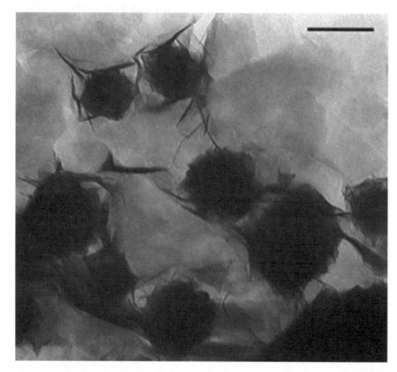

Figure 3.4   Cryo-TEM image (scale bar: 1 µm) of a PAAm latex in the presence of hydrophobized MMT (MMT20) from inverse emulsion polymerization. A water-soluble azo compound, VA-086, was used as the free-radical initiator.

was generated when clay encapsulation was attempted. Furthermore, the miniemulsion polymerization technique has proven to be another versatile route towards encapsulated materials. In the past decade, about 100 papers per year were published [excluding the many papers on clay nanocomposites produced using techniques other than (mini)emulsion polymerization].

Usually, 'maximum' properties can be obtained when the inorganic particles are distributed evenly and as single (primary) particles in the matrix. This means that in the steps towards obtaining the final product, keeping the particles well dispersed is of major importance. Initially the particles should be well dispersed in the aqueous phase and (partial) coagulation during the emulsion polymerization must be avoided because this leads to irreversible fixation of the coagulates.

Depending on the conditions one can distinguish three different mechanisms (Figure 3.5):

- Formation of a *hemimicelle* on the surface of the inorganic particle, where after swelling with monomer the encapsulating polymer is formed.
- *Precipitation* of polymer chains initiated in the aqueous phase.

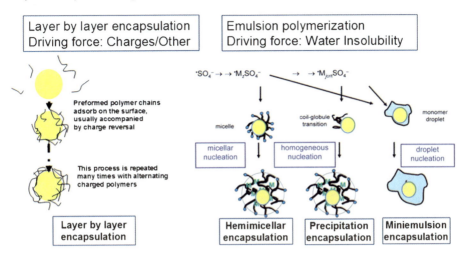

Figure 3.5 Schematic representation of encapsulation of inorganic (submicron) particles through different approaches.

- Formation of polymer inside the particle containing monomer droplets (*miniemulsion* polymerization approach).

Each of these mechanisms can be combined with controlled radical polymerization.

Examples of the *hemimicelle approach* include some early work to encapsulate TiO$_2$ particles.[28–30] Normally the so-called coupling agents were applied; these usually contain at least one alkoxy group, which can react with hydroxyl groups on the surface of particles. Other functional groups are also available that can interact physically or chemically with the surrounding polymer matrix, thereby aiding the dispersion of the particles.

Other additives that have been used to make the particle surface hydrophobic include a combination of methacrylic acid and aluminum nitrate as a coupling agent[31] or compounds such as stearic acid. Without chemical modification of the surface, one can also use bilayers of adsorbed surfactant as a locus of polymerization.[32] The first step is the formation of a so-called hemimicelle, a bilayer of the surfactant molecules at the solid–aqueous interface (Figure 3.5). A hemimicelle may be considered to be the surface analogue of a micelle and therefore this approach can be compared to the emulsion polymerization approach.

Some early examples of the *precipitation approach* include the aqueous solution polymerizations reported by Chaimberg *et al.*[33] for the graft polymerization of polyvinylpyrrolidone on to silica. The nonporous silica particles were modified with vinyltriethoxysilane in xylene, isolated and dispersed in an aqueous solution of vinylpyrrolidone. The polymerization was performed at 70 °C and initiated by hydrogen peroxide, after which precipitation on the surface occurred, leading to encapsulation. Nagai *et al.*[34] in 1989 reported on the aqueous polymerization of the quaternary salt of dimethylaminoethyl methacrylate with lauryl bromide, a surface-active monomer, on silica gel. Although the aim was to polymerize only on the surface, separate latex particles were also formed.

The *miniemulsion approach* is extensively discussed in Chapter 10.

The following strategies can be used to help improve the particle encapsulation efficiencies: using small particles at a high concentration, a hydrophobic initiator, low surfactant concentrations and monomers with low water solubility (added semi-continuously). By using less water-soluble monomers in combination with a nonionic initiator, the formation of surface-active oligomers in the aqueous phase can be minimized, thus increasing the efficiency of particle encapsulation.

In addition to emulsion polymerization with a separate monomer phase, emulsions consisting of diluted monomer droplets together with inorganic particles have also been polymerized by Park and Ruckenstein[35] They created emulsions of decane, a monomer and silica in an aqueous solution of surfactant, which were then polymerized to latexes containing fairly uniformly distributed inorganic particle clusters of submicrometer size.

3.4.2 Encapsulation of Clay by (Mini)emulsion Polymerization

In the past few years, many groups have tried to encapsulate clay platelets inside latex particles. This encapsulation poses some extra challenges because of the tendency of the clay platelets to form stacks and house-of-cards structures in an aqueous medium. Furthermore, the clay platelets also tend to reside at the water–polymer interface, as discussed previously in this chapter. This means that most efforts would lead to clay platelets situated mainly at the outside of the latex particles. In this section, a few examples are shown in which clay has been proven to be located inside the latex particles. Another challenge in this respect is to confirm where the clay platelets reside. It is very difficult to visualize natural clay platelets when they do not form stacks. Individual clay platelets are too thin to see by TEM; they only become visible when their basal planes are oriented more or less parallel to the electron beam. A (cryo)-TEM instrument with a stage that can rotate can help verify the location of an exfoliated clay platelet, *i.e.* whether it is inside or on the surface of a latex particle.[9,17]

Many of the papers described emulsion or miniemulsion polymerization in the presence of unmodified or modified clays (often MMT). The modification can be surface modification, edge modification or both. Ianchis *et al.*[36] claimed that they obtained both clay platelets (silylated MMT) inside the latex particles and on the surface of latex particles. Although the clay was not easily visible in the TEM pictures presented, they indirectly inferred the presence of MMT platelets inside the latex particles by looking at the shape of the latex particles. Snowman morphologies were associated with the encapsulated clay. Recently, reversible addition–fragmentation chain transfer (RAFT)-mediated miniemulsion polymerization was applied to encapsulate clays by Hartmann *et al.*,[7] in which a RAFT agent was first anchored on the clay surface.

Most of the attempts resulted in so-called armored latex particles with clay platelets located on the surface of the latex. Recently, natural and synthetic clays were successfully encapsulated.[9,10] The anisotropy of the clay platelets resulted in non-spherical latex particles, in the shape of either peanut-like particles (Figure 3.6) containing MMT[9] or flat latex particles (Figure 3.7) containing synthetic gibbsite platelets,[10] It has been demonstrated that, for successful MMT encapsulation, it is crucial to modify the clay edge covalently with polymerizable moieties and to starve-feed monomer during the surfactant-free emulsion polymerization.[9] With this approach, about 60–75 wt% of MMT platelets were encapsulated inside latex particles. The location of MMT (inside the PMMA particles) was confirmed by cryo-TEM analysis through examining micrographs at several tilt angles of the cryo-stage between –45° and +45°.[9] The absence of MMT at the PMMA particle surface was further corroborated by the very smooth particle surface in the SEM image,[9] as opposed to the rugged surface for the clay-armored particles.

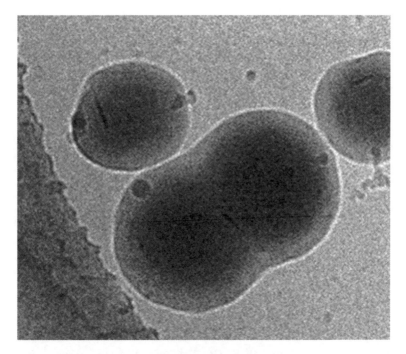

Figure 3.6 Cryo-TEM picture of PMMA latex particles with encapsulated MMT: the clay was edge-modified with a reactive titanate and the surrounding polymer was created by a starve-fed emulsion polymerization of methyl methacrylate.[9] The image size is ∼500×440 nm.

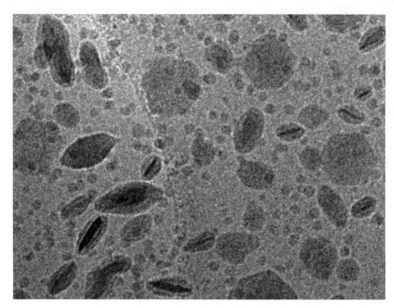

Figure 3.7 Cryo-TEM image of encapsulated gibbsite platelets (typical thickness: 10 nm). The polymerization was performed from the surface with amphiphilic RAFT agents. Note that some of the flat latex particles were seen from the face and some from the side. Also, secondary nucleation was visible in this image.[10] The image size is ~1200×960 nm.

References

1. A. M. van Herk and A. L. German, in *Microspheres, Microcapsules and Liposomes, Vol 1: Preparation and Chemical Applications*, ed. R. Arshady, Citus Books, London, 1998 pp. 457–486.
2. E. Bourgeat-Lami, *J. Nanosci. Nanotechnol.*, 2002, **2**, 1.
3. R. Kroker and K. Hamann, *Angew. Makromol. Chem.*, 1978, **13**, 1.
4. R. Laible and K. Hamann, *Adv. Colloid Interface Sci.*, 1980, **13**, 65.
5. T. Ono, *Prog. Org. Coat.*, 1986, **18**, 279.
6. W. Xie, J. M. Hwu, G. J. Jiang, T. M. Buthelezi and W. P. Pan, *Polym. Eng. Sci.*, 2003, **43**, 214.
7. A. Sakamande, R. D. Sanderson and P. C. Hartmann, *J. Polym. Sci., Part A: Polym. Chem.*, 2008, **46**, 7114.
8. P. Rittigstein and J. M. Torkelson, *J. Polym. Sci., Part B: Polym. Phys.*, 2006, **44**, 2935.
9. D. J. Voorn, W. Ming and A. M. Van Herk, *Macromolecules*, 2006, **39**, 4654.
10. S. I. Ali, J. P. A. Heuts, B. S. Hawkett and A. M. van Herk, *Langmuir*, 2009, **25**, 10523.
11. J. Zhang, K. Chen and H. Zhao, *J. Polym. Sci., Part A: Polym. Chem.*, 2008, **46**, 2632.

12. B. P. Binks and S. O. Lumsdon, *Langmuir*, 2000, **16**, 2539.
13. R. P. Moraes, A. M. Santos, P. C. Oliveira, F. C. T. Souza, M. do Amaral, T. S. Valera and N. R. Demarquette, *Macromol. Symp.*, 2006, **245**, 106.
14. A. A. Gusev and H. R. Lusti, *Adv. Mater.*, 2001, **13**, 1641.
15. H. R. Lusti, A. A. Gusev and O. Guseva, *Model. Simul. Mater. Sci. Eng.*, 2004, **12**, 1201.
16. C. Lu and Y.-W. Mai, *Comput. Sci. Technol.*, 2007, **67**, 2895.
17. D. J. Voorn, W. Ming and A. M. van Herk, *Macromolecules*, 2006, **39**, 2137.
18. S. U. Pickering, *J. Chem. Soc.*, 1907, **91**, 2001.
19. S. Cauvin, P. J. Colver and S. A. F. Bon, *Macromolecules*, 2005, **38**, 7887.
20. N. P. Ashby and B. P. Binks, *Phys. Chem. Chem. Phys.*, 2000, **2**, 5640.
21. S. A. F. Bon and P. J. Colver, *Langmuir*, 2007, **23**, 8316.
22. B. zu Putlitz, K. Landfester, H. Fischer and M. Antonietti, *Adv. Mater.*, 2001, **13**, 500.
23. N. N. Herrera, J. M. Letoffe, J. L. Putaux, L. David and E. Bourgeat-Lami, *Langmuir*, 2004, **20**, 1564.
24. G. Diaconu, M. Paulis and J. R. Leiza, *Polymer*, 2008, **49**, 2444.
25. C. H. Lee, A. T. Chien and M. H. Yen *et al.*, *J. Polym. Res.*, 2008, **15**, 331.
26. J. Faucheu, C. Gauthier and L. Chazeau *et al.*, *Polymer*, 2010, **51**, 6.
27. N. Greesh, P. C. Hartmann, V. A. Cloete and R. D. Sanderson, *J. Polym. Sci., Part A: Polym. Chem.*, 2008, **46**, 3619.
28. C. H. M. Caris, L. P. M. van Elven, A. M. van Herk and A. L. German, *Br. Polym. J.*, 1989, **21**, 133.
29. C. H. M. Caris, R. P. M. Kuijpers, A. M. van Herk and A. L. German, *Makromol. Chem. Makromol. Symp.*, 1990, **35/36**, 535.
30. R. Q. F Janssen, A. M. van Herk and A. L. German, *J. Oil Colour Chem. Assoc.*, 1993, **11**, 455.
31. J. P. Lorimer, T. J. Mason, D. Kershaw, I. Livsey and R. Templeton-Knight, *Colloid Polym. Sci.*, 1991, **269**, 392.
32. K. Meguro, T. Yabe, S. Ishioka, K. Kato and K. Esumi, *Bull. Chem. Soc. Jpn.*, 1986, **59**, 3019.
33. M. Chaimberg, R. Parnas and Y. Cohen, *J. Appl. Polym. Sci.*, 1989, **37**, 2921.
34. K. Nagai, Y. Ohishi, K. Ishiyama and N. Kuramoto, *J. Appl. Polym. Sci.*, 1989, **38**, 2183.
35. J. S. Park and E. Ruckenstein, *Polymer*, 1990, **31**, 175.
36. R. Ianchis, D. Donescu, C. Petcu, M. Ghiurea, D. F. Anghel, G. Stanga and A. Marcu, *Appl. Clay Sci.*, 2009, **45**, 164.

CHAPTER 4

PMMA-based Montmorillonite Nanocomposites by Soap-free Emulsion Polymerization

KING-FU LIN[a, b] AND KENG-JEN LIN[b]

[a] Department of Materials Science and Engineering, National Taiwan University, Taipei, Taiwan; [b] Institute of Polymer Science and Engineering, National Taiwan University, Taipei, Taiwan

4.1 Introduction

Poly(methyl methacrylate) (PMMA) is a transparent solid with a glass transition temperature (T_g) of $\sim 100\,^\circ$C. It possesses many useful properties such as high strength, rigidity, good optical properties and weathering resistance. It was developed in 1928 and was introduced commercially in 1933 by the German company Rohm and Haas.[1] In 1936, the first commercially viable production of acrylic safety glass began. Since then, PMMA has been widely applied in daily life products and is commonly called acrylic glass or simply acrylic. Recently, owing to its good biocompatibility and low immune response, PMMA has also been applied in biomedical products.[2–5]

Ever since PMMA was invented, its copolymers and composites have frequently been investigated by polymer researchers in order to expand their properties and applications. Notably, PMMA nanocomposites with a variety of nanoscale dispersive phases often exhibit unusual physical and chemical properties that are dramatically different from those of conventional composites. The nanoscale dispersive phases that have been used for PMMA nanocomposites in the literature include nanoparticles (TiO_2, ZnO, SiO_2, F_3O_4, *etc.*),[6–15] single- and multi-walled carbon nanotubes,[16–22] clays

RSC Nanoscience & Nanotechnology No. 16
Polymer Nanocomposites by Emulsion and Suspension Polymerization
Edited by Vikas Mittal
Published by the Royal Society of Chemistry, www.rsc.org

(montmorillonite, laponite, mica, hectorite, *etc.*),[23–33] and graphenes,[34–37] which individually impart specific properties to the nanocomposites. It is noteworthy that the polymer nanocomposites with montmorillonite (MMT) as a nanophase are generally considered as one of the most successful nano-technologies even applied to polymers. Numerous publications[38–53] have appeared following the pioneering work of the Toyota research group.[54–57]

MMT is a smectic clay composed of several ~ 1 nm thick crystalline nano-platelets, which are formed by sandwiching an edge-shared octahedral layer of alumina between two tetrahedral silica layers;[58] a schematic diagram is shown in Figure 4.1. Stacking of the nanoplatelets creates a van der Waals gap. Small fractions of the tetrahedral silicon atoms substituted by aluminum ions and the octahedron aluminum ions substituted by magnesium ions give each sandwich layer an overall negative charge, which is counterbalanced by some external cations, such as sodium ions in the van der Waals gap. Because of the weak interactions between the stacking nanoplatelets, the cations inside the gallery can be easily exchanged with other cations.[59] Researchers usually organify MMT by using alkylammonium salts to exchange with cations for accessing the intercalated or exfoliated nanocomposites.[60–65]

Although the conventional processing methods such as extrusion, injection and thermoforming have often been used to manufacture MMT–PMMA nanocomposites,[66–71] the dispersion of organified MMT into the PMMA matrix on the nanoscale is still a major concern to polymer researchers. To improve the dispersion of MMT, the emulsion polymerization of MMA in the

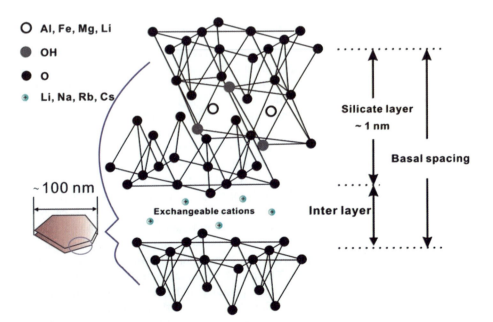

Figure 4.1 Schematic diagram of structural units in MMT crystal.

presence of organified MMT has been used to fabricate partially exfoliated or intercalated MMT–PMMA nanocomposite latexes following the conventional methods to process these nanocomposite latex particles.[72–75] However, the surfactant used for emulsion polymerization and alkylammonium salts for organization of the MMT cannot be removed from the fabricated nanocomposites, creating interfacial problems between the PMMA matrix and MMT.

Recently, exfoliated MMT–PMMA nanocomposite latex has been successfully fabricated by soap-free emulsion polymerization without any pretreatment or organization of MMT.[76,77] As PMMA was copolymerized with MA to lower the glass transition temperature, the resulting exfoliated MMT–P(MMA-*co*-MA) nanocomposite latex can be cast into film directly.[78–80] Many unusual properties of the cast films have been reported, including low permeability, high flame retardation and superior tensile properties. The last was attributed to the grafting of P(MA-*co*-MMA) chain segments on to the exfoliated MMT nanoplatelets.[80] Owing to the implausibly easy processing method to achieve these good properties of PMMA-based MMT nanocomposites, this chapter will elucidate the exfoliation mechanism of MMT and how the P(MA-*co*-MMA) chain segments are grafted on the exfoliated MMT nanoplatelets during soap-free emulsion polymerization in the presence of MMT.

4.2 Fabrication of PMMA Nanocomposites Through Soap-free Emulsion Polymerization

4.2.1 Soap-free Emulsion Polymerization

The conventional emulsion polymerization requires the surfactant to form micelles in water, where the monomers diffuse inside for polymerization. A water-soluble initiator such as potassium persulfate (KPS) is often used and decomposes in water to form radicals, which also diffuse inside the micelles to initiate or terminate the polymerization. Thus, the products fabricated by conventional emulsion polymerization contain the surfactant, which cannot be removed and may adversely affect their properties. In soap-free emulsion polymerization,[81–83] the polymerization mechanism is similar to that in conventional emulsion polymerization except that no surfactant is involved. The typical initiator is also KPS. It decomposes and initiates the polymerization of MMA monomers in water first to form oligomeric ion radicals carrying sulfate anions with potassium cations as a counterion. Because MMA monomers only slightly dissolve in water, the polymerization rate is slow in this stage. As the oligomeric ion radicals grow to a certain length, they will precipitate in water to form primary particles. The primary particles tend to aggregate into a secondary particle as the number of primary particles is above the critical micelle concentration. The secondary particles with the oligomeric sulfate ions facing the water phase and chain segments facing inside the particle are similar to the micelles formed by the ionic surfactants in conventional

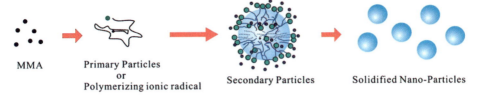

Figure 4.2 Schematic diagram of soap-free emulsion polymerization to fabricate PMMA latex particles.

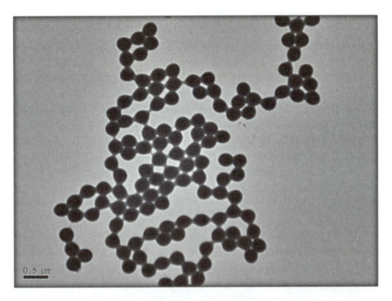

Figure 4.3 Typical TEM image of PMMA latex particles fabricated by soap-free emulsion polymerization.

emulsion polymerization. As the secondary particles are formed, the MMA monomers and sulfate ion radicals from the decomposed KPS will begin to diffuse into the particles for polymerization. The polymerization rate increases significantly in this stage until the MMA droplets in water run out. The mechanism of soap-free emulsion polymerization is illustrated schematically in Figure 4.2.

When the polymerization is complete, PMMA latex particles will form in a uniform size. Figure 4.3 shows the typical TEM images of PMMA latex particles fabricated by soap-free emulsion polymerization. The particle size is about 250 nm on average. The sulfate ionic groups on the surface of PMMA latex particles prevent them from aggregation. Because the T_g of PMMA is $\sim 100\,^{\circ}$C, they will not form a film after casting and drying. However, the T_g of PMMA can be lowered by copolymerization with PMA so that cast films are able to form.[78]

4.2.2 Exfoliated MMT–PMMA Nanocomposites by Soap-free Emulsion Polymerization

MMT is a commercial product but can be extracted from bentonite. Figure 4.4 shows a TEM image of MMT that was extracted from bentonite quarried from the mountain side of Taitung in the east of Taiwan.[76] It clearly reveals a layer structure from the edge of the MMT. The interlayer distance of MMT measured from the X-ray diffraction pattern is 1.23 nm, as shown in Figure 4.5. Because the thickness of the exfoliated MMT nanoplatelets is ~1 nm on average, the thickness of the gallery in MMT would be ~0.23 nm, which can be increased by swelling in water on the basis of the ionic osmotic pressure. To begin the soap-free emulsion polymerization of MMA in the presence of MMT to fabricate exfoliated MMT–PMMA nanocomposite latex, MMA monomers were added to the aqueous suspension solution of MMT at 70 °C that had previously been mixed with KPS by stirring at room temperature for at least 24 h.

When the polymerization had proceeded for 5 min, some of the polymerizing PMMA chains were found to insert into the gallery of MMT and aggregate into a disk-like configuration, as shown in Figure 4.6a and b.[76] The number of disk-like domains was increased as more polymerizing PMMA chains were adsorbed and eventually exfoliated MMT. The formed disk-like domains and exfoliated MMTs adhered together and aggregated into a particulate form (Figure 4.6c).

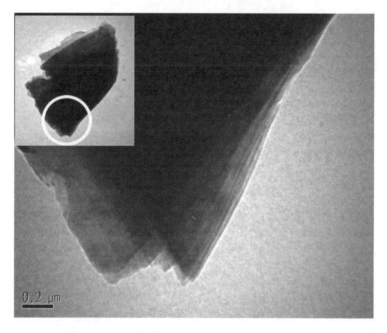

Figure 4.4   TEM image of MMT that was extracted from bentonite quarried from the mountain side of Taitung in the east of Taiwan. Reproduced from reference 76 with permission from Wiley.

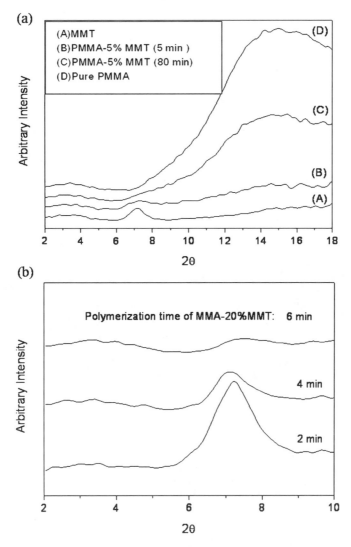

Figure 4.5 X-ray diffraction patterns of (a) the indicated powder samples and (b) 20% MMT–PMMA powder samples after soap-free emulsion polymerization for the indicated times. Reproduced from reference 76 with permission from Wiley.

Fully grown latex particles of the exfoliated MMT–PMMA nanocomposite with a size of 300–600 nm were not as uniform as those prepared by the typical soap-free emulsion polymerization of MMA. Some of them aggregated together and were either embedded or covered with the same exfoliated MMT nanoplatelets (Figure 4.6d).

The exfoliation of MMT during soap-free emulsion polymerization of MMA has been confirmed by X-ray diffraction.[76] When MMA was polymerized for

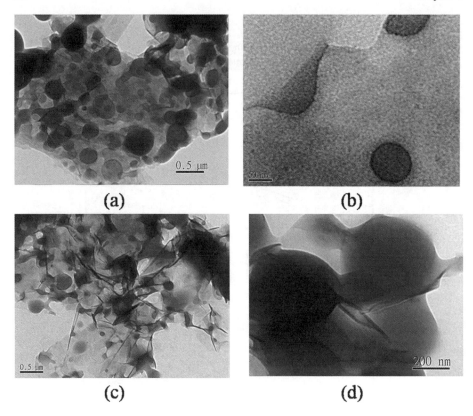

(a)   (b)

(c)   (d)

Figure 4.6   TEM images representing the different stages of formation of the latex
particles during the emulsion polymerization of MMA in the presence of
MMT: (a) micelle formation as a disk form inserting into the interlayer
regions of MMT; (b) exfoliation of the MMT nanosheet; (c) growth of the
micelles after the exfoliation of MMT; (d) fully grown latex particles of
exfoliated MMT–PMMA nanocomposites. Reproduced from reference 76
with permission from Wiley.

5 min at 70 °C with 5 wt% MMT, its X-ray diffraction peak at $2\theta = 7.28°$,
corresponding to the interlayer spacing of MMT, separated into two peaks (see
Figure 4.5a). The peak at the larger angle had the same position as that of
pristine MMT, whereas the peak at the smaller angle ($2\theta = 6.658°$) corresponded to an interlayer spacing of 1.33 nm, 0.1 nm more than that of pristine
MMT. Referring to the TEM image shown in Figure 4.6b, the latter peak might
be associated with the intercalation of MMT by the insertion of the disk-like
micelles. According to the X-ray results, only some of the MMTs were intercalated at this stage. After polymerization for 80 min, the peak at the smaller
angle diminished, whereas the peak at the larger angle overlapped with the
diffraction peak of PMMA.

However, it might be argued that the concentration of MMT in 5% MMT–
PMMA was too low to provide comprehensible evidence for the exfoliation of

MMT from the X-ray diffraction data. Therefore, the MMT concentration was increased to 20 wt%. The 20% MMT–PMMA powder samples for X-ray diffraction measurements after soap-free emulsion polymerization for 2, 4 and 6 min were prepared with the following considerations:[76] at 2 min, the micellation stage had not yet been reached; at 4 min, the micellation stage had just been reached; and at 6 min, the micellation stage was presumably over. Their X-ray diffraction patterns shown in Figure 4.5b revealed that the intensity of the peak at $2\theta = 7.28°$ decreased significantly and that the peak position shifted slightly to a smaller angle as the polymerization reached the micellation stage. As the polymerization passed through the micellation stage, the peak at $2\theta = 7.28°$ diminished significantly and shifted back to the original position. Apparently, MMT was rapidly intercalated and exfoliated during the micellation stage and this confirmed the TEM results. Because of the larger quantity of MMT in 20% MMT–PMMA compared with that in 5% MMT–PMMA, the extent of intercalation by the insertion of disk-like micelles was less for the former than for the latter, so the peak at $2\theta = 7.28°$ shifted less to a smaller angle. However, once the MMT had been intercalated by the disk-like micelles, it readily exfoliated.

On the basis of these observations, a brief mechanism for the formation of latex particles of exfoliated MMT–PMMA nanocomposites was suggested, as illustrated schematically in Figure 4.7. When the primary particles with the polymerizing PMMA chains carrying a sulfate ionic group at one end and free radical at the other end form during soap-free emulsion polymerization, they tend to diffuse into the gallery of MMT. The polymerizing chains will aggregate into micelles in a disk form resulting from the confined space in the gallery. As more and more micelles in disk form appear in the gallery, they will trigger the exfoliation of MMT. Once the MMTs have been exfoliated, the PMMA disks have more space to grow into a particulate form (see Figure 4.6c). In this stage, the disks may aggregate together when they collide with one another. Consequently, the latex particles contain several exfoliated MMT nanoplatelets, as illustrated in Figure 4.6d.

Notably, this may now raise the questions of why MMA monomers do not diffuse into the gallery of MMT for polymerization and how the polymerizing chains that carry an anionic group can diffuse into the gallery. To answer the first question, the adsorption of MMA by MMT in aqueous solution was investigated by the following method:[77] 3 g of MMA and 0.15 g of MMT in 30 ml of water were mixed at room temperature with stirring for 3 days. Subsequently, the solution was centrifuged to obtain an MMT–MMA slurry, which was then dried to a powder form at 50 °C. The FTIR spectrum of MMA–MMT powder did not detect any substantial adsorption of MMA by MMT, as shown in Figure 4.8. The absorption band at 2924 cm^{-1} contributed by the $-CH_2-$ and $-CH_3$ stretching of MMA and the band at 1265 cm^{-1} contributed by the $-C-O-$ stretching of MMA did not appear in the spectrum of MMA–MMT powder. It is conceivable that MMA is not a hydrophilic monomer so that it cannot be adsorbed into the gallery of MMT.

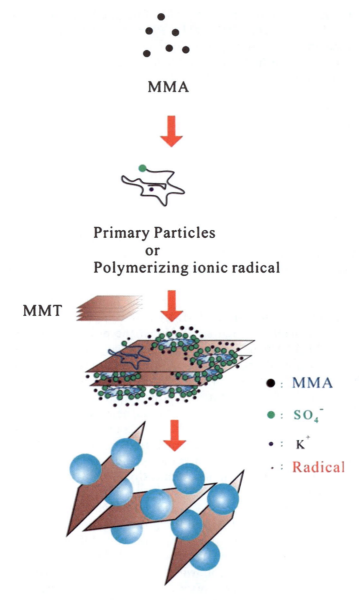

Figure 4.7 Schematic diagram of the formation mechanism of exfoliated MMT–
PMMA nanocomposite latexes via soap-free emulsion polymerization.

To answer the second question, sodium dodecyl sulfate (SDS), which carries
a similar sulfate ionic group as a model compound, was used to investigate its
intercalation with MMT.[77] After mixing MMT with 0.5 and 3 cationic
exchange capacity (CEC) of SDS, it was indeed observed that SDS surfactant

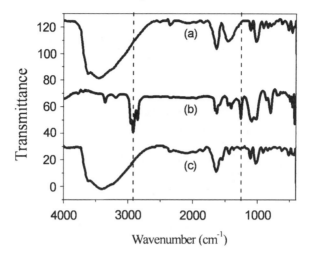

Figure 4.8 FTIR spectra of (a) pristine MMT, (b) MMA and (c) MMT–MMA. Reproduced from reference 77 with permission from Wiley.

had diffused into the gallery of MMT and formed two-dimensional lamellae or vesicle-like configurations, as shown in Figure 4.9a and b. The domain size was much smaller than that aggregated by polymerizing PMMA chains (see Figure 4.6a), because of the shorter chain length of SDS ($\sim 2 \, \text{nm}$)[84] than the contour length of PMMA chains. The X-ray diffraction pattern of SDS–MMT powder containing 1 CEC of SDS shown in Figure 4.10 also revealed that the characterization peak of MMT has been shifted from $2\theta = 7.2$ to $6.2°$. As the amount of SDS was increased to 3 CEC, not only was the peak further shifted to $6.0°$ but also the peak intensity was drastically increased. The results supported the fact that SDS was able to diffuse into the gallery of MMT, although it did not exfoliate MMT due to the small size of the SDS molecules. Similar intercalation of MMT by anionic surfactants has also been reported.[85–88] However, the basic difference between the cationic surfactant-intercalated MMT and anionic surfactant-intercalated MMT is that the former is hydrophobic and usually referred to as an organified MMT whereas the latter is still hydrophilic and cannot be directly dispersed in polymer. Nevertheless, the anionic surfactant-intercalated MMT has indeed been employed to fabricate the exfoliated MMT–PMMA nanocomposites by emulsion polymerization.[88]

It is believed that the driving force for the PMMA polymerizing chains that carry an anionic group to diffuse into the gallery of MMT is the ionic osmotic pressure. The system is more like the ionic networks in aqueous solution containing small ions from the dissociated external electrolytes.[89] Thus, water tends to diffuse into MMT along with potassium counterions and the polymerizing chains that carry an anionic group act as ion pairs. As more and more polymerizing chains that carry an anionic group diffuse into the gallery of MMT, they form micelles in disk form and exfoliate the MMT.

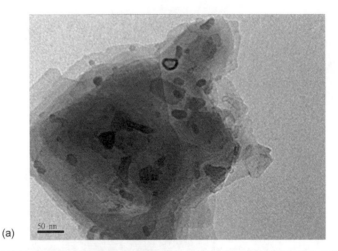

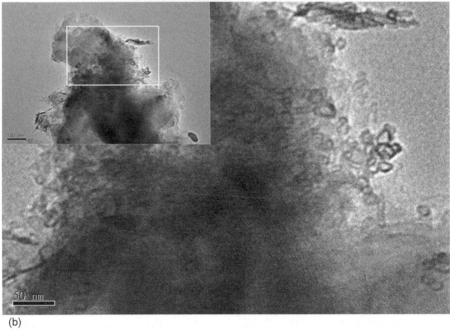

Figure 4.9   TEM images of SDS–MMT prepared by mixing MMT with (a) 0.5 CEC
and (b) 3 CEC of SDS for 3 days. Reproduced from reference 77 with
permission from Wiley.

4.3   Structure and Mechanical Properties of Exfoliated MMT–P(MA-*co*-MMA) Films

The exfoliated MMT–PMMA latex cannot be cast into film because the T_g of
PMMA is well above room temperature. To lower the T_g of nanocomposites,

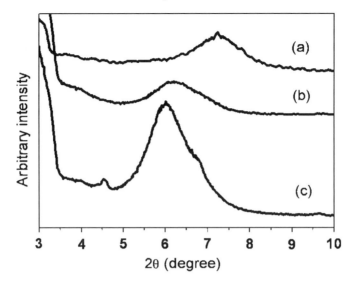

Figure 4.10 X-ray diffraction patterns of (a) MMT and of SDS–MMT powders prepared by mixing MMT with (b) 1 CEC and (c) 3 CEC of SDS. Reproduced from reference 77 with permission from Wiley.

MA monomers were employed to copolymerize with MMA in the presence of MMT during soap-free emulsion polymerization. The fabricated exfoliated MMT–P(MA-*co*-MMA) nanocomposite latexes could be readily cast into films, which were found to have improved mechanical properties resulting from the MgO components of the exfoliated MMT nanoplatelets being grafted by P(MA-*co*-MMA) chains, as verified by FTIR and XPS spectra.[80] Accordingly, the relationship between the structure and mechanical properties of exfoliated MMT–P(MA-*co*-MMA) nanocomposite films is discussed below.

4.3.1 Grafting of P(MA-*co*-MMA) Chains on to Exfoliated MMT Nanoplatelets

First, 1 wt% MMT–P(MA-*co*-MMA) nanocomposite film was fabricated by the following method.[80] To a three-necked glass flask that has been loaded with 500 ml of water were added 1.541 g of KPS and 0.384 g of MMT. After being stirred at room temperature for 3 days, the solution was heated to 70 °C under a nitrogen atmosphere and then 32.24 ml of MA and 7.712 ml of MMA were added for copolymerization. After 24 h, the resulting solution was filtered through an 80-mesh sieve to remove large aggregates and then poured into a 15×30 cm aluminum foil rectangular mold. After drying at room temperature, the cast film was removed from the mold and further dried in an oven for 24 h at 50 °C. The final form of the nanocomposite films had a thickness of 0.25 mm.

The exfoliated MMT–P(MA-*co*-MMA) nanocomposite film was then dissolved in toluene. By adding water to the solution until phase separation, the

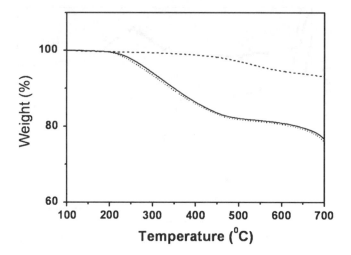

Figure 4.11 TGA plots of (---) pristine MMT and of exfoliated MMT nanoplatelets after purification with toluene (—) five and (· · ·) seven times. Reproduced from reference 80 with permission from Wiley.

water phase which contained the exfoliated MMT was collected. To purify the MMT nanoplatelets further, the collected water phase was further treated with clean toluene until it separated into two phases again. The aqueous phase was collected and the same purification process was repeated until the thermogravimetric analysis (TGA) of the collected exfoliated MMT nanoplatelets showed no difference in the weight change, as illustrated in Figure 4.11.[80] The remaining P(MA-*co*-MMT) matrix on the exfoliated MMT nanoplatelets, which was not dissolvable in toluene but decomposed after heating to 500 °C, was estimated to be about 15 wt%. The image of the remaining P(MA-*co*-MMT) matrix could be clearly seen from the three-dimensional atomic force microscope (AFM) image of the exfoliated MMT nanoplatelets, as shown in Figure 4.12.[80]

The grafting of P(MA-*co*-MMT) chains on the MMT nanoplatelets was then investigated by FTIR spectroscopy and XPS. The FTIR spectrum of the MMT nanoplatelets after purification with toluene is presented in Figure 4.13 along with those of the pristine MMT and neat P(MA-*co*-MMT) for comparison.[80] First, the absorption peak at $1035\,cm^{-1}$ representing the out-of plane Si–O stretching mode became narrower as the MMT was exfoliated into nanoplatelets. Narrowing of this peak associated with the intercalation and exfoliation of MMT has been well studied by Cole[90] and Yan *et al.*[91] The peak at $3020\,cm^{-1}$ representing the =C–H– alkene proton group stretching mode appeared in the FTIR spectrum of exfoliated MMT nanoplatelets, implying that the radicals in the PMMA chain ends transformed into the alkene groups grafting into the MMT nanoplatelets. It has been suggested that the chain transfer of polymerizing radicals to MMT should occur during soap-free emulsion polymerization of MMA in the presence of MMT because the molecular weight of

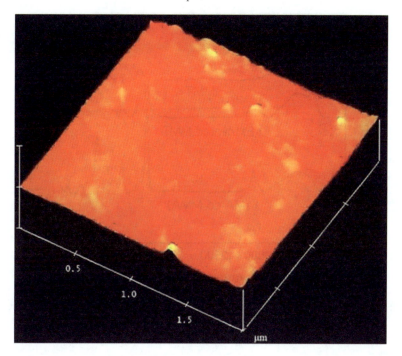

Figure 4.12 Typical AFM image of exfoliated MMT nanoplatelets after purification five times with toluene. Reproduced from reference 80 with permission from Wiley.

the polymerized PMMA matrix did not increase with the observed substantial increase in polymerization rate.[77] Moreover, the peak at 1481 cm^{-1} that has been assigned to the stretching mode of the MgO components[92] was diminished for the exfoliated MMT nanoplatelets compared with the pristine MMT.

The MgO components of exfoliated MMT nanoplatelets grafted by the polymerizing chains were further confirmed by the XPS results.[80] Figure 4.14 presents the XPS spectra of Si 2p, Al 2p, Mg 1s and Mg 2p bonding energies for pristine MMT and exfoliated MMT nanoplatelets. The Si 2p bonding energy peak at 102.8 eV and the Al 2p peak at 74.8 eV contributed by the SiO$_2$ and Al$_2$O$_3$ components[93] of MMT did not change in position but became slightly broader as MMT was exfoliated. However, the Mg 1s bonding energy peak at 1305.8 eV contributed by MgO components[93] of MMT were shifted to 1305.1 eV, whereas the Mg 2p peak at 50.0 eV shifted slightly to 49.8 eV. The decreased bonding energy of MgO components suggested that they had reacted or at least associated with the electron donor groups. Hence it is plausible to support the FTIR data that the P(MA-*co*-MMA) chains have grafted on to the MMT nanoplatelets through chain transfer to the MgO components. The polymerizing chain ends that originally carried free radicals transferred electrons to the MgO groups of MMT nanoplatelets while being transformed themselves into alkene groups.

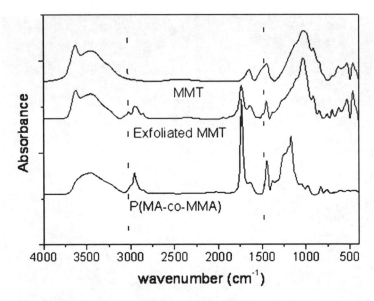

Figure 4.13 FTIR spectra of pristine MMT, exfoliated MMT nanoplatelets after
purification with toluene and neat P(MA-*co*-MMT). Reproduced from
reference 80 with permission from Wiley.

4.3.2 Mechanical Properties of Exfoliated MMT–P(MA-*co*-MMA) Nanocomposite Film

Grafting of P(MA-*co*-MMA) on to the exfoliated MMT nanoplatelets
significantly enhanced the mechanical properties of nanocomposite films.
Figure 4.15 illustrates typical curves for neat P(MA-*co*-MMA) and its nanocomposite films containing 1 wt% MMT.[80] Their average Young's modulus,
yield strength at 0.2% offset strain and ultimate tensile strength and strain are
listed in Table 4.1. Surprisingly, not only was the ultimate tensile strength
doubled but also the ultimate strain increased from 1140% for the neat
copolymer to 1210% for the nanocomposite film. Although increased Young's
modulus and yield strength for the MMT–polymer nanocomposites with
increase in the content of MMT have frequently been reported,[94–97] there have
been very few studies that have reported that the ultimate strain of an MMT–
polymer nanocomposite was greater than that of the neat polymer. Notably,
strain hardening after 400% strain for the nanocomposite film as a result of
plastic deformation was also observed but not for neat P(MA-*co*-MMA), as
revealed in Figure 4.15.[80] This might be due to the fact that the MMT nanoplatelets were aligned in the straining direction during high elongation of the
specimen, leading to an increase in the modulus in stretching. In addition, a
much larger strain recovery ratio after creep for 1 wt% MMT–P(MA-*co*-
MMA) nanocomposite film than that for the neat P(MA-*co*-MMA) was
observed, as shown in Figure 4.16. On applying a constant stress of 1 MPa for

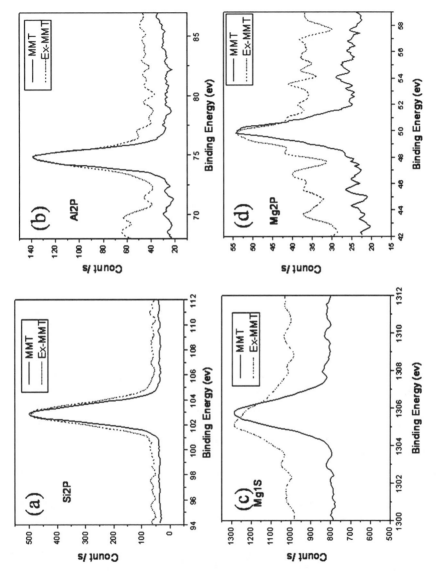

Figure 4.14 Normalized XPS spectra of (a) Si 2p, (b) Al 2p, (c) Mg 1s and (d) Mg 2p for (——) pristine MMT and (···) exfoliated MMT nanoplatelets after purification with toluene. Reproduced from reference 80 with permission from Wiley.

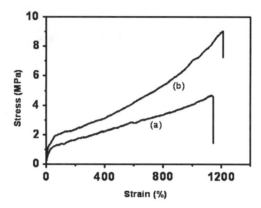

Figure 4.15   Stress–strain curves for (a) neat P(MA-*co*-MMA) and (b) 1 wt% MMT–P(MA-*co*-MMA) nanocomposite films. Reproduced from reference 80 with permission from Wiley.

Table 4.1   Properties of P(MA-*co*-MMA) and 1 wt% MMT–P(MA-*co*-MMA) nanocomposite films.

Film	Young's modulus (MPa)	Yield strength (MPa)	Ultimate strength (MPa)	Ultimate strain (%)
P(MA-*co*-MMA)	36.6 (±7.1)	1.23 (±0.04)	4.95 (±0.29)	1000 (±93)
1 wt% MMT–P(MA-*co*-MMA)	56.6 (±7.7)	1.97 (±0.16)	9.97 (±0.79)	1202 (±78)

60 min, the neat copolymer film only crept to 23% strain whereas 1 wt% MMT–P(MA-*co*-MMA) nanocomposite crept to 180%. Because the MMT nanoplatelets will not deform, the excess creep for the nanocomposite film indicates that the P(MA-*co*-MMA) matrix was much easier to deform than the neat P(MA-*co*-MMA). This might be attributable to the slightly lower T_g of the P(MA-*co*-MMA) matrix compared with the neat P(MA-*co*-MMA) as measured by differential scanning calorimetry (DSC) (see Table 4.2). After the stress had been released for 180 min, the neat P(MA-*co*-MMA) film recovered to 14% strain whereas the nanocomposite also recovered to 14%. Interestingly, 157% excess strain for the crept nanocomposite film was fully recovered in 180 min.

The grafted MMT nanoplatelets can retard the crack propagation of nanocomposite films. To investigate this effect, a TEM sample was prepared by the following method.[80] First, a small piece of 1 wt% MMT–P(MA-*co*-MMA) film was dissolved in tetrahydrofuran. After it had completely dissolved, a carbon film-coated copper grid was dipped into the solution and then immediately withdrawn and dried in a desiccator. When it was fully dried, the specimen was ripped with tweezers and subjected to TEM investigation.

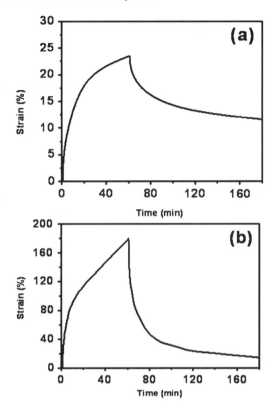

Figure 4.16 Creep and recovery curves for (a) neat P(MA-*co*-MMA) and (b) 1 wt%
MMT–P(MA-*co*-MMA) nanocomposite films by applying a constant
stress of 1 MPa for 60 min and then releasing the stress. Reproduced
from reference 80 with permission from Wiley.

Table 4.2 Properties of exfoliated MMT–P(MA-*co*-MMA) nanocomposite
films with various MMT contents.

MMT (wt%)	Young's modulus (MPa)	Yield strength (MPa)	Ultimate tensile strain (%)	T_g (°C) By DSC	T_g (°C) By DMA
0	36.6 ($\pm$7.1)	1.23 ($\pm$0.04)	1000 ($\pm$93.2)	19.2	22.7
1	56.6 ($\pm$7.7)	1.97 ($\pm$0.16)	1202 ($\pm$78.2)	17.5	26.4
5	165.4 ($\pm$29.5)	4.07 ($\pm$0.10)	484 ($\pm$67.9)	18.3	27.2
10	200.4 ($\pm$32.4)	6.12 ($\pm$0.12)	184 ($\pm$45.5)	17.5	36.5
15	657.4 ($\pm$77.4)	21.7 ($\pm$2.17)	12 ($\pm$2.5)	16.3	32.9
20	920.6 ($\pm$69.2)	24.7 ($\pm$0.24)	3 ($\pm$1.7)	17.2	32.3

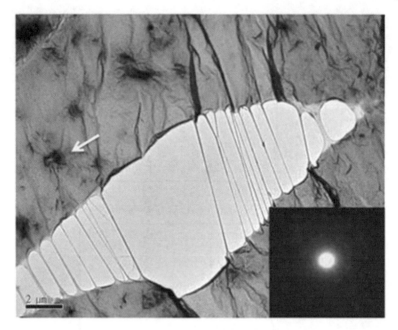

Figure 4.17   TEM image of 1 wt% MMT–P(MA-*co*-MMA) nanocomposite film after ripping of the carbon-coated copper grid to generate cracks. The arrow indicates the area where the diffraction pattern of MMT was taken. Reproduced from reference 80 with permission from Wiley.

Figure 4.17 shows the TEM image of the film with a crack. Microfibrils formed by crazing inside the crack can be clearly seen. The crack developed within the polymer matrix did not pass through the interfacial region between the polymer matrix and MMT nanoplatelets. Further, the polymer matrix adjacent to the crack crinkled and tended to withdraw towards the MMT nanoplatelets, which were identified by the electron diffraction pattern as shown in the inset. The results indicate that the interfacial strength between P(MA-*co*-MMA) and its grafted MMT nanoplatelets was higher than the cohesion strength of the P(MA-*co*-MMA) matrix and the well-dispersed MMT nanoplatelets acted like a fortifier for the nanocomposite film.

4.3.3   Structure–Thermomechanical Property Relationship of Exfoliated MMT–P(MA-*co*-MMA) Nanocomposite Films

The exfoliated MMT nanoplatelets in the P(MA-*co*-MMA) nanocomposite films cast from their latexes tended to restack with each other when the content of MMT was more than 10 wt%, which greatly affected the thermomechanical properties and fracture behavior of the nanocomposite films. Figure 4.18a shows the XRD patterns of neat MMT and its prepared MMT–P(MA-*co*-MMA) nanocomposite latex particles with different contents of MMT.[79] The

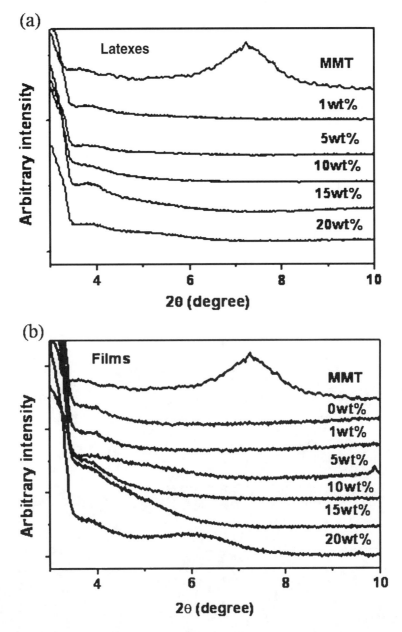

Figure 4.18 X-ray diffraction patterns of MMT–P(MA-*co*-MMA) nanocomposite (a) latex particles and (b) films with the indicated contents of MMT. The X-ray diffraction pattern of pristine MMT is also included for comparison. Reproduced from reference 79.

characteristic diffraction peak of neat MMT disappeared and no additional peak was observable for any of the MMT–P(MA-*co*-MMA) nanocomposite latex particles, suggesting that the MMTs had been fully exfoliated during soap-free emulsion polymerization. As the exfoliated MMT–P(MA-*co*-MMA) nanocomposite latex solutions were cast into films, their XRD patterns were barely changed until the content of MMT was more than 10 wt% (Figure 4.18b). A broad shoulder at $2\theta = 3.5$–$6°$ in the diffraction pattern appeared when the content was increased to 15 wt%. It shifted to a larger angle and formed a broad peak at $2\theta = 6.2°$ when the content was further increased to 20 wt%, implying that the exfoliated MMT nanoplatelets tended to restack with each other during film formation to generate the diffraction peak.

Figure 4.19a and b show the TEM images of 15 wt% MMT–P(MA-*co*-MMA) and 20 wt% MMT–P(MA-*co*-MMA) nanocomposite dried latexes, respectively.[79] Because the T_g of the P(MA-*co*-MMA) matrix measured by DSC was only 17 °C (see Table 4.2), the exfoliated MMT–P(MA-*co*-MMA) latex particle had no well-defined texture. When the 15 wt% MMT–P(MA-*co*-MMA) and 20 wt% MMT–P(MA-*co*-MMA) nanocomposite latexes were cast

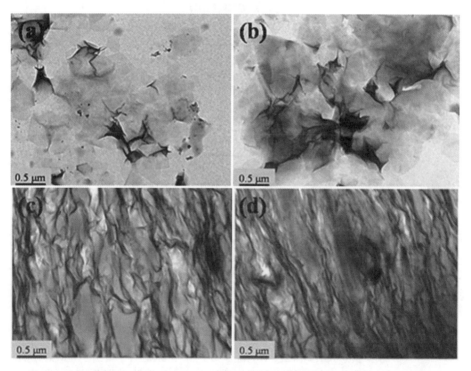

Figure 4.19   TEM images of (a) 15 wt% MMT–P(MA-*co*-MMA) and (b) 20 wt% MMT–P(MA-*co*-MMA) nanocomposite dried latexes after dilution and cross-sections of their respective cast films: (c) 15 wt% MMT–P(MA-*co*-MMA) and (d) 20 wt% MMT–P(MA-*co*-MMA). Reproduced from reference 79.

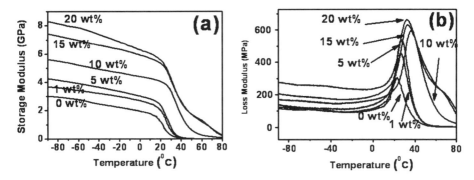

Figure 4.20 (a) Storage and (b) loss modulus spectra of neat P(MA-*co*-MMA) and its exfoliated MMT nanocomposite films with the indicated contents of MMT. Reproduced from reference 79.

into films, the TEM images of cross-sectioned samples revealed that the exfoliated MMT nanoplatelets had restacked with each other (Figure 4.19c and d).[79] Further, the 20 wt% MMT–P(MA-*co*-MMA) nanocomposite film exhibited a higher packing density and shorter packing distance than the 15 wt% MMT–P(MA-*co*-MMA), which is consistent with the WAXS results (Figure 4.18b).

Figure 4.20 shows the dynamic mechanical properties of neat P(MA-*co*-MMA) and its exfoliated MMT nanocomposite films with different MMT contents.[79] Not only did the storage modulus of the nanocomposite increase with increase in MMT content, but also the T_g measured from the loss modulus peak increased until 10 wt% MMT was reached. Further, the T_g peak became broader along with an increase in intensity as the content of MMT increased, indicating that the incorporation of MMT nanoplatelets increased the potential energy for the local motion of polymer segments in glass transition.[98,99] Notably, the T_g measured by DSC decreased slightly with increase in the content of MMT, as shown in Figure 4.21, also suggesting that the increase in T_g measured by DMA resulted from the hindered segmental motion of P(MA-*co*-MMA) chains by MMT nanoplatelets.[79] T_g data measured by DSC and DMA are both listed in Table 4.2 for comparison. Interestingly, although the T_g of 15 wt% MMT–P(MA-*co*-MMA) nanocomposite film measured by DMA is slightly lower than that of 10 wt% MMT–P(MA-*co*-MMA) nanocomposite film, its loss modulus peak contains an additional shoulder located at $\sim 70\,^\circ$C, which corresponds to the restacking of MMT nanoplatelets as observed from its XRD pattern and TEM. This implies that the shoulder was contributed by the excessively blocked local motion of polymer segments sandwiched in between the restacking MMT nanoplatelets. It is noteworthy that the decreased T_g for the nanocomposite compared with the neat copolymer measured by DSC might be due to the 'nanoconfinement' of MMT nanoplatelets on the P(MA-*co*-MMA) matrix, a phenomenon which has been well studied by Li and Simon.[100] They suggested that the magnitude of the T_g

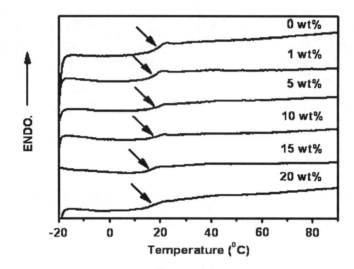

Figure 4.21   DSC curves for neat P(MA-*co*-MMA) and its exfoliated MMT nanocomposite films with the indicated contents of MMT. The T_g values for
the samples are indicated by arrows. Reproduced from reference 79.

depression measured by DSC is probably related to the size of the polymer
segments being confined relative to the confinement size.

The typical stress–strain curves for the neat P(MA-*co*-MMA) and its
exfoliated nanocomposite films with different MMT contents are shown in
Figure 4.22.[79] Their average Young's modulus, yield strength at 0.2% offset
strain and ultimate tensile strain are included in Table 4.2. Both Young's
modulus and yield strength were substantially increased with increase in the
content of MMT. The increase was much more significant when the MMT
content was increased from 10 to 15 wt%. When the MMT content was more
than 1 wt%, the ultimate strain of the nanocomposite films decreased with
increase in the content of MMT, as shown in Figure 4.22 and Table 4.2.
Nevertheless, 10 wt% MMT–P(MA-*co*-MMA) nanocomposite film still has
231% strain, behaving like a ductile material, whereas the strain of the film
containing 15 wt% MMT decreased to 12%, behaving like a brittle material.
With further increase in MMT content to 20 wt%, the strain decreased to 3%.
Obviously, restacking of the exfoliated MMT nanoplatelets significantly
affected the tensile properties of the nanocomposite films.

Notably, the ductile films of neat P(MA-*co*-MMA) and exfoliated MMT–
P(MA-*co*-MMA) nanocomposites could be stretched to show visible whitening
owing to crazing or void formation. Figure 4.23 shows the SEM images of neat
P(MA-*co*-MMA) and 1 and 10 wt% MMT–P(MA-*co*-MMA) nanocomposite
films being stretched to show the appearance of whitening.[79] It is surprising that
the neat P(MA-*co*-MMA) film showed cracks with a close to circular shape,
whereas the cracks in the 1 wt% MMT–P(MA-*co*-MMA) nanocomposite film

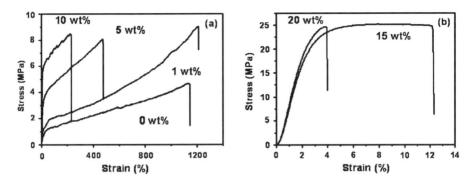

Figure 4.22   Stress–strain curves for neat P(MA-*co*-MMA) and its exfoliated MMT nanocomposite films with the indicated contents of MMT. Reproduced from reference 79.

exhibited an elongated elliptical shape. According to the fracture mechanics, the specimen with elongated elliptical cracks can withstand higher stress, σ, than that with the circular cracks, as derived by[101]

$$\sigma_y = \sigma\left(1 + \frac{2a}{b}\right) \qquad (4.1)$$

where σ_y is the yield stress on the crack tip and a and b are the length from the crack center to the edges in the axes perpendicular and parallel to the stress direction respectively. So, if the deformation energy is not considered, once the applied stress is higher than one-third of σ_y, the circular crack will propagate. For the elongated elliptical crack with $2a/b = \frac{1}{2}$ as for the roughly estimated crack profile in the stretched 1 wt% MMT–P(MA-*co*-MMA) nanocomposite films illustrated in Figure 4.23b, the crack propagates only when the applied stress is higher than two-thirds of σ_y. This plausibly explains why 1 wt% MMT–P(MA-*co*-MMA) nanocomposite film has a higher yield strength and ultimate tensile strain than the neat copolymer. The formation of elongated elliptical cracks supported the view that the crack propagation in the polymer matrix has been hindered by the exfoliated MMTs, as shown in the TEM image of Figure 4.17.

However, when the content of MMT reaches 10 wt%, the surface of the stretched nanocomposite film shows line cracks with the long axis perpendicular to the stretching direction. It is noteworthy that for the 10 wt% MMT–P(MA-*co*-MMA) nanocomposite film, the tensile strength and Young's modulus were increased about fivefold whereas the ultimate tensile strain was reduced to about one-fifth of that of the neat P(MA-*co*-MMA) film. When the MMT content was increased to more than 10%, no whitening appeared by stretching to fracture. Obviously, restacking of the MMT nanoplatelets in the nanocomposite films changed the fracture behavior from ductile to brittle nature.

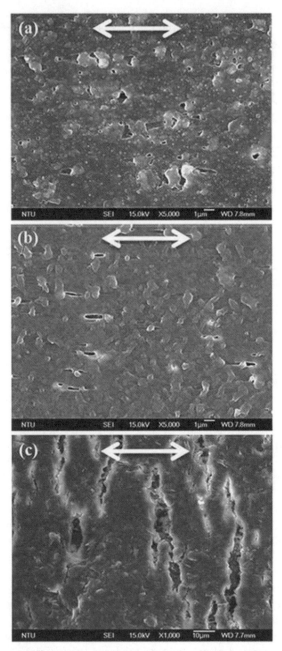

Figure 4.23   SEM images of (a) neat P(MA-*co*-MMA), (b) 1 wt% MMT–P(MA-*co*-MMA) nanocomposite and (c) 10 wt% MMT/ P(MA-*co*-MMA) nano-composite films after stretching so whitening occurs. Reproduced from reference 79.

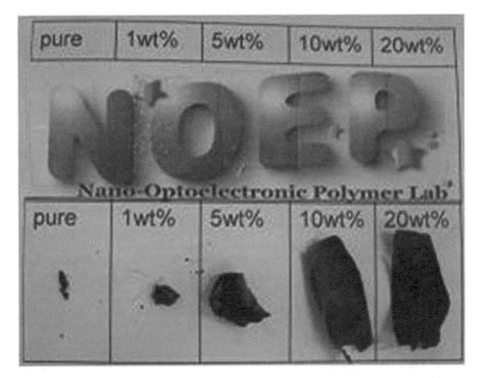

Figure 4.24 Photograph of the MMT–P(MA-*co*-MMA) nanocomposite films prepared with a 3:1 weight ratio of MA to MMA with the indicated MMT contents (top) and their residuals after combustion (bottom). Reproduced from reference 78 with permission from Springer.

4.4 Physical Properties of Exfoliated MMT–PMMA Nanocomposite Films

The cast exfoliated MMT–P(MA-*co*-MMA) nanocomposite films are transparent even though they become blurred when the content of MMT exceeds 5 wt% owing to the light scattering from exfoliated MMT nanoplatelets.[78] Figure 4.24 illustrates typical photographs of the cast films. Their absolute transmittance measured by UV/VIS spectrophotometry is given in Figure 4.25, showing no absorption peak in the visible region. Interestingly, when all the cast films were subjected to combustion in a fire, the char residuals of P(MA-*co*-MMA)/MMT nanocomposite films with MMT content higher than 5 wt% preserved their original film profile.[78] However, the neat copolymer films were immediately burned to ash as soon as they became close to the flame. Apparently, the films containing MMT could impede the burning. If the amount of exfoliated MMT nanoplatelets is sufficient to form a continue phase inside the cast film, they will act like a scaffold to preserve the film profile after burning. Conceivably, as soon as the combustion occurs, this scaffold would have the potential to block the flame propagation.

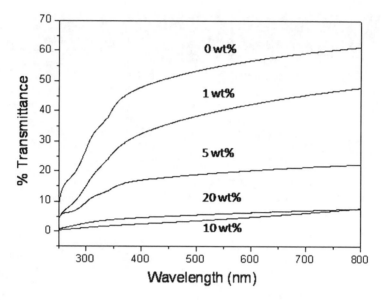

Figure 4.25   Transmittance of MMT–P(MA-*co*-MMA) nanocomposite films prepared with a 3:1 weight ratio of MA to MMA with the indicated MMT
contents. Reproduced from reference 78 with permission from Springer.

The water vapor permeability of exfoliated MMT–P(MA-*co*-MMA) nanocomposite films was investigated at 40 °C following the test method of ASTM
E96.[78] The samples were dried at 80 °C to a constant weight before testing.
First, ~ 150 ml of deionized water was placed in a cylindrical cup. The mouth
of the cup was sealed with a dried film specimen and then the cup was placed in
an oven maintained at 40 °C and 39% relative humidity. The weight was
recorded periodically until the change of weight with time was linear. The slope
of the linear region was used to estimate the water vapor transmission δ, which
was defined as the weight loss per unit time and unit area of the film specimen.
Then, the permeability coefficient K_V of water vapor was estimated by using the
following equation:

$$K_V = \frac{\delta d}{P_s(1 - R_h)} \tag{4.2}$$

where d is the thickness of film specimen, P_s is the saturated water vapor
pressure at the test temperature and R_h is the relative humidity outside the test
cup expressed as a fraction. The results shown in Figure 4.26 indicate that the
permeability decreased with decrease in the weight ratio of MA to MMA and
increase in the content of MMT. However, the extent of decreased permeability
for the exfoliated MMT–P(MA-*co*-MMA) nanocomposite films compared with
the neat copolymers was not as great as the data reported for cast exfoliated
MMT–PVAc nanocomposite films.[102,103] This might be due to the fact that T_g

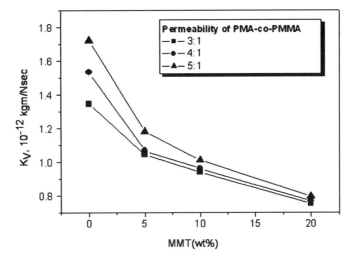

Figure 4.26 Permeability coefficient K_V of MMT–P(MA-*co*-MMA) nanocomposite films as a function of the MMT content, prepared with (■) 3:1, (●) 4:1 and (▲) 5:1 weight ratios of MA to MMA. Reproduced from reference 78 with permission from Springer.

of the MMT–P(MA-*co*-MMA) nanocomposite films is lower than room temperature. Further, the T_g of the P(MA-*co*-MMA) matrix in nanocomposite films measured by DSC was also lower than that of neat copolymers (see Table 4.2), creating excess free volume for the water vapor to be transported in the polymer matrix.

4.5 Conclusion

This chapter introduced the method of fabrication of exfoliated MMT–PMMA nanocomposite latex through soap-free emulsion polymerization, by which the polymer chains were able to graft on to the exfoliated MMT nanoplatelets. For the exfoliation mechanism of MMT it has been demonstrated that during micellation of PMMA in the presence of MMT, the polymerizing chains were able to aggregate into a disk domain in the interlayer regions of MMT and execute the exfoliation. However, the fabricated MMT–PMMA nanocomposite latex particles are too rigid to be cast into a film. To lower the T_g of nanocomposites, MA monomers were used to copolymerize with MMA and the resulting exfoliated MMT–P(MA-*co*-MMA) nanocomposite latexes could be readily cast into films, which exhibited many unusual properties, such as low permeability, high flame retardation and superior tensile properties.

Notably, this fabrication method can also be applied to other polymers, such as PVAc and poly(*n*-isopropylacrylamide) (PNIPAm).[102–105] The unusual phase transition behavior of the fabricated MMT–PNIPAm nanocomposites has been reported to be due to the hydrogen bonding between the isopropylacrylamide side-groups of PNIPAm chain segments and the anionic groups of the exfoliated MMT nanoplatelets.[104] The exfoliated MMT–PNIPAm nanocomposite latex particles have also been applied to gel the liquid electrolyte system for fabrication of gel-type dye sensitized solar cells.[105] Further, the exfoliated MMT nanoplatelets can be collected by removing the polymer matrix from the nanocomposite latex with solvent. They were found to be well dispersed in water and performed like a two-dimensional electrolyte. Their ionic conductivity roughly follows Manning's limiting law[106] for the conduction of a polyelectrolyte. The dissociated MMT nanoplatelets that carry negative charges in water were able to adsorb cations, such as tris(2,2′-bipyridyl)ruthenium(II) [$Ru(bpy)_3^{2+}$] and methylene blue (MB^+) rapidly and recover into a smectic configuration.[107] Since MMT is a bio-benign material, there has been growing interest in including MMT as a potential component in biomedical applications such as tissue engineering.[108,109] Hence exfoliated MMT nanoplatelets may have a chance to be applied as a nanoencapsulate for drug-releasing agents. Overall, the potential applications of exfoliated MMT nanocomposite latexes fabricated by soap-free emulsion polymerization are versatile.

References

1. J. A. Brydson, *Plastics Materials*, Butterworth-Heinemann, Oxford, 1999.
2. G. Kong, R. Braun and M. Dewhirst, *Cancer Res.*, 2000, **60**, 4440.
3. M. T. Koesdjojo, Y. H. Tennico and V. T. Remcho, *Anal. Chem.*, 2008, **80**, 2311.
4. M. Crne, J. O. Park and M. Srinivasarao, *Macromolecules*, 2009, **42**, 4353.
5. J. B. Park and R. S. Lakes, *Biomaterials: an Introduction*, Plenum Press, New York, 1992.
6. N. Daraboina and G. Madras, *Ind. Eng. Chem. Res.*, 2008, **47**, 6828.
7. A. Chatterjee, *J. Appl. Polym. Sci.*, 2010, **116**, 3396.
8. J. Wang and X. Ni, *J. Appl. Polym. Sci.*, 2008, **108**, 3552.
9. A. Convertino, G. Leo, M. Striccoli, G. D. Marco and M. L. Curri, *Polymer*, 2008, **49**, 5526.
10. S. Hess, M. M. Demir, V. Yakutkin, S. Baluschev and G. Wegner, *Macromol. Rapid Commun.*, 2009, **30**, 394.
11. R. Y. Hong, J. Z. Qian and J. X. Cao, *Powder Technol.*, 2006, **163**, 160.
12. T. C. Chang, Y. T. Wang, Y. S. Hong and Y. S. Chiu, *J. Polym. Sci., Part A: Polym. Chem.*, 2000, **38**, 1972.
13. J. Zheng, R. Zhu, Z. He, G. Cheng, H. Wang and K. Yao, *J. Appl. Polym. Sci.*, 2010, **115**, 1975.
14. P. C. Wang, W. Y. Chiu and T. H. Young, *J. Appl. Polym. Sci.*, 2006, **100**, 4925.
15. Z. Cao, W. Jiang, X. Ye and X. Gong, *J. Magn. Magn. Mater.*, 2008, **320**, 1499.
16. M. Vaysse, M. K. Khan and P. Sundararajan, *Langmuir*, 2009, **25**, 7042.

17. R. B. Mathur, S. Pande, B. P. Singh and T. L. Dhami, *Polym. Compos.*, 2008, **29**, 717.

18. S. M. Yuen, C. C. M. Ma, C. Y. Chuang, K. C. Yu, S. Y. Wu, C. C. Yang and M. H. Wei, *Compos. Sci. Technol.*, 2008, **68**, 963.

19. C. Zeng, N. Hossieny, C. Zhang and B. Wang, *Polymer*, 2010, **51**, 655.

20. J. Liu, A. Rasheed, M. L. Minus and S. Kumar, *J. Appl. Polym. Sci.*, 2009, **112**, 142.

21. S. J. Park, M. S. Cho, S. T. Lim, H. J. Choi and M. S. Jhon, *Macromol. Rapid Commun.*, 2003, **24**, 1070.

22. L. Cui, N. H. Tarte and S. I. Woo, *Macromolecules*, 2009, **42**, 8649.

23. S. S. Hou and K. S. Rohr, *Chem. Mater.*, 2003, **15**, 1938.

24. P. A. Wheeler, J. Wang and L. J. Mathias, *Chem. Mater.*, 2006, **18**, 3937.

25. J. M. Yeh, S. J. Liou, C. Y. Lin, C. Y. Cheng, Y. W. Chang and K. R. Lee, *Chem. Mater.*, 2002, **14**, 154.

26. M. Si, T. Araki, H. Ade, A. L. D. Kilcoyne, R. Fisher, J. C. Sokolov and M. H. Rafailovich, *Macromolecules*, 2006, **39**, 4793.

27. H. Muenstedt, N. Katsikis and J. Kaschta, *Macromolecules*, 2008, **41**, 9777.

28. G. B. Rossi, G. Beaucage, T. D. Dang and R. A. Vaia, *Nano Lett.*, 2002, **2**, 319.

29. A. S. Zerda, T. C. Caskey and A. J. Lesser, *Macromolecules*, 2003, **36**, 1603.

30. S. Mallick, A. K. Dhibar and B. B. Khatua, *J. Appl. Polym. Sci.*, 2010, **116**, 1010.

31. Y. Wang and J. Y. Guo, *Polym. Compos.*, 2010, **31**, 596.

32. D. Lerari, S. Peeterbroeck, S. Benali, A. Benaboura and P. Dubois, *Polym. Int.*, 2010, **59**, 71.

33. L. M. Stadtmueller, K. R. Ratinac and S. P. Ringer, *Polymer*, 2005, **46**, 9574.

34. K. P. Pramoda, N. T. T. Linh, P. S. Tang, W. C. Tjiu, S. H. Goh and C. B. He, *Compos. Sci. Technol.*, 2010, **70**, 578.

35. G. Chen, W. Weng, D. Wu and C. Wu, *Eur. Polym. J.*, 2003, **39**, 2329.

36. J. Y. Jang, M. S. Kim, H. M. Jeong and C. M. Shin, *Compos. Sci. Technol.*, 2009, **69**, 186.

37. T. Ramanathan, S. Stankovich, D. A. Dikin, H. Liu, H. Shen, S. T. Nguyen and L. C. Brinson, *J. Polym. Sci., Part B: Polym. Phys.*, 2007, **45**, 2097.

38. J. Zhu, A. B. Morgan, F. J. Lamelas and C. A. Wilkie, *Chem. Mater.*, 2001, **13**, 3774.

39. X. Hu and J. Meng, *J. Polym. Sci., Part A: Polym. Chem.*, 2005, **43**, 994.

40. S. Ray, G. Galgali, A. Lele and S. J. Sivaram, *J. Polym. Sci., Part A: Polym. Chem.*, 2005, **43**, 304.

41. M. W. Weimer, H. Chen, E. P. Giannelis and D. Y. Sogah, *J. Am. Chem. Soc.*, 1999, **121**, 1615.

42. H. Zhao, S. D. Argoti, B. P. Farrell and D. A Shipp, *J. Polym. Sci., Part A: Polym. Chem.*, 2004, **42**, 916.

43. D. Yebassa, S. Balakrishnan, E. Feresenbet, D. Raghavan, P. R. Start and S. D. Hudson, *J. Polym. Sci., Part A: Polym. Chem.*, 2004, **42**, 1310.
44. J. Ma, Z. Z. Yu, Q. X. Zhang, X. L. Xie, Y. W. Mai and I. Luck, *Chem. Mater.*, 2004, **16**, 757.
45. X. Fan, C. Xia and R. C. Advincula, *Langmuir*, 2005, **21**, 2537.
46. Y. S. Choi, H. T. Ham and I. Chung, *J. Polym.*, 2003, **44**, 8147.
47. Y. S. Choi and I. Chung, *J. Polym.*, 2004, **45**, 3827.
48. X. Huang and W. J. Brittain, *Macromolecules*, 2001, **34**, 3255.
49. S. K. Choudhari and M. Y. Kariduraganavar, *J. Colloid Interface Sci.*, 2009, **157**, 44.
50. M. S. Kim, J. K. Jun and H. M. Jeong, *Compos. Sci. Technol.*, 2008, **68**, 1919.
51. C. I. W. Calcagno, C. M. Mariani, S. R. Teixeira and R. S. Mauler, *Compos. Sci. Technol.*, 2008, **68**, 2193.
52. T. Pojanavaraphan and R. Magaraphan, *Polymer*, 2010, **51**, 1111.
53. L. Qiu, W. Chen and B. Qu, *Polymer*, 2006, **47**, 922.
54. Y. Kojima, A. Usuki, M. Kawasumi, A. Okada, Y. Fukushima, T. Kurauchi and O. Kamigaito, *J. Polym. Sci., Part A: Polym. Chem.*, 1993, **31**, 1755.
55. Y. Kojima, A. Usuki, M. Kawasumi, A. Okada, Y. Fukushima, T. Kurauchi and O. Kamigaito, *J. Mater. Res.*, 1993, **8**, 1185.
56. A. Usuki, Y. Kojima, M. Kawasumi, A. Okada, Y. Fukushima, T. Kurauchi and O. Kamigaito, *J. Mater. Res.*, 1993, **8**, 1179.
57. Y. Kojima, A. Usuki, M. Kawasumi, A. Okada, T. Kurauchi and O. Kamigaito, *J. Polym. Sci., Part A: Polym. Chem.*, 1993, **31**, 983.
58. A. Usuki, N. Hasegawa, H. Kadoura and T. Okamoto, *Nano Lett.*, 2001, **1**, 271.
59. Z. Zhao, T. Tang, Y. Qin and B. Huang, *Langmuir*, 2003, **19**, 9260.
60. T. Lan, P. D. Kaviratna and T. J. Pinnavaia, *J. Phys. Chem. Solids*, 1996, **57**, 1005.
61. H. Z. Shi, T. Lan and T. Pinnavaia, *J. Chem. Mater.*, 1996, **8**, 1584.
62. J. J. Lin, Y. C. Hsu and K. L. Wei, *Macromolecules*, 2007, **40**, 1579.
63. X. Kornmann, H. Lindberg and L. A. Berglund, *Polymer*, 2001, **42**, 1303.
64. E. P. Giannelis, R. Krishnamoorti and E. Manias, *Adv. Polym. Sci.*, 1999, **138**, 107.
65. K. Yano, A. Usuki and A. Okada, *J. Polym. Sci., Part A: Polym. Chem.*, 1997, **35**, 2289.
66. Z. Shen, Y. B. Cheng and G. P. Simon, *Macromolecules*, 2005, **38**, 1744.
67. L. Cui, N. H. Tarte and S. I. Woo, *Macromolecules*, 2008, **41**, 4268.
68. N. Moussaif and G. Groeninckx, *Polymer*, 2003, **44**, 7899.
69. J. H. Wang, T. H. Young, D. J. Lin, M. K. Sun, H. S. Huag and L. P. Cheng, *Macromol. Mater. Eng.*, 2006, **291**, 661.
70. M. Lewin, *Polym. Adv. Technol.*, 2006, **17**, 758.
71. M. Hernandez, B. Sixou, J. Duchet and H. Sautereau, *Polymer*, 2007, **48**, 4075.
72. D. C. Lee and L. W. Jang, *J. Appl. Polym. Sci.*, 1996, **61**, 1117.

73. F. Tiarks, K. Landfester and M. Antonietti, *Langmuir*, 2001, **17**, 5775.
74. J. L. L. Xavier, A. Guyot and E. B. Lami, *J. Colloid Interface Sci.*, 2002, **250**, 82.
75. P. Meneghetti and S. Qutubuddin, *Langmuir*, 2004, **20**, 3424.
76. K. F. Lin, S. C. Lin, A. T. Chien, C. C. Hsieh, M. H. Yen, C. H. Lee, C. S. Lin, W. Y. Chen and Y. H. Lee, *J. Polym. Sci., Part A: Polym. Chem.*, 2006, **44**, 5572.
77. K. J. Lin, C. A. Dai and K. F. Lin, *J. Polym. Sci., Part A: Polym. Chem.*, 2009, **47**, 459.
78. C. H. Lee, A. T. Chien, M. H. Yen and K. F. Lin, *J. Polym. Res.*, 2008, **15**, 331.
79. K. J. Lin, C. H. Lee and K. F. Lin, *J. Polym. Sci., Part B: Polym. Phys.*, 2010, **48**, 1064.
80. K. J. Lin, H. W. Ting, C. H. Lee and K. F. Lin, *J. Polym. Sci., Part A: Polym. Chem.*, 2009, **47**, 5891.
81. Y. C. Chen, C. F. Lee and W. Y. Chiu, *J. Appl. Polym. Sci.*, 1996, **61**, 2235.
82. K. F. Lin and Y. D. Shieh, *J. Appl. Polym. Sci.*, 1998, **69**, 2069.
83. K. F. Lin and Y. D. Shieh, *J. Appl. Polym. Sci.*, 1998, **70**, 2313.
84. P. C. Hiemenz, *Principles of Colloid and Surface Chemistry*, Wiley, New York, 1986, Chapter 8.
85. M. Mravěáková, M. Omastová, K. Olejníková B. Pukánszky and M. M. Chehimi, *Synth. Met.*, 2007, **157**, 347.
86. Y. S. Choi, M. Xu and I. J. Chung, *Polymer*, 2005, **46**, 531.
87. C. C. Chou, M. L. Chiang and J. J. Lin, *Macromol. Rapid Commun.*, 2001, **26**, 1841.
88. Y. S. Choi, M. H. Choi, K. H. Wang, S. O. Kim, Y. K. Kim and I. J. Chung, *Macromolecules*, 2001, **34**, 8978.
89. P. J. Flory, *Principles of Polymer Chemistry*, Cornell University Press, Ithaca, NY, 1953, Chapter 8.
90. K. C. Cole, *Macromolecules*, 2008, **41**, 834.
91. L. Yan, C. B. Roth and P. F. Low, *Langmuir*, 1996, **12**, 4421.
92. R. A. Niquist and R. O. Kagel, *Infrared Spectra of Inorganic Compounds*, Academic Press, New York, 1971.
93. D. Briggs and M. P. Seah, *Practical Surface Analysis by Auger and X-ray Photoelectron Spectroscopy*, Wiley, New York, 1983.
94. C. Tang, L. Xiang, J. Su, K. Wang, C. Yang, Q. Zhang and Q. Fu, *J. Phys. Chem. B*, 2008, **112**, 3876.
95. F. D. C. Braganca, L. F. Valadares, C. A. D. P. Leite and F. Galembeck, *Chem. Mater.*, 2007, **19**, 3334.
96. T. D. Fornes and D. R. Paul, *Macromolecules*, 2004, **37**, 7698.
97. C. He, T. Liu, W. C. Tjiu, H. J. Sue and A. F. Yee, *Macromolecules*, 2008, **41**, 193.
98. T. Mcnally, W. R. Murphy, C. Lew, R. Turner and G. Brennan, *Polymer*, 2003, **44**, 2761.
99. J. Mijoviæ and K. F. Lin, *J. Appl. Polym. Sci.*, 1985, **30**, 2527.

100. Q. Li and S. L. Simon, *Macromolecules*, 2008, **41**, 1310.
101. J. G. Williams, *Fracture Mechanics of Polymers*, Ellis Horwood, Chichester, 1984.
102. A. T. Chien and K. F. Lin, *J. Polym. Sci., Part A: Polym. Chem.*, 2007, **45**, 5583.
103. A. T. Chien, Y. H. Lee and K. F. Lin, *J. Appl. Polym. Sci.*, 2008, **109**, 355.
104. M. H. Yen and K. F. Lin, *J. Polym. Sci., Part B: Polym. Phys.*, 2009, **47**, 524.
105. C. W. Tu, K. Y. Liu, A. T. Chien, M. H. Yen, T. H. Weng, K. C. Ho and K. F. Lin, *J. Polym. Sci., Part A: Polym. Chem.*, 2008, **46**, 47.
106. G. S. Manning, *J. Phys. Chem.*, 1981, **85**, 1506.
107. C. H. Lee, T. H. Weng, K. Y. Liu, K. J. Lin and K.F. Lin, *J. Appl. Polym. Sci.*, early view on web, DOI: 10.1002/app.32390.
108. K. F. Lin, C. Y. Hsu, T. S. Huang, W.-Y. Chiu, Y.-H. Lee and T. H. Young, *J. Appl. Polym. Sci.*, 2005, **98**, 2042.
109. K. S. Katti, D. R. Katti and R. Dash, *Biomed. Mater.*, 2008, **3**, 034122.

CHAPTER 5

Acrylic–Clay Nanocomposites by Suspension and Emulsion Polymerization

URŠKA ŠEBENIK AND MATJAŽ KRAJNC

Faculty of Chemistry and Chemical Technology, University of Ljubljana, Aškerčeva cesta 5, Ljubljana 1000, Slovenia

5.1 Introduction

Acrylic–clay nanocomposite materials have attracted intensive academic activity and extensive industrial attention lately due to their peculiar and unexpected properties and their unique applications in the commercial sector. Acrylic–clay nanocomposites comprise dispersions of nano-clay platelets throughout an acrylic polymer matrix. The clay disk-like platelets or layers have a thickness < 1 nm and extend laterally up to 1 µm. A complete dispersion of clay nanolayers in a polymer matrix maximizes the number of available reinforcing elements and, owing to the high surface area ($\sim 760 \, \mathrm{m^2 \, g^{-1}}$)[1] of clay particles, materials with improved properties can be synthesized with much smaller amounts of clay compared with traditionally used fillers. Often these materials possess novel properties and exhibit a balance of previous antagonistic properties.[2] Improvements in properties such as high moduli, increased heat and strength resistance, higher chemical resistance, lower solvent uptake, enhanced barrier characteristics, reduced gas permeability, reduced flammability and increased biodegradability of biodegradable polymers have been observed.[1–6]

In recent years, the majority of research activity in the field of acrylic–clay nanocomposite materials has gradually shifted from organic-based to

RSC Nanoscience & Nanotechnology No. 16
Polymer Nanocomposites by Emulsion and Suspension Polymerization
Edited by Vikas Mittal
© Royal Society of Chemistry 2011
Published by the Royal Society of Chemistry, www.rsc.org

water-based systems because of environmental and health issues. Therefore, suspension and emulsion polymerization, both used in industry to produce a large variety of acrylic polymers, which can be used as paints, adhesives, coatings, binders, *etc.*, represent a promising approach to synthesize water-based acrylic–clay nanocomposites. Acrylic–clay nanocomposites can be synthesized *in situ*, by carrying out an emulsion or suspension (co)polymerization of a chosen monomer or monomers in the presence of clay. When hydrophilic (pristine) clay is employed, *in situ* emulsion polymerization is the best suited alternative. However, if for the production of waterborne nanocomposites organophilic clays (pristine clays where the naturally occurring cations have been replaced with long alkylammonium salts) are used, suspension polymerization is usually the best choice.[7–9]

Up to now, poly(methyl methacrylate)[10–24] and methyl methacrylate copolymers (*e.g.* with styrene, butyl acrylate and dodecyl methacrylate)[7,8,25–29] have been the most widely used acrylic polymers for nanocomposite preparation by emulsion and suspension polymerization. Less research has been based on other acrylic polymers, such as polyacrylonitrile,[30–33] poly(butyl acrylate),[34–36] poly(butyl methacrylate),[37] poly(2-ethylhexyl acrylate),[38] poly(2-hydroxyethyl methacrylate),[39] polyacrylamide,[39] poly(lauryl acrylate),[40] poly(butyl acrylate-*co*-styrene),[41–43] poly(acrylonitrile-*co*-styrene),[44,45] poly(acrylonitrile-*co*-methacrylate),[46] poly(ethyl acrylate-*co*-2-ethylhexyl acrylate)[47] and poly(2-ethylhexyl acrylate-*co*-acrylic acid),[48] and sometimes small amounts of hydophilic acrylic monomers, such as hydroxyethyl methacrylate, methacrylic acid and acrylic acid, have been used as comonomers.[34,49] Therefore, it may be stated that, so far, the preparation of acrylic–clay nanocomposites has been based mainly on high glass transition temperature polymers, although nanocomposite materials with lower glass transition temperatures with improved or novel properties, which exhibit a balance of previous antagonistic properties, can also be achieved and are very desirable. Regarding nanocomposites of low glass transition temperature polymers, such as poly(butyl acrylate), poly(ethyl acrylate) and poly(2-ethylhexyl acrylate), which have been utilized as the main components of acrylic pressure-sensitive adhesives, little information is available.

This chapter is focused on the synthesis and properties of acrylic–clay nanocomposite pressure-sensitive adhesives, which are an example of the use of low glass transition temperature acrylic–clay nanocomposite materials.

5.2   Pressure-sensitive Adhesives Reinforced with Clays

Pressure sensitive adhesives (PSAs) are defined as materials which in dry form are aggressively and permanently tacky at room temperature and adhere firmly to a variety of dissimilar surfaces upon mere contact without the need for more than finger pressure.[50] Due to saturated nature of the polymer and its resulting resistance to oxidation, acrylic polymers are one of the most widely used materials for the production of PSAs.[51] Acrylic PSAs are available as solutions, aqueous emulsions and suspensions, hot melts and 100% reactive solids. Commercially solution polymerization was first, but interest in suspension and emulsion polymers has increased with the need to solve solvent emission problems and produce more environmentally friendly materials.[51] When polymer latex particles are employed, each individual particle can be considered to be a 'building block' of the PSA film, offering structural control at the nanoscale.[40] Structures within the particles can be translated into the film structure[52] and influence its macroscale mechanical and adhesive properties.[53] The most distinct difference between suspension and emulsion PSAs is their particle size and particle size distribution (PSD). The PSD has a great influence on the rheological characteristics of PSA dispersions. These are very important for the coating machinery, while the rheological characteristics of dried PSA dispersions are important for the performance of the adhesive systems. The products of suspension polymerization are commonly referred to as microspheres (or microbeads) and consist of spherical polymer particles in the 10–250 μm range. In emulsion systems, fine adhesive particles of diameter less than 2 μm, when dried, coalesce to form a continuous film. This smooth adhesive film in prolonged contact with a surface will increase in contact and peel strength with time. On the other hand, in the case of the microsphere adhesive, the adhesion arises from the physical structure. The film formed is discontinuous and spheres are in contact with the surface only at their tops, which causes a lower peel strength, but also improves their removability.[54–57] The properties of a PSA depend on balance of three basic applied properties: tack (the ability to adhere quickly), peel strength (the ability to resist removal by peeling) and shear resistance (the ability to resist flow when shear forces are applied). The balance between these three properties is adjusted according to the specific end use of the PSA.[50,51] Tack, peel and shear are influenced by the properties of the base polymer. Inherent properties, such as copolymer composition and microstructure, molecular weight and distribution, are among the most influential parameters affecting the PSA properties directly, and also indirectly through their influence on the physical properties (*e.g.* glass transition temperature, T_g) and thus also the rheological properties of polymer (*e.g.* viscoelastic behavior, moduli).

The incorporation of a nanofiller into latex films offers a simple and effective means of modifying its viscoelastic properties. Therefore, with the aim of optimizing the adhesive properties, efforts have been made recently to synthesize waterborne PSA nanocomposites. In most cases,[34,38,40,47,48] layered silicates (clays) have been used as reinforcing nanomaterial. It was observed that acrylic–clay nanocomposite PSAs exhibited excellent pressure-sensitive adhesion and that their adhesive properties were strongly influenced by the type and amount of clay added.[47] The incorporation of clay in PSA formulations resulted in significant improvements in storage modulus,[34,47] dissipation energy,[40] cohesive strength,[34] shear strength,[47] tack adhesion,[40] peel strength,[38] thermal stability,[38,48] superabsorbency[38,48] and biodegradability.[38,48] Moreover, it was demonstrated that the enrichment of the surface of the composite PSA film by the surfactant was depressed.[34]

However, when designing a nanocomposite PSA, its composition and nanostructure, the theoretical background of how the adhesion performance depends on material properties, must be understood and taken into account to avoid some risks, such as too high an increase in elastic modulus[40] and decrease in optical clarity[34] due to the too high amount of nanofiller used. PSAs are soft adhesives, which are able to wet surfaces and to achieve close contact (under application of light pressure and even on rough surfaces) as a result of their relatively low elastic modulus. To achieve a high adhesion energy, the polymer must be able to flow and thus be drawn into fibrils when under high strain in confinement. If the adhesive is too stiff, wetting is poor and fibril formation is restricted. Fibrils that are created will detach under small strains because the stress to deform them is greater than the adhesive force on the adherend. As a result, tackiness is lost. On the other hand, if the adhesive is too liquid-like, the resistance to creep under shear stress will be low. Hence, to be an effective PSA, the material needs a balance of elasticity and viscosity,[40] which, of course, is changed according to the specific end use of the PSA. It was demonstrated that the viscoelastic properties can be balanced by choosing an appropriate nanocomposite synthesis method by which a suitable type and amount of clay are incorporated in the polymer matrix.[47]

5.3   Synthesis and Structure Characterization of Acrylic–Clay Nanocomposite Pressure-sensitive Adhesives

Waterborne acrylic–clay nanocomposite PSAs have been synthesized by suspension polymerization,[47,48] emulsion polymerization[34,38] and Pickering miniemulsion polymerization of a 'soft' acrylate monomer in the presence of clay, which was followed by blending the polymer–clay nanocomposite particles with a standard acrylate latex for application as a waterborne PSA.[40]

Kajtna and Šebenik[47] synthesized poly(ethyl acrylate-*co*-2-ethylhexyl acrylate) [P(EA-*co*-2-EHA)]–clay nanocomposite PSAs in order to improve the poor shear resistance of their adhesive, whose adhesive properties had previously been optimized through molecular weight and level of crosslinking control. Different types and amounts (0.1, 0.25, 0.5, 1 and 2 wt%) of modified and unmodified montmorillonite (MMT) clays were dispersed in an ethyl acrylate (EA)–2-ethylhexyl acrylate (2-EHA) monomer mixture, which was then polymerized using a suspension polymerization technique, where an initiator (dibenzoyl peroxide), surface-active agents (modified esters of sulfo-carboxylic acid and ethoxylated oleyl alcohol), chain transfer agent (*n*-dodecanethiol), multifunctional acrylic monomer (butanediol diacrylate) and suspension stabilizer [poly(vinyl alcohol)] were used. Polymerization was monitored in-line using attenuated total reflectance Fourier transform infrared (ATR-FTIR) spectroscopy. The results showed that the kinetics of suspension polymerization were independent of the addition of MMT clays. An attempt was also made[47] to prepare a nanocomposite adhesive with 5 wt% of MMT clay, but on adding more than 2 wt% of MMT clay the adhesive suspension became unstable and subsequently batch coagulation occurred.

Li *et al.*[34] introduced 1, 3 and 5 wt% of pristine sodium montmorillonite (Na-MMT) into a poly(butyl acrylate) (PBA) PSA matrix by seeded semi-batch emulsion polymerization to study the effect of reinforcement on the properties of PSAs. First, the seeds were prepared using batch soap-free emulsion polymerization by polymerizing butyl acrylate (BA) and smaller amounts of strongly polar co-monomers [acrylic acid (AA) and 2-acrylamido-2-methyl-1-propanesulfonic acid (AMPS)] in the Na-MMT dispersion in water. The strongly polar co-monomers were employed with the consideration that strong interaction of carbonyl and amido moieties with Na-MMT might make the polymer end-tethered on silicate layers. Likewise, AMPS can widen the clay interlayer spacing and facilitate insertion of monomers into the clay galleries. To increase the stability of the Na-MMT aqueous dispersion, a small amount of sodium diphosphate was added, whose multivalent anions can be strongly adsorbed on the clay silicate layers and thus increase their negative charge density. As initiator system an aqueous solution of *tert*-butyl hydroperoxide and ascorbic acid was applied. After the seed preparation was completed, monomer emulsion (containing BA, AA, water and sodium dodecyl sulfate) and initiator system solution were added semi-batchwise. The monomer conversions obtained at the end of polymerization were between 96 and 93%. On the basis of X-ray diffraction (XRD) and transmission electron microscopy (TEM) results, it was concluded[34] that both intercalated and exfoliated, structures of Na-MMT coexisted in the composites.

Rana *et al.*[38] prepared P(2-EHA)–silicate nanocomposites by batch emulsion polymerization of 2-EHA in the presence of sorbitol, using potassium persulfate as initiator. The nanocomposites so prepared were shown to be intercalated by XRD, TEM and IR spectroscopy. However, according to the authors,[48] the nanocomposite[38] failed to satisfy the desired application in transdermal drug delivery. Consequently, they improved their product by preparing a P(2-EHA-*co*-AA)–silicate nanocomposite. AA was chosen for its polarity and known bioadhesive properties and 2-EHA for its hydrophobicity and softness. Copolymers with different molar ratios of the monomers were prepared. Sorbitol was used as emulsifier and benzoyl peroxide as initiator; therefore, the synthesis process[48] may be classified as suspension copolymerization. The drug cloxacillin sodium was added to the reaction mixture after 2 h of polymerization and 1 h before the reaction was terminated. The nanocomposites so prepared were shown to be intercalated by XRD and TEM. NMR and IR spectroscopy confirmed the copolymer formation and the absence of monomer impurities in the product.

Wang *et al.*[40] succeeded in preparing waterborne PSAs with nanostructured features that can build in an energy dissipation mechanism without stiffening the structure too much. In their study, 'soft–hard' polymer particles having Laponite clay armor were synthesized by the Pickering miniemulsion polymerization of *n*-lauryl acrylate (LA). The resulting poly(lauryl acrylate) (PLA)–Laponite hybrid particles were then blended at various low concentrations with a standard poly(butyl acrylate-*co*-acrylic acid) (PBA) latex, prepared by emulsion polymerization with a BA:AA molar ratio of 99:1, for application as waterborne PSAs. Dynamic light scattering, atomic force microscopy and TEM were used[39] for the structure characterization of clay-armored PLA hybrid particles and films of their blends with PBA. The average diameter of the Laponite-armored PLA hybrid particles was found to be 286 nm. Their surface appeared to be uniformly covered with the clay disks, with just a few gaps between them. The blend of PLA and PBA latexes formed a film with Laponite-armored PLA hybrid particles dispersed in a continuous PBA matrix.

5.4 Properties of Acrylic–Clay Nanocomposite Pressure-sensitive Adhesives

As mentioned above, the properties of a PSA depend on the balance between three adhesive properties: tack, peel strength and shear resistance. Depending on the adhesive properties of the starting PSA formulation, attempts have been made to improve a selected property (shear strength, peel strength or tack) by adding clay to the formulation.

Li *et al.*[34] achieved a large improvement in shear strength by incorporation of 1–3 wt% of clay in the polymer matrix during emulsion polymerization. Moreover, they observed that the failure mode changed from cohesive to adhesive. The increase in shear resistance was derived from the significant increase in the storage modulus of the adhesive in the rubbery plateau, which occurred at room temperature. However, when a 5 wt% clay loading was used, very good shear strength was achieved at the cost of tack and peel strength, so it was concluded that a 5 wt% clay loading was too high. Further, it was demonstrated, by X-ray photoelectron spectroscopy, that enrichment of the PSA surface by the surfactant during drying of emulsion PSAs was depressed by the layered silicate. This is important because the enrichment of the surface by the surfactants, due to incompatibility of conventional surfactants with the polymer matrix and resulting surfactant migration to the polymer surface, creates a weak boundary layer when the PSA layer is applied to an adherend and, thus, results in a decrease in tack and peel strength.

Kajtna and Šebenik,[47] who prepared PSA nanocomposites with pristine and differently modified MMT clays by suspension polymerization, succeeded in raising the shear resistance of their adhesives by adding clay to their PSA formulation. They observed similar trends to Li *et al.*[34] regarding changes in adhesive properties (shear strength, peel strength and tack) with the amount of clay used. The results of the determination of adhesive properties are shown in Figures 5.1–5.3, and the main characteristics of the clays used (Cloisite MMT clays, Southern Clay Products, Gonzalez, TX, USA) are presented in Table 5.1. The FTM 8 test method (FINAT test method 8), FTM 1 test method and probe test (Polyken test) were used for the determination of shear resistance, peel adhesion at 180° and tack, respectively.

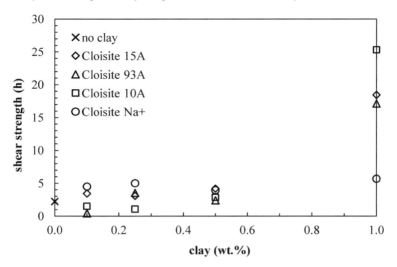

Figure 5.1 Influence of type and amount of clay added on the shear strength of adhesives. The FTM 8 test method was used to determine the ability of an adhesive to withstand static forces applied in the same plane as the coated adhesive. Resistance to shear surface is defined as the time required for a standard area of pressure-sensitive coated material to slide from a standard flat surface in a direction parallel to the surface.

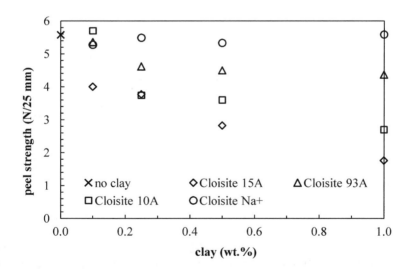

Figure 5.2 Influence of type and amount of clay added on the peel strength of adhesives. The FTM 1 test method was used. Peel adhesion is defined as the force required to remove pressure-sensitive coated material that has been applied to a standard test plate at an angle of 180° and a speed of 300 mm min^{-1}.

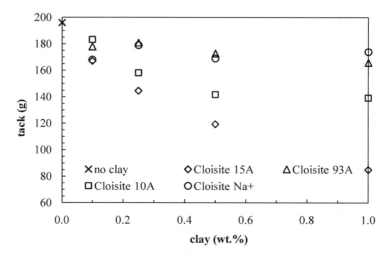

Figure 5.3 Influence of type and amount of clay added on the tack of adhesives. The probe test (Polyken test) was used. The device measures the force required to separate a stainless-steel probe (0.5 cm diameter) from the adhesive under low pressure and at short contact times. Tack was tested with a $100\,g\,cm^{-1}$ contact pressure, 1 s dwell time and at a $1\,cm\,s^{-1}$ test rate.

Table 5.1 Characteristics of MMT clays.[47]

MMT type	*Organic modifier*[a]	*Modifier concentration (mequiv. per 100 g clay)*
Cloisite 15A	2M2HT	125
Cloisite 93A	M2HT	90
Cloisite 10A	2MBHT	125
Cloisite Na$^+$	None	None

[a]2M2HT, dimethyl dehydrogenated tallow ($\sim$65% C_{18}, $\sim$30% C_{16}, $\sim$5% C_{14}), quaternary ammonium; M2HT, methyl dehydrogenated tallow ($\sim$65% C_{18}, $\sim$30% C_{16}, $\sim$5% C_{14}), quaternary ammonium; 2MBHT, dimethylbenzyl hydrogenated tallow ($\sim$65% C_{18}, $\sim$30% C_{16}, $\sim$5% C_{14}), quaternary ammonium.

It can be seen in Figure 5.1 that the shear strength increased substantially when 1 wt% of modified clays was added. In comparison, the shear strength for a nanocomposite containing an equivalent amount of pristine clay did not lead to an increase of the same magnitude. Therefore, the incorporation of modified clays must have stiffened the PSA material and enhanced the resistance of the adhesive to creep under shear stress. DMA of PSAs showed how the storage modulus increased after the addition of clay at room temperature in a frequency range from 0.1 to 100 Hz. The adhesive with the highest storage modulus also exhibited the highest shear strength. The results of DMA measurements in the shear mode are shown in Table 5.2. It can be seen that the values of tanδ for nanocomposites were lower than that of the adhesive without clay. Tanδ is defined as the ratio of loss modulus (G'') to storage modulus (G')

Table 5.2 Values of storage modulus (G') and loss factor (tanδ) for adhesives without and with 1 wt% of clay measured in shear mode at different frequencies.

Frequency (Hz)	No clay		Cloisite 15A		Cloisite 93A		Cloisite 10A		Cloisite Na$^+$	
	G' (kPa)	Tanδ	G' (kPa)	Tanδ	G' (kPa)	Tanδ	$G\delta$ (kPa)	Tanδ	G' (kPa)	Tanδ
0.1	35.5	0.30	41.2	0.29	43.1	0.25	53.6	0.25	41.0	0.29
1	51.4	0.27	60.8	0.26	60.6	0.26	72.3	0.22	58.2	0.27
10	73.2	0.33	84.9	0.34	84.9	0.31	100	0.31	82.3	0.36
100	140	0.66	140	0.64	140	0.62	150	0.62	140	0.67

and also correlates well with adhesive cohesive strength. An adhesive with high cohesion and hence high shear strength has a low value of tanδ. In contrast, an adhesive with a high value of tanδ has high adhesive strength. In the latter case, the adhesive dissipates energy through its own deformation.

Therefore, the decreasing peel values with increasing amount of clay in Figure 5.2 may be attributed to the improved cohesive strength of the adhesive, which has an influence on the degree of deformation and hence results in a lower force needed for the debonding process (adhesive failure was observed on debonding). The decrease in tack (Figure 5.3) with increasing amount of MMT may be explained by the ratio of the loss factor to the storage modulus, which is a good predictor of the tack energy.[39,58–60] The lower the ratio is, then the more PSA can be drawn without detaching it from the substrate and the higher is the tack. However, a decrease in tack might also be caused by the distribution of clay particles, which have negligible tack, on the adhesive surface. However, since the hydrophobic clays lowered tack more than the pristine clay, it is not likely that this was the key reason for the decrease in tack. However, since the microsphere adhesives studied by Kajtna and Šebenik[47] were used for removable applications, a decrease of peel and tack to this extent was more or less desirable. Further, it was shown that the degree of clay modification influenced the adhesive properties. In cases when modified clays were used, all adhesive properties (tack, peel and shear strength) changed similarly on varying the amount of clay, but there was a noticeable difference in peel strength and tack values between adhesives prepared by using differently modified (type and amount of modifier used for the modification) MMT clays. When an unmodified type of MMT clay (the most hydrophilic) was used, the distribution of clay between the continuous (water) and dispersed phase (monomers) played an important role in adhesive properties, as can be seen in Figures 5.1–5.3. Although a minor decrease in tack and peel was measured, the values were relatively independent of the amount of added clay. An overall improvement of shear strength was determined, but the values were independent of the amount of pristine clay. The authors speculated that only a certain amount of clay was dispersed in the monomer phase, which caused minor changes in adhesion properties.

In the study of Kajtna and Šebenik,[47] the amount of PSAs' gel phase was also determined, and was found to decrease with increasing amount of clay in the adhesive formulation. Since the acrylate chain-growth kinetics are complicated by the inter- and intramolecular transfer to polymer,[50,57,61–65] the gel is formed when intermolecular chain transfer to the polymer is followed by termination by combination. The relative amounts of the soluble and gel polymer phase and also the molar mass distribution of the sol fraction and the crosslinking density of the gel fraction are among the most important factors that influence the adhesive properties. In addition to the crosslinking of polymer molecules, the amount of gel phase is also dependent on entanglement of polymer molecules. Therefore, the authors speculated that the presence of clay particles in the reaction mixture hindered the entanglement of polymer molecules in the bulk, which resulted in a smaller amount of gel phase with increasing amount of clay. When they used a chain transfer agent and multifunctional acrylic monomer together with clays, the amount of gel also decreased with increasing amount of clay. However, in this case the amount of gel phase was considerably higher as compared with the adhesives without a chain transfer agent and multifunctional acrylic monomer but with the same amount of clay. This was due to the crosslinking reaction of polymer molecules as a consequence of two C=C double bonds present in the multifunctional acrylic monomer molecule. However, the shear strength of adhesives synthesized with a combination of chain transfer agent and multifunctional acrylic monomer was very poor regardless of the amount of clay added, which was attributed to the formation of crosslinked polymer structures and subsequent hindering of the intercalation process.

On the other hand, Wang et al.[40] showed that the use of clay-armored latex particles with a soft PLA core as a nanocomposite filler in a standard PBA waterborne PSA led to marked enhancement of the mechanical properties. These nanocomposite particles prepared *via* Pickering miniemulsion polymerization led to an increase in the tack energy, by raising the plateau stress and increasing the strain at failure. In comparison, the tack energy for nanocomposites containing an equivalent amount of non-armored PLA, clay disks or both did not lead to increases of the same magnitude. DMA analysis showed that the clay-armored latex particles did not increase the PSA storage modulus; therefore, they did not stiffen the adhesive, but did increase the loss tangent, which is a key requirement for increasing the tack energy. An optimum concentration of nanocomposite filler was found for the maximum tack energy increase, above which the tack began to decrease. The authors suspected that at higher concentrations the excess of nanosized clay, which existed in the latex serum after the Pickering polymerization and could not be avoided, saturated the air interface of the adhesive and consequently caused a decrease in tack.

5.5 Conclusion

It has been shown that acrylic–clay nanocomposites can exhibit physical, chemical and mechanical properties that are dramatically different from those of polymers or of their microcomposites. Therefore, the nanocomposites offer great potential for new and improved commercial applications. Undoubtedly, for this purpose, suspension and emulsion polymerization are very promising production methods, since they are relatively simple and their products are environmentally friendly. The preparation of acrylic–clay nanocomposites has been mainly based on 'hard' or high glass transition temperature polymers, even though it has been demonstrated that the 'soft' (low glass transition temperature) polymer–clay nanocomposites can also show dramatically improved viscoelastic behavior and novel properties. However, to design a nanocomposite material with desired superior functional properties, the correlation between features of the synthesis method, nanostructure of the material obtained and the resulting physical, chemical and mechanical properties of the material must be understood in detail. Nanocomposite pressure-sensitive adhesives are an excellent example of how material properties, *i.e.* shear strength, tack and peel strength, can be controlled and designed through optimizing the balance of elasticity and viscosity. This balance is changed according to the specific end use of the adhesive and it was demonstrated that it can be adjusted by choosing an appropriate nanocomposite synthesis method and by using the proper type and amount of clay.

Acknowledgment

Financial support from the Slovenian Ministry of Higher Education, Science and Technology (Grant P2-0191) is gratefully acknowledged.

References

1. P. C. LeBaron, Z. Wang and T. J. Pinnavaia, *Appl. Clay Sci.*, 1999, **15**, 11.
2. S. S. Ray and M. Okamoto, *Prog. Polym. Sci.*, 2003, **28**, 1539.
3. M. Biswas and S. S. Ray, *Adv. Polym. Sci.*, 2001, **55**, 167.
4. E. P. Giannelis, *Appl. Organomet. Chem.*, 1998, **12**, 675.
5. R. Xu, E. Manias, A. J. Snyder and J. Runt, *Macromolecules*, 2001, **34**, 337.
6. S. S. Ray, K. Yamada, M. Okamoto and K. Ueda, *Nano Lett.*, 2002, **2**, 1093.
7. G. Diaconu, M. Paulis and J. R. Leiza, *Polymer*, 2008, **49**, 2444.
8. G. Diaconu, M. Paulis and J. R. Leiza, *Macromol. React. Eng.*, 2008, **2**, 80.
9. Q. Sung, Y. Deng and Z. L. Wang, *Macromol. Mater. Eng.*, 2004, **289**, 288.
10. D. C. Lee and L. W. Jang, *J. Appl. Polym. Sci.*, 1996, **61**, 1117.
11. M. Okamoto, S. Morita, H. Taguchi, Y. H. Kim, T. Kotaka and H. Tatayama, *Polymer*, 2000, **41**, 3887.
12. M. Okamoto, S. Morita, Y. H. Kim, T. Kotaka and H. Tateyama, *Polymer*, 2001, **42**, 1201.
13. Y. S. Choi, M. H. Choi, K. H. Wang, S. O. Kim, Y. K. Kim and I. J. Chung, *Macromolecules*, 2001, **34**, 8978.
14. X. Huang and W. J. Brittain, *Macromolecules*, 2001, **34**, 3255.

15. N. Clarke, L. R. Hutchings, I. Robinson, J. A. Elder and S. A. Collins, *J. Appl. Polym. Sci.*, 2001, **113**, 1307.
16. Y. Li, B. Zhao, S. B. Xie and S. M. Zhang, *Polym. Int.*, 2003, **52**, 892.
17. H. Essawy, A. Badran, A. Youssef and A. E. Abd El-Hakim, *Polym. Bull.*, 2004, **53**, 9.
18. P. Meneghetti and S. Qutubuddin, *Langmuir*, 2004, **20**, 3424.
19. P. Meneghetti and S. Qutubuddin, *Thermochim. Acta*, 2006, **442**, 74.
20. D. J. Voorn, W. Ming and A. M. van Herk, *Macromolecules*, 2006, **39**, 4654.
21. K. F. Lin, S. C. Lin, A. T. Chien, C. C. Hsieh, M. H. Yen, C. H. Lee, C. S. Lin, W. Y. Chiu and Y. H. Lee, *J. Polym. Sci., Part A: Polym. Chem.*, 2006, **44**, 5572.
22. Q. H. Kong, Y. Hu, L. Yang, W. C. Fan and Z. Y. Chen, *Polym. Compos.*, 2006, **27**, 49.
23. P. K. Sahoo and R. Samal, *Polym. Degrad. Stabil.*, 2007, **92**, 1700.
24. H. Essawy, A. M. Youssef, A. E. Abd El-Hakim and A. M. Rabie, *Polym.-Plast. Technol.*, 2009, **48**, 177.
25. C. Zheng and L. J. Lee, *Macromolecules*, 2001, **34**, 4098.
26. Z. Zhang, N. Zhao, W. Wei, D. Wu and Y. Sun, *Int. J. Nanosci. Ser.*, 2006, **5**(Nos. 2 & 3), 291.
27. G. Diaconu, M. Mièušík, A. Bonnefond, M. Paulis and J. R. Leiza, *Macromolecules*, 2009, **42**, 3316.
28. G. Gonzáles, E. Colmenar, G. Diaconu, F. Alarcia, M. Manea, M. Paulis, M. J. Barandiaran, J. R. Leiza, J. C. de la cal and J. M. Asua, *Macromol. React. Eng.*, 2009, **3**, 233.
29. F. Dietche, Y. Thomann, R. Thomann and R. Mülhaupt, *J. Appl. Polym. Sci.*, 2000, **75**, 396.
30. Y. S. Choi, K. H. Wang, M. Xu and J. Chung, *Chem. Mater.*, 2002, **14**, 2936.
31. T. S. Yu, J. P. Lin, J. F. Xu, T. Chen and S. L. Lin, *Polymer*, 2005, **46**, 5695.
32. T. S. Yu, J. P. Lin, J. F. Xu, T. Chen, S. L. Lin and X. H. Tian, *Compos. Sci. Technol.*, 2007, **67**, 3219.
33. B. Samal, P. K. Rana and P. K. Sahoo, *Polym. Compos.*, 2008, **29**, 1203.
34. H. Li, Y. Yang and Y. Yu, *J. Adhes. Sci. Technol.*, 2004, **18**, 1759.
35. P. K. Sahoo, R. Samal, S. K. Swain and P. K. Rana, *Eur. Polym. J.*, 2008, **44**, 3522.
36. R. Samal, S. K. Swain, P. K. Rana and P. K. Sahoo, *J. Polym. Mater.*, 2008, **25**, 397.
37. R. Samal, P. K. Rana, G. P. Mishra and P. K. Sahoo, *Polym. Compos.*, 2008, **29**, 173.
38. P. K. Rana, S. K. Swain and P. K. Sahoo, *J. Appl. Polym. Sci.*, 2004, **93**, 1007.
39. D. J. Voorn, W. Ming and A. M. van Herk, *Macromolecules*, 2006, **39**, 2137.

40. T. Wang, P. J. Colver, S. A. F. Bon and J. L. Keddie, *Soft Matter*, 2009, **5**, 3842.
41. N. Negrette-Herrera, J. L. Putaux, L. David and E. Bourgeat-Lami, *Macromolecules*, 2006, **39**, 9177.
42. N. N. Herrera, S. Persoz, J. L. Putaux, L. David and E. Bourgeat-Lami, *J. Nanosci. Nanotechnol.*, 2006, **6**, 421.
43. N. Negrete-Herrera, J. L. Putaux, L. David, F. D. Haas and E. Bourgeat-Lami, *Macromol.Rapid. Commun.*, 2007, **28**, 1567.
44. M. H. Noh, L. W. Jang and D. C. Lee, *J. Appl. Polym. Sci.*, 1999, **74**, 179.
45. M. H. Noh and D. C. Lee, *J. Appl. Polym. Sci.*, 1999, **74**, 2811.
46. B. Samal, P. K. Rana and P. K. Sahoo, *Polym. Polym. Compos.*, 2009, **17**, 385.
47. J. Kajtna and U. Šebenik, *Int. J. Adhes. Adhes.*, 2009, **29**, 543.
48. P. K. Rana and P. K. Sahoo, *J. Appl. Polym. Sci.*, 2007, **106**, 3915.
49. F. A. Zhang, L. Chen and J. Q. Ma, *Polym. Adv. Technol.*, 2009, **20**, 589.
50. I. Benedek and L. J. Heymans, in *Pressure-sensitive Adhesives Technology*, Marcel Dekker, New York, 1996.
51. D. Satas, *Handbook of Pressure Sensitive Adhesive Technology*, Van Nostrand Reinhold, New York, 1989.
52. F. D. Dos Santos and L. Leiber, *J. Polym. Sci., Part B: Polym. Phys.*, 2003, **41**, 224.
53. F. Deplace, C. Carelli, T. Yamaguchi, A. Foster, M. A. Rabjohns, P. A. Lovell, C. H. Lei, J. L. Keddie, K. Ouzineb and C. Creton, *Soft Matter*, 2009, **7**, 1440.
54. J. Kajtna, M. Krajnc and J. Golob, *Macromol. Symp.*, 2006, **243**, 132.
55. J. Kajtna, U. Šebenik, M. Krajnc and J. Golob, *Drying Technol.*, 2008, **26**, 323.
56. J. Kajtna, B. Likozar, J. Golob and M. Krajnc, *Int. J. Adhes. Adhes.*, 2008, **28**, 382.
57. J. Kajtna, J. Golob and M. Krajnc, *Int. J. Adhes. Adhes.*, 2009, **29**, 186.
58. T. Wang, C.-H. Lei, A. B. Dalton, C. Creton, Y. Lin, K. A. S. Fernando, Y.-P. Sun, M. Manea, J. M. Asua and J. L. Keddie, *Adv. Mater.*, 2006, **18**, 2730.
59. J. Nase, A. Lindner and C. Creton, *Phys. Rev. Lett.*, 2008, **101**, 074503.
60. C. Carelli, F. Deplace, L. Boissonnet and C. Creton, *J. Adhes*, 2007, **83**, 491.
61. O. Elizalde, G. Arzamendi, J. R. Leiza and J. M. Asua, *Ind. Eng. Chem. Res.*, 2004, **43**, 7401.
62. R. Jovanovič and M. A. Dube, *Ind. Eng. Chem. Res.*, 2005, **44**, 6668.
63. I. Gonzales, J. R. Leiza and J. M. Asua, *Macromolecules*, 2006, **39**, 5015.
64. M. Van der Brink, M. Pepers, A. M. Van Herk and A. L. German, *Polym. React. Eng.*, 2001, **9**, 101.
65. O. Kammona, E. G. Chatzi and C. Kiparissides, *J. Macromol. Sci.*, 1999, **C39**, 57.

CHAPTER 6

Biodegradable Polymer–Clay Nanocomposite Fire Retardants via Emulsifier-free Emulsion Polymerization

PRAFULLA KUMAR SAHOO

Department of Chemistry, Utkal University, Vani Vihar, Bhubaneswar 751 004, India

6.1 Introduction

Polymer nanocomposites combine two concepts, *i.e.* composites and nanometer sized materials. Polymer nanocomposites,[1–3] developed in the late 1980s, represent a new class of materials with enhanced performance and have great academic and industrial interest compared with conventional microcomposites.[4,5] The main goal of the polymer nanocomposite research is to enhance the mechanical strength and toughness of the polymeric components using molecular or nanoscale fillers. Fillers have important roles in modifying the properties of various polymers. A novel class of fillers is anisotropic layered silicates of the montmorillonite (MMT) or sodium silicate (SS) type; classed as organoclays, organophilic modification affords compatibility between fillers and polymers.

In recent years, MMT has attracted great academic and commercial interest because of its high aspect ratio of silicate nanolayers, high surface area and wide applications in polymer materials.[6,7] Much attention has been paid to polymer nanocomposites, essentially polymer–MMT or polymer–SS nanocomposites, which exhibit physical and chemical properties that are dramatically

RSC Nanoscience & Nanotechnology No. 16
Polymer Nanocomposites by Emulsion and Suspension Polymerization
Edited by Vikas Mittal
© Royal Society of Chemistry 2011
Published by the Royal Society of Chemistry, www.rsc.org

different from those of conventional filled polymers. The polymer–layered silicate nanocomposites (PLSNs) can exhibit increased modulus,[8–10] decreased thermal coefficient, reduced gas permeability,[11–13] increased solvent resistance,[14] enhanced ionic conductivity[15] and improved flame retardant properties[16–21] compared with the polymers alone. In general, two idealized polymer–layered silicate structures are possible, intercalated and exfoliated. The intercalation of polymer chains results in the finite expansion of silicate layers without the ordered structure of the silicate layers being disturbed. Therefore, an intercalated structure consists of well-ordered multilayers of polymer and silicate layers, a few nanometers thick. Exfoliated structures, however, are formed because of the extensive penetration of the polymers into the silicate layers, resulting in delamination of the silicate layers and, therefore, a loss of ordered structure. In practice, however, partially intercalated and exfoliated structures are usually obtained. For the development of fundamental research and commercial requirements, highly exfoliated polymer–MMT nanocomposites obtained in a direct low-loss way are becoming increasingly attractive.[22,23] Accordingly, the processes for and mechanisms of highly exfoliated polymer–MMT nanocomposites have been a focus in fundamental and application research.[24–30] There are three dominant ways to achieve exfoliated nanocomposites, *i.e.* solution intercalation, melt intercalation and *in situ* intercalative polymerization.[12,31,32]

Out of these, the author's group has been successful in utilizing the *in situ* emulsion intercalative polymer method for the synthesis of a number of polymer–clay nanocomposites for possible use as superabsorbents, pressure-sensitive adhesives and flame retardants, along with the study of their biodegradability. However, *in situ* polymerization is more versatile due to the variety of polymerization conditions and clay treatment and is therefore a promising way to produce tailored nanocomposites with controlled morphology. Among them, heterogeneous polymerizations, *i.e.* emulsion and dispersion methods, present the advantage of easier removal of the resulting product from the reactor compared with bulk polymerization. Finally, considering environmental and ecological aspects, heterophase polymerization in particular has attracted increasing interest, and in the presence of nanoclays appears to be a suitable way to produce polymer–clay nanocomposites. Emulsion polymerization is used for nanocomposite synthesis. The first composite synthesis *via* emulsion polymerization was reported in 1996 by Lee and Jang.[33]

Despite the considerable number of studies concerned with the preparation of various types of polymer–silicate nanocomposites[19,34,35] by the conventional emulsion technique, there has been no report on syntheses of novel polymer–silicate nanocomposites by the emulsifier-free emulsion method. The advantage of this method of polymerization is that a high concentration of surfactant/emulsifier is considered to be one of the main drawbacks of emulsion polymerization due to difficulties in removing the residual emulsifier from the nanoparticle surface.[36,37] The recently developed unconventional emulsifier-free emulsion polymerization enables the latter problem to be resolved and composite nanoparticles with tailored morphology to be synthesized.

The present review of nanocomposites synthesized by *in situ* polymerization highlights an increasing number of studies on emulsifier-free emulsion polymerization. In our laboratory, using the *in situ* emulsifier-free emulsion technique, we have prepared a number of polymer nanoparticles and nanocomposites, such as PMMA[38] and PAN[39] nanoparticles, PEHA–SS nanocomposite[40] used for biodegradable superabsorbents, pressure-sensitive adhesives (PSAs), PMMA–MMT,[15] PBA–SS,[41] PBMA–SS-Mg(OH)$_2$[16] nanocomposites used for fire retardants, PAN–SS[42] used as a water absorbent, poly(EHA-*co*-AA)–SS nanocomposite to act as a PSA in transdermal drug delivery (TDD)[43] and poly(AA-*co*-AM)–MBA nanohydrogel for colon-specific drug delivery.[44]

Another very significant outcome of these studies is the conversion of non-biodegradable hydrophobic homopolymers such as PAN and PMMA into biodegradable hydrophilic material *via* the application of nanotechnology. Hence different biodegradation tests such as soil burial activated sludge and cultured media methods are carried out for the commercialization and ecologically friendly nature of these materials.

Since this chapter is focused on the development of different polymer nanocomposite fire retardants, some aspects of fire/flame retardants need to be discussed. Current research in the area of condensed fire retardants for polymers usually builds upon existing techniques. These include metal hydroxides (alumina, magnesium hydroxide) and phosphorus,[45–47] bromine,[48–50] antimony,[51] silicon,[52,53] nitrogen[54] and graphite[55] based materials. However, these materials tend to weaken mechanical properties while improving flammability resistance. However, polymer–clay nanocomposite flame retardants have generated a great deal of interest primarily due to improved mechanical and thermal properties and also they are green materials. Morgan and Tour[56] studied the flammability of polymer nanocomposites by considering two types of polymer–clay nanocomposites which have greatly reduced heat release rates and also they have observed that polymers which normally do not char or leave any carbonaceous residue upon burning do produce char in the presence of clay. Generally, fire retardants improve flammability but reduce the mechanical properties of polymers. However, in this study it was observed that polymer–clay nanocomposites reduced the flammability of polymers with improvement in the mechanical properties. The most important difficulty in the development of polymer–clay nanocomposites with the purpose of enhancing fire retardancy is that the most efficient structure for such enhancement may not result in the best mechanical properties. However, polymer–layered silicate nanocomposites offer effective fire retardancy while avoiding environmental pollution in terms of combustion, recycling and the disposal of the end-products. This is the most successful approach developed so far to produce environmental friendly fire retardant polymers.

In this chapter, detailed studies on the synthesis, particularly, of PBA–SS nanocomposite *via* the *in situ* emulsifier-free emulsion technique are reported along with their properties such as thermal, fire retardancy and biodegradability.

6.2 Experimental

6.2.1 Materials

The monomer, butyl acrylate (BA), was purchased from SRL India Ltd and sodium silicate (SS) was a gift of a solid granular sample (CAS 1344-098, Batch No. 2023BB H2O-1) from PQ Corporation, Maastricht, The Netherlands. Copper sulfate, glycine and ammonium persulfate (APS) were of analytical grade and used as such. All solutions were prepared using doubly distilled water.

6.2.2 Preparation of Nanocomposite

We prepared PBA–SS nanocomposite *via* the *in situ* emulsion polymerization strategy described in our previous reports on PEHA–SS[20] and the emulsion process[38] by taking distilled BA dispersed in deionized water by stirring with *in situ*-developed $CuSO_4$ (0.1 M)–glycine (0.1 M) complex. Silicate solution, prepared on a weight percentage basis, was added and the mixture was slowly heated to 50 °C followed by the addition of APS (0.1 M) dissolved in water. Polymerization was carried out with stirring at 400–600 rpm for 3 h. Polymerization was terminated by addition of a 0.1 M solution of ferrous ammonium sulfate solution. The coagulated products were purified. The variations of different components along with conversions are presented in Table 6.1.

6.2.3 Characterization

IR spectra of PBA (degree of polymerization ∼2600) and PBA–SS samples in the form of (0.5 g) KBr pellets were recorded in a Perkin-Elmer Paragon 500 FTIR spectrophotometer in the range 400–4000 cm^{-1}. Using a Rigaku X-ray machine operated at 40 kV and 150 mA, X-ray diffraction (XRD) patterns were obtained to determine the mean interlayer spacing of the (001) plane (d_{001}) for the PBA–silicate. The insertion of PBA into the silicate layer was confirmed by using an XRD monitoring diffraction angle 2θ up to 10°.

The nanoscale structure of PBA–SS was investigated by means of TEM (Hitachi H-700 instrument), operated at an accelerating voltage of 100 kV. An ultrathin section (the edge of the sample sheet perpendicular to the compression mold) obtained using a diamond knife with a thickness of 100 nm was micro-tomed at 80 °C.

Thermal properties were measured by using a Shimadzu DTA-500 system in air from 30 to 600 °C at a heating rate of 10 °C min^{-1}. Tensile bars were obtained on a Van Dorn 55 HPS 2.8 F mini injection molding machine under the following processing conditions: a melt temperature of 150 °C, a mold temperature of 25 °C, an injection speed of 40 mm s^{-1}, an injection pressure of 10 MPa and a holding time of 2 s, with a total cycle time of 30 s. Tensile measurements on injection molded samples of nanocomposites were performed

Table 6.1   Effect of variation of concentration of BA, SS and APS and time and temperature on the percentage conversion. Reproduced from reference 41 with permission from Elsevier

Sample code	[BA] (mol dm^{-3})	$[SS] \times 10^{-2}$ (mol dm^{-3})	$[APS] \times 10^{-2}$ (mol dm^{-3})	Time (min)	Temperature (°C)	Conversion (%)
S_0	7.0	0	2.0	180	50	65.45
S_1	1.75	1.0	1.0	180	50	37.19
S_2	3.5	1.0	1.0	180	50	40.08
S_3	7.0	1.0	1.0	180	50	44.54
S_4	10.5	1.0	1.0	180	50	40.83
S_5	14.0	1.0	1.0	180	50	39.53
S_6	17.5	1.0	1.0	180	50	33.43
S_7	21.0	1.0	1.0	180	50	28.77
S_8	7.0	0.25	1.0	180	50	25.05
S_9	7.0	0.5	1.0	180	50	31.18
S_{10}	7.0	1.0	1.0	180	50	44.54
S_{11}	7.0	1.5	1.0	180	50	35.63
S_{12}	7.0	2.0	1.0	180	50	30.06
S_{13}	7.0	2.5	1.0	180	50	19.49
S_{14}	7.0	3.0	1.0	180	50	13.36
S_{15}	7.0	1.0	0.5	180	50	40.09
S_{16}	7.0	1.0	1.0	180	50	44.54
S_{17}	7.0	1.0	1.5	180	50	52.89
S_{18}	7.0	1.0	2.0	180	50	79.06
S_{19}	7.0	1.0	2.5	180	50	54.00
S_{20}	7.0	1.0	3.0	180	50	36.75
S_{21}	7.0	1.0	3.5	180	50	28.95
S_{22}	7.0	1.0	2.0	60	50	33.41
S_{23}	7.0	1.0	2.0	90	50	42.31
S_{24}	7.0	1.0	2.0	120	50	46.77
S_{25}	7.0	1.0	2.0	150	50	49.55
S_{26}	7.0	1.0	2.0	180	50	79.06
S_{27}	7.0	1.0	2.0	210	50	79.07
S_{28}	7.0	1.0	2.0	240	50	79.08
S_{29}	7.0	1.0	2.0	300	50	79.09
S_{30}	7.0	1.0	2.0	360	50	79.09
S_{31}	7.0	1.0	2.0	180	35	15.59
S_{32}	7.0	1.0	2.0	180	40	19.49
S_{33}	7.0	1.0	2.0	180	45	49.55
S_{34}	7.0	1.0	2.0	180	50	79.06
S_{35}	7.0	1.0	2.0	180	55	59.57
S_{36}	7.0	1.0	2.0	180	60	45.65
S_{37}	7.0	1.0	2.0	180	65	29.51

according to ASTM D-638-00 using an Instron Model 5567 test machine. Tests were carried out at a crosshead speed of $50\,mm\,min^{-1}$ with a 1 kN load cell without the use of an extensiometer. All tests were performed at room temperature and the results were the average of five measurements. The highest value of the standard deviation was 15%.

6.2.4 Flame Retardancy

The flame retardancy of the nanocomposites was assessed by the limiting oxygen index (LOI) test according to ASTM D-2863, using apparatus from Fire Instrument Research Equipment with a digital readout of oxygen concentration to $\pm 0.1\%$. The specimens used had the dimensions $100 \times 6.5 \times 3$ mm. The LOI value corresponds to the minimum concentration of oxygen in an oxygen–nitrogen mixture necessary to burn the sample during 3 min or over a length of 80 mm. The flammability properties of PBA–SS nanocomposite, under fire-like conditions, were determined using a cone calorimeter[25] in accordance with the procedure outlined in AS/NZS 3837:1998, which is based on ISO 5660-1:1993, with dimensions $100 \times 100 \times 4$ mm at a heat flux of 35 kW m^{-2}.[26] The cone calorimeter measures fire-relevant properties such as heat release rate (HRR), mass loss rate and smoke production rate, among others. The HRR, in particular the peak HRR (PHRR), has been found to be the most important parameter to evaluate fire safety.

6.2.5 Biodegradation by Activated Sludge

Activated sludge water was collected from tank areas receiving toilet and domestic wastewater. In most areas of in India, waste materials after use are usually dumped near the sludge and the sludge water contains many microorganisms (bacteria, fungi, yeasts, *etc.*) responsible for the biodegradation of waste materials. The sludge was collected[57] in a polypropylene container, which was filled completely and then tightly closed, then brought to the laboratory immediately. After settling for about 1 h, the total solid concentration was increased to 5000 mg l^{-1}. The activated sludge water and a polymer sample (0.2 g) were incubated together in a sterilized vessel at room temperature (28 ± 2 °C). Duplicate samples were removed at intervals for biodegradation study *via* weight loss measurements. Vessels containing polymer samples without sludge water were treated as controls.[58]

6.3 Results and Discussion

From the series of experiments, it was found that the PBA was intercalated into the gallery structure of silicate by the catalytic action of the Cu(II)–glycine complex. The complex initiating system helps to stabilize the emulsion latex to a high conversion level in the absence of added emulsifier (Table 6.1). The initiation is a surface catalysis with adequate energy transfer from the complex to the initiator (APS), resulting in a complex initiation mechanism deviating from a simple path of decomposition. The concerted generation–consumption criterion is the driving force in complex-catalyzed peroxide and vinyl polymerization.[59] This explanation is similar to that for an earlier study of PEHA–SS nanocomposite.[40]

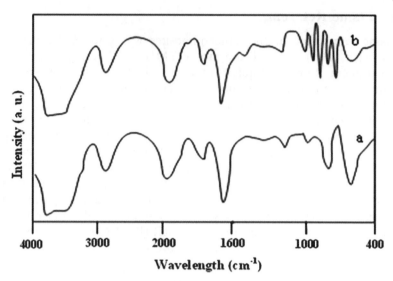

Figure 6.1 FTIR spectra of (a) PBA and (b) PBA–SS (5%, w/v) nanocomposite. Reproduced from reference 41 with permission from Elsevier.

6.3.1 FTIR Spectroscopy

The formation of the polymer PBA and its nanocomposite with SS was investigated by FTIR spectroscopy as shown in Figure 6.1. The absorption bands in the region 2650–3000 cm^{-1}, namely at 2980, 2840 and 2640 cm^{-1}, were due to C–H stretching vibrations of the methylene (CH_2), methyl (CH_3) and butyl (C_4H_9) groups. In Figure 6.1a, the carbonyl group peak for PBA appears at 1630 cm^{-1}. In Figure 6.1b, the Si–O–Si bond stretching shows up as a peak at 900–1050 cm^{-1}, which is absent in Figure 6.1a, indicating the presence of silicate in the PBA–SS nanocomposite matrix.

6.3.2 XRD Analysis

The systematic arrangement of the silicate layers of the intercalated composites was elucidated by using XRD to determine the interlayer spacing with the help of Bragg's equation. Due to the intercalation of PBA into the galleries of silicate *via* emulsion polymerization, the *d*-spacing of the PBA–SS nanocomposites increased with a shift of 2θ to lower values. The XRD pattern of PBA–SS nanocomposites before and after combustion, shown in Figure 6.2, indicates the complete disappearance of the clay peak, which confirmed the better dispersion and partial exfoliation of silicate layers over the PBA matrix. This conclusion was further evidenced with the TEM results in Figure 6.3b.

6.3.3 TEM Analysis

The internal structure of the PBA–SS nanocomposites was further conformed by TEM, which directly visualized the expanded layer structure in the

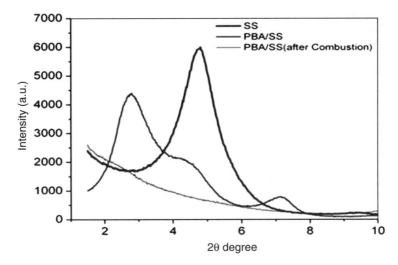

Figure 6.2 XRD of SS and PBA–SS (5%, w/v) before and after combustion. Reproduced from reference 41 with permission from Elsevier.

nanocomposites and partial dispersion of silicate layers in the polymer matrix.[60] Figure 6.3a demonstrates a mixed nanostructure with well-dispersed layers of polymer blend nanocomposites showing compatibility between PBA and silicate. The interlayer spacing in the figure reveals the intercalation of the PBA matrix with silicate layers[61] as evidenced by XRD. Figure 6.3b shows the TEM image of the nanocomposite after combustion, which is in agreement with the result obtained in the XRD study of PBA–SS nanocomposite after combustion. In Figure 6.3b, the silicate platelets are well separated and disordered as compared with Figure 6.3a, indicating better dispersion of nanolayers.

6.3.4 Thermal Analysis

The thermal properties of the nanocomposite materials were evaluated by TGA as shown in Figure 6.4. In contrast to PBA, the onset of decomposition for the PBA–SS nanocomposite is shifted towards higher temperature with inclusion of SS, indicating an enhancement of the thermal stability upon intercalation. The PBA–SS nanocomposite exhibited higher thermal stability due to the higher decomposition onset temperature than that of PBA, which can be attributed to the nanoscale silicate layers preventing diffusion of the volatile decomposition product. On the other hand, since the inorganic part (silicate) of the nanocomposite film hardly lost weight during the heating period, the shift of weight loss to higher temperatures might occur simply because the nanocomposite films possessed a relatively small amount (about 4 wt%) of organic polymer that contributed to the weight loss.

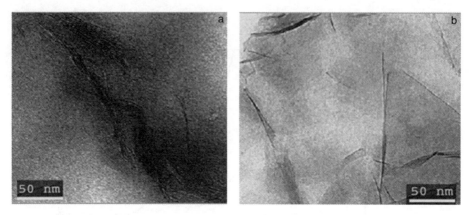

Figure 6.3 TEM image of PBA–SS (5%, w/v) nanocomposite film (a) before and (b) after combustion. Reproduced from reference 41 with permission from Elsevier.

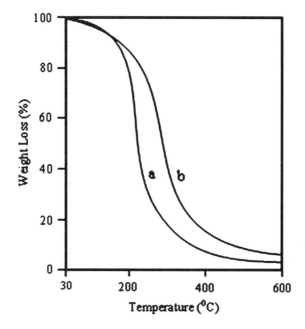

Figure 6.4 TGA: weight loss *versus* temperature for (a) PBA and (b) PBA–SS (5%, w/v). Reproduced from reference 41 with permission from Elsevier.

6.3.5 Mechanical Properties

The mechanical properties, including Young's modulus, elongation at break, toughness, yield stress and yield strain, of all the nanocomposites prepared in this study, together with the corresponding values for the virgin polymer, are

Table 6.2 Comparative data on mechanical properties ($\pm$ error) of PBA and PBA–SS nanocomposites. Reproduced from reference 41 with permission from Elsevier

Sample	Young's modulus (MPa)	Elongation at break (%)	Toughness (MPa)	Yield strain (%)	Yield stress (MPa)
S_0	218 ± 11	708 ± 40	142.7 ± 6	24.8 ± 1.8	18.1 ± 1.5
S_9	264 ± 13	659 ± 29	82.4 ± 34	23.6 ± 2.0	17.5 ± 1.2
S_{10}	363 ± 12	587 ± 32	52.1 ± 29	22.1 ± 1.0	17.1 ± 1.4
S_{11}	382 ± 17	493 ± 28	33.9 ± 15	21.6 ± 1.4	16.4 ± 1.3
S_{12}	413 ± 23	367 ± 19	20.4 ± 12	20.2 ± 0.9	15.6 ± 1.0
S_{14}	429 ± 21	346 ± 14	16.6 ± 20	19.2 ± 0.7	12.4 ± 0.9

given in Table 6.2. The Young's modulus of the nanocomposites increased significantly with increase in silicate concentration, whereas the yield stress and strain decreased monotonically with increase in silicate content. Due to their rigidity, silicate filler particles cannot be deformed by external stress in the specimen but act only as stress concentrators during the deformation process.[62] The elongation at break and toughness of the nanocomposites decreased tremendously with increase in silicate content, which is in accordance with the results obtained earlier.[63] Therefore, the emulsion process played a vital role in the dispersion of silicate in PBA, creating strong interfacial adhesion with the matrix.

6.3.6 Flame Retardancy

6.3.6.1 LOI Test

For evaluation of the efficiency of flame retardation of the synthesized nanocomposites, we measured the burning rates of a virgin PBA sample and the PBA–SS nanocomposite samples. The virgin PBA starts to burn very slowly at 20% of oxygen and an increase in the oxygen concentration increases dramatically the burning rate of PBA–SS nanocomposites, as can be seen in Figure 6.5. This observation is in good agreement with the results of Wang *et al.*[14] and Randoux *et al.*,[64] who studied the flame retardancy of nanocomposites by a conventional testing evaluation method using the LOI test.

SEM (Figure 6.6) revealed that the localized silicate humps on the surface of the nanocomposite were ruptured after combustion, as shown in Figure 6.6b, which indicates that charring of the silicate prevented further combustion of the composite. This result is evidenced by the dripping-melt and carbonaceous residue of the polymer and its nanocomposite as shown in Figure 6.7. The amount of carbonaceous residue is greater in the case of the nanocomposite in Figure 6.7b than for the virgin polymer in Figure 6.7a. Figure 6.7a and b represent digital figures for virgin PBA and PBA–SS nanocomposite burning

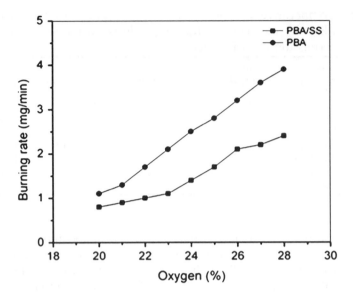

Figure 6.5   Burning of PBA and PBA–SS (5%, w/v) (S₅) *versus* % oxygen. Reproduced from reference 41 with permission from Elsevier.

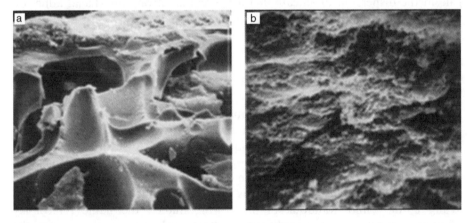

Figure 6.6   SEM images of PBA–SS (5%, w/v) (a) before and (b) after combustion in
the LOI test. Reproduced from reference 41 with permission from
Elsevier.

in air with dimensions of $20 \times 5 \times 2$ mm. These observations demonstrate that
the virgin PBA burns with dripping melt but the PBA–SS nanocomposite
produces a char residue after some time. Hence this preliminary study implies
the PBA–SS nanocomposites have a greater tendency for fire retardation than
the virgin PBA latex.

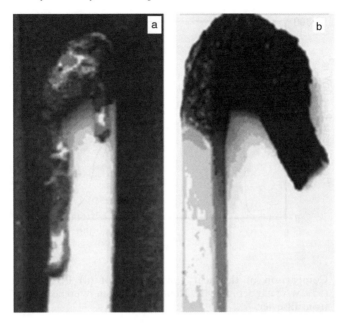

Figure 6.7 (a) Dripping-melt of PBA and (b) carbonaceous residue of PBA–SS (5%, w/v) nanocomposite. Reproduced from reference 41 with permission from Elsevier.

6.3.6.2 Cone Calorimetry

Cone calorimetry is another most effective bench-scale method for studying the flammability properties of materials. The cone calorimeter measures fire-relevant properties such as HRR, mass loss rate (MLR) and smoke yield, among others.

Of these, the HRR, in particular the PHRR, has been found to be the most important parameter to evaluate fire safety.[18,65] Substances having a higher PHRR spread the flame more, whereas the flame spread is hindered with substances having a lower PHRR value. From cone calorimetry, it was found that the nanocomposite has a lower PHRR than virgin PBA, as shown in Figure 6.8.

The flammability properties of a solid-phase flame retardant can be the result of the formation of carbon residue with silicate layer present in it. The resulting incomplete combustion is reflected in a lower specific heat of combustion and higher CO yield. The primary parameter responsible for the lower HRR of the nanocomposite is the mass loss rate (MLR) during combustion. as shown in Figure 6.9. The MLR of the nanocomposite is significantly reduced from the values observed for the virgin polymer, and also the smoke production rate (SPR), as shown in Figure 6.10.

The curves for the HRR, MLR and SPR *versus* time for virgin PBA and PBA–SS nanocomposite are shown in Figures 6.8–6.10, from which it can be

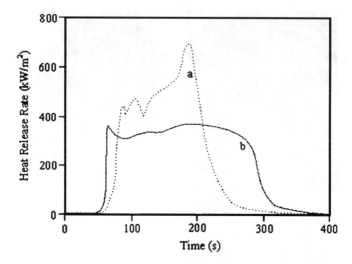

Figure 6.8   Comparison of the heat release rates of (a) PBA and (b) PBA–SS (5%, w/v) nanocomposite. Reproduced from reference 41 with permission from Elsevier.

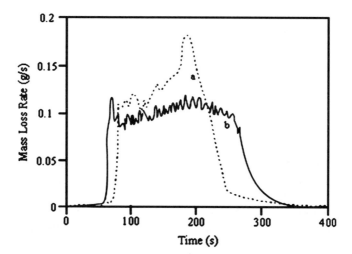

Figure 6.9   Comparison of the mass loss curves of (a) PBA and (b) PBA–SS (5%, w/v) nanocomposite. Reproduced from reference 41 with permission from Elsevier.

concluded that the addition of SS to PBA matrices always causes a decrease in the time-to-ignition (TTI) due to the first decomposition of silicate present in nanocomposites.[66] After the decomposition of SS in the nanocomposite, it produces a char residue which acts as a barrier to protect the loss of mass and heat transfer. Since this is similar to what is seen for the clay systems, one may

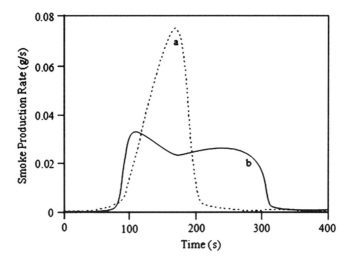

Figure 6.10 Smoke production rate of (a) PBA and (b) PBA–SS (5%, w/v) nano-composite. Reproduced from reference 41 with permission from Elsevier.

suspect that a similar process, the barrier process, is operative for the SS systems.

An alternative interpretation of these results is that two peaks, one at short time and another at longer time, are seen. A similar result is seen for intumescent systems in which the first peak corresponds to the formation of the barrier and the second to its destruction. In the case of these SS systems, the first peak may be attributed to the degradation whereas the second may be suggestive of some stabilization of the char formation. The interpretation of these results is somewhat speculative.

One may conjecture that the SS absorbs and re-irradiates the heat to cause rapid charring of the polymer, which gives rise to the rapid rise seen in the HRR curve. This charred polymer then undergoes thermal degradation but more slowly than does the virgin polymer since the SS now functions as a barrier to mass transport and partly insulates the underlying polymer from the heat source. Since a barrier process, similar to that seen for polymer–clay systems, is implied by the cone data, it is important to determine if a barrier does form and its composition.

6.3.7 Biodegradation

With regard to its commercialization and ecologically friendly nature, biodegradation tests are carried out. Biodegradation by activated sludge is less in the case of PBA than for PBA–SS nanocomposites. The biodegradation is increased with increasing percentage of silicate (Table 6.3). From this observation, it is confirmed that the growth of microorganisms is greater because of

Table 6.3 Biodegradation: % weight loss by active sludge of PBA and PBA–
SS nanocomposites. Reproduced from reference 41 with permission
from Elsevier.

Time (days)	S_0	S_9	S_{10}	S_{11}	S_{12}	S_{14}
8	2	13.04	15.12	15.52	16.15	17.18
15	5	20.28	21.29	23.36	23.41	24.4
21	20	27.44	28.48	29.52	30.62	31.65
30	25	34.58	35.62	36.68	37.74	38.84
180	32	41.70	42.73	43.78	44.80	45.84

the presence of more hydrophilic silicate and more water absorbance resulting
in more degradation, *i.e.* a greater percentage weight loss of nanocomposite
samples. Again, due to the hydrophobic nature of PBA, the water absorbance is
less, which results in it reducing the biodegradation.

6.4   Conclusion

PBA–SS nanocomposites were synthesized by a novel, cost-effective, unconventional emulsifier-free emulsion technique. The XRD results showed a disordered layer structure due to intercalation of polymer into the layered silicate.
These composites were further characterized by TGA, TEM and their burning
rates and exhibited improved properties of thermal stability with good fire
retardant properties. It is found that the fire retardant nanocomposites have
average biodegradability and have promising future prospects.

6.5   Abbreviations

APS	ammonium persulfate
BA	butyl acrylate
FTIR	Fourier transform infrared
HRR	heat release rate
LOI	limiting oxygen index
MBA	*N*,*N*-methylenebisacrylamide
MLR	mass loss rate
MMT	montmorillonite
PAA	poly(acrylic acid)
PAM	polyacrylamide
PAN	polyacrylonitrile
PBA	poly(butyl acrylate)
PBMA	poly(butyl methacrylate)
PEHA	poly(ethylhexyl acrylate)
PHEMA	poly(2-hydroxyethyl methacrylate)
PHRR	peak heat release rate
PLSN	polymer layered silicate nanocomposite

PMMA   poly(methyl methacrylate)
SEM   scanning electron microscopy
SPR   smoke production rate
SS   sodium silicate
TDD   transdermal drug delivery
TEM   transmission electron microscopy
TGA   thermogravimetric analysis
TTI   time-to-ignition
XRD   X-ray diffraction

References

1. X. Huang and W. J. Bittain, *Macromolecules*, 2001, **34**, 3255.
2. J. Zhing and C. A. Wilkie, *Polym. Degrad. Stabil.*, 2004, **83**, 301.
3. B. K. Kim, J. W. Seo and H. M. Jeong, *Eur. Polym. J.*, 2003, **39**, 85.
4. B. M. Navak, *Adv. Mater.*, 1993, **85**, 422.
5. S. Komarnenl, *J. Mater. Chem.*, 1992, **2**, 1219.
6. B. Lepoittevin, N. Pantoustier, M. Devalckenaere, M. Alexandre, D. Kubies, C. Calberg, R. Jerome and P. Dubois, *Macromolecules*, 2002, **35**, 8385.
7. S. Morlat-Therias, B. Mailhot, G. Gonzalez and J. Gardette, *Chem. Mater.*, 2005, **17**, 1072.
8. A. Usuki, Y. Kojima, M. Kawasmi, A. Okada, Y. Fukusima and T. Kunuanchi, *J. Mater. Res.*, 1993, **8**, 1179.
9. K. Yano, A. Usuki, T. Kuruanchi and O. Kamigaino, *J. Polym. Sci., Part A: Polym. Chem.*, 1993, **31**, 2493.
10. P. B. Messermith and E. P. Giannelis, *Chem. Mater.*, 1999, **46**, 1719.
11. P. B. Messermith and E. P. Giannelis, *J. Polym. Sci., Part A: Polym. Chem.*, 1995, **33**, 1047.
12. S. D. Burnside and E. P. Giannelis, *Chem. Mater.*, 1995, **7**, 1597.
13. R. A. Vaia, S. Vasudevan, W. Krawiec, L. G. Scanlon and E. P. Giannelis, *Adv. Mater.*, 1995, **7**, 154.
14. S. Wang, Y. Hu, R. Zong, Y. Tang, Z. Chen and W. Fan, *Appl. Clay Sci.*, 2004, **25**, 49.
15. P. K. Sahoo and R. Samal, *Polym. Degrad. Stabil.*, 2007, **92**, 1700.
16. R. Samal, P. K. Rana, G. P. Mishra and P. K. Sahoo, *Polym. Compos.*, 2008, **29**, 173.
17. G. Y. Marosi, P. Anna and A. Szabo, *Polym. Mater. Sci. Eng.*, 2004, **91**, 29.
18. C. M. L. Preston, G. Amarasinghe, R. A. Shanks, J. L. Hopewell and Z. Mathys, *Polym. Degrad. Stabil.*, 2004, **84**, 533.
19. G. J. Marosi, *Express Polym. Lett.*, 2007, **1**, 545.
20. P. K. Rana, S. K. Swain and P. K. Sahoo, *J. Appl. Polym. Sci.*, 2004, **93**, 1007.

21. S. S. Ray, O. A. Kampala, M. Okiamoto and M. Ueda, *Macromol. Rapid Commun.*, 2002, **23**, 943.
22. C. Chou and J. Lin, *Macromolecules*, 2005, **38**, 230.
23. K. Wang, L. Wang, J. Wu, L. Chen and C. He, *Langmuir*, 2005, **21**, 3613.
24. C. J. G. Plummer, L. Garamszegi, Y. Leterrier, M. Rodlert and J. E. Manson, *Chem. Mater.*, 2002, **14**, 486.
25. Y. S. Choi, M. H. Choi, K. H. Wang, S. O. Kim, Y. K. Kim and I. J. Chung, *Macromolecules*, 2001, **34**, 8978.
26. P. Viville, R. Lazzaroni, E. Pollet, M. Alexandre, P. Dubois, G. Borcia and J. Pireaux, *Langmuir*, 2003, **19**, 9425.
27. J. H. Park and S. C. Jana, *Macromolecules*, 2003, **36**, 2758.
28. W. R. Mariott and E. Y.-X. Chen, *J. Am. Chem. Soc.*, 2003, **125**, 15726.
29. D. R. Robello, N. Yamaguchi, T. Blanton and C. Barnes, *J. Am. Chem. Soc.*, 2004, **126**, 8118.
30. Z. Zhang, L. Zhang, Y. Li and H. Xu, *Polymer*, 2005, **46**, 129.
31. S. S. Ray and M. Okamoto, *Prog. Polym. Sci.*, 2003, **28**, 1539.
32. K. E. Strawhecher and E. Manias, *Chem. Mater.*, 2000, **12**, 2943.
33. D. C. Lee and L. W. Jang, *J. Appl. Polym. Sci.*, 1996, **61**, 1117.
34. A. B. Morgan and J. W. Gilman, *J. Appl. Polym. Sci.*, 2003, **87**, 1329.
35. T. K. Kim, L. W. Jang, D. C. Lee, H. J. Choi and M. S. Jhon, *Macromol. Rapid Commun.*, 2002, **23**, 191.
36. P. K. Sahoo, B. Samal and S. K. Swain, *J. Appl. Polym. Sci.*, 2004, **91**, 3120.
37. G. Darid, F. Ozer, B. C. Simionescu, H. Zareir and E. Piskin, *Eur. Polym. J.*, 2002, **38**, 73.
38. P. K. Sahoo and R. Mahapatra, *Eur. Polym. J.*, 2003, **39**, 1838.
39. P. K. Sahoo and B. Samal, *Chin. J Polym Sci.*, 2007, **25**, 145.
40. P. K. Rana, S. K. Swain and P. K. Sahoo, *J. Appl. Polym. Sci.*, 2004, **93**, 1007.
41. P. K. Sahoo, R. Samal, S. K. Swain and P. K. Rana, *Eur. Polym. J.*, 2008, **44**, 3522.
42. B. Samal, P. K. Rana and P. K. Sahoo, *Polym. Compos.*, 2008, **29**, 1203.
43. P. K. Rana and P. K. Sahoo, *J. Appl. Polym. Sci.*, 2007, **106**, 3915.
44. D. Ray, D. K. Mohapatra, R. Mohapatra, G. P. Mohanta and P. K. Sahoo, *J. Biomater. Sci. Polym. Ed.*, 2008, **19**, 1487.
45. R. Allcock, T. J. Hartle, J. P. Taylor and N. J. Sunderland, *Macromolecules*, 2001, **34**, 3896.
46. D. Price, K. J. Bullett, L. K. Cunliffe, T. R. Hull, G. J. Milnes, J. R. Ebdon, B. J. Hunt and P. Joseph, *Polym. Degrad. Stabil.*, 2005, **88**, 74.
47. D. Price, K. Pyrah, T. R. Hull, G. J. Milnes, J. R. Ebdon, B. J. Hunt and P. Joseph, *Polym. Degrad. Stabil.*, 2002, **77**, 227.
48. F. Barontini, V. Cozzani and L. Petarca, *Ind. Eng. Chem. Res.*, 2001, **40**, 3270.
49. G. Soderstrom and S. Marklund, *Environ. Sci. Technol.*, 2004, **38**, 825.
50. A. Sjodin, H. Carlsson, K. Thuresson, S. Sjolin, A. Bergman and C. Ostman, *Environ. Sci. Technol.*, 2000, **35**, 448.

51. M. Zanetti, G. Camino, D. Canavese, A. B. Morgan, F. J. Lamelas and C. A. Wilkie, *Chem. Mater.*, 2002, **14**, 189.
52. Z. Wang, E. Han and W. Ke, *Polym. Degrad. Stabil.*, 2006, **91**, 1937.
53. K. Zhang, L. Zheng, X. Zhang, X. Chen and B. Yang, *Colloids Surf. A.*, 2006, **277**, 145.
54. H. Horacek and R. Grabner, *Polym. Degrad. Stabil.*, 1996, **54**, 205.
55. J. Wang and Z. Han, *Polym. Adv. Technol.*, 2006, **17**, 335.
56. A. B. Morgan and J. M. Tour, *Macromolecules*, 1998, **31**, 2857.
57. T. W. Federle, M. A. Barlaz, C. A. Pettugrew, K. M. Kerr, J. J. Kemper and B. A. Nuck, *Biomacromolecules*, 2002, **3**, 838.
58. Y. Liu, C. Hsu and K. Hsu, *Polymer*, 2005, **46**, 1851.
59. P. K. Sahoo, M. Dey and S. K. Swain, *J. Appl. Polym. Sci.*, 1999, **74**, 2784.
60. P. Maiti, P. H. Nam, M. Okamoto, N. Hasegawa and A. Usuki, *Macromolecules*, 2002, **35**, 2042.
61. O. Becker, Y. B. Change, R. J. Verley and G. P. Simon, *Macromolecules*, 2003, **36**, 1616.
62. K. H. Wang, M. H. Choi, C. M. Koo, M. Xu, J. Chung and M. C. Jang, *J. Polym. Sci., Part. B: Polym. Phys.*, 2002, **40**, 1454.
63. S. K. Swain and A. I. Isayev, *Polymer*, 2007, **48**, 281.
64. T. Randoux, J. C. Vanovervelt, H. V. Bergen and G. Camino, *Prog. Org. Coat.*, 2002, **45**, 281.
65. E. P. Giannelis, C. A. Wilkie and E. Manias, in *Chemistry and Technology of Polymer Additives*, ed. S. Al-Malaika, A. Golovoy and C. A. Wilkie, Blackwell Scientific, Oxford, 1999, p. 249.
66. J. Wang and Z. Han, *ACS Symp. Ser.*, 2005, **922**, 172.

CHAPTER 7

Polymer Nanocomposites Prepared by Suspension Polymerization of Inverse Emulsion

JINTAO YANG,[a] BIN ZHU[b] AND L. JAMES LEE[b]

[a] College of Chemical Engineering and Material, Zhejiang University of Technology, Hangzhou 310014, P.R. China; [b] Department of Chemical and Biomolecular Engineering, Ohio State University, Koffolt Laboratories, 140 W. 19th Avenue, Columbus, OH 43210, USA

7.1 Introduction

The application of inorganic nanoparticles in polymers and the development of high-performance functional polymer nanocomposites have been intensively investigated in the last twenty years. Some milestone developments in this area include the following: Iijima's paper published in *Nature* boosted high interest in carbon nanotubes (CNTs),[1] a study on nylon-6–montmorillonite (MMT) nanocomposite[2] was reported by Toyota Central Research Laboratories in Japan and, more recently, the extraction of graphene and its application in polymer nanocomposites were reported.[3] Additionally, there are many good reviews on the performance and physical properties of various nanoparticles and nanocomposite manufacturing processes.[4–7]

Nanoparticles refer to small particles that have at least one dimension in the 1–100 nm range. According to the number of dimensions at the nanoscale, nanoparticles can be classified as planar nanosheets, nanofibers/tubes and nanospheres.[8] A variant of nanospheres is nanoporous microparticles. Whereas the

RSC Nanoscience & Nanotechnology No. 16
Polymer Nanocomposites by Emulsion and Suspension Polymerization
Edited by Vikas Mittal
© Royal Society of Chemistry 2011
Published by the Royal Society of Chemistry, www.rsc.org

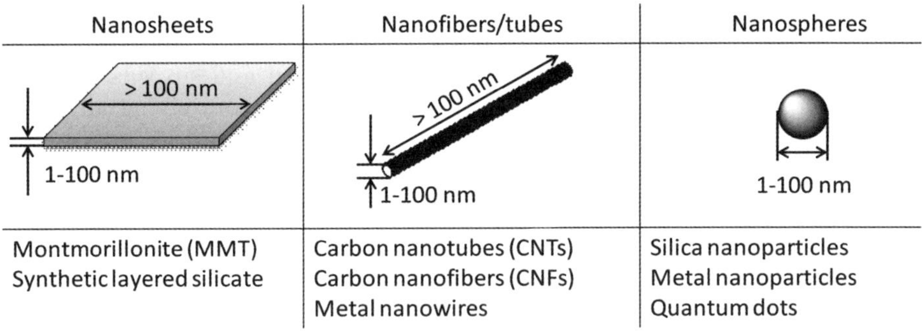

Nanosheets	Nanofibers/tubes	Nanospheres
> 100 nm 1-100 nm	>100 nm ↑1-100 nm	↔ 1-100 nm
Montmorillonite (MMT) Synthetic layered silicate	Carbon nanotubes (CNTs) Carbon nanofibers (CNFs) Metal nanowires	Silica nanoparticles Metal nanoparticles Quantum dots

Figure 7.1 Different nanoparticle shapes and examples.

diameter of the particle may be in the order of microns, the pore sizes are in the order of nanometers. A diagram of different nanoparticle shapes and examples of widely investigated nanoparticles are shown in Figure 7.1. Obvious features of nanoparticles are their large specific surface area, high aspect ratio and functionality.

When nanoparticles are used as inorganic fillers dispersed in a polymer matrix, the hybrid material is defined as a polymer nanocomposite. Exfoliated nanoparticles, especially those with high aspect ratio, can significantly affect the rheological behavior and morphology of polymers.[9–15] Some nanoparticles, such as CNTs and graphene, can bring functionality into the polymer matrix, achieving the percolation threshold at a much lower concentration than conventional inorganic fillers.[16–19] Properties of the polymer matrix that can be reinforced by adding nanoparticles include modulus,[9,20–23] strength,[23–25] thermal stability,[12] heat resistance,[26–28] electrical conductivity,[16,17] reduced gas permeability[28] and low flammability.[29–33] In addition to dispersion, the orientation and location of nanoparticles in the polymer matrix are also important issues in the preparation of polymer nanocomposites, but they are not discussed in this chapter.

In the last twenty years, many polymers have been used to make polymer nanocomposites. Thermoplastic polymers include nylon,[2,20,21] polyaniline (PANI),[34–37] poly(ε-caprolactone),[38] polycarbonate (PC), polyether ether ketone (PEEK),[39] polyethylene (PE), poly(ethyl acrylate) (PEA), polyisoprene (PI), polylactide (PLA),[26] poly(methyl methacrylate) (PMMA),[12,40–47] polypropylene (PP),[30] polypyrrole (PPy),[48] polystyrene (PS),[10,14–17,27,30,49–64] poly(vinyl acetate) (PVAc),[65,66] poly(vinyl alcohol) (PVA), poly(vinyl chloride) (PVC)[32,67] and thermoplastic polyurethane (TPU),[9] and thermosets include Bakelite, butadiene rubber, epoxy,[19,22,68] polydimethylsiloxane (PDMS), polyurethane (PU),[69] styrene–butadiene rubber (SBR) and unsaturated polyester resin.

Since most inorganic particles are not very compatible with polymers, mechanical agitation and ultrasonication are usually applied to facilitate particle dispersion. Particle surface modification and surfactants are also used to increase the compatibility between nanoparticles and the polymer matrix. Based on the original medium where the nanoparticles are dispersed, the manufacturing process for making polymer nanocomposites includes melt blending, solution mixing and *in situ* polymerization. The *in situ* polymerization can be further divided into *in situ* bulk, solution, emulsion and suspension polymerization and a combination of *in situ* emulsion and suspension polymerization, which is the main focus of this chapter. Before giving an introduction to the preparation of polymer nanocomposites *via* suspension polymerization of inverse emulsion, we first review recent developments in preparing polymer nanocomposites using either inverse emulsion polymerization or suspension polymerization.

7.2 Preparation of Polymer Nanocomposites *via* Inverse (Mini)emulsion Polymerization

Emulsion polymerization is a type of radical polymerization starting in an emulsion system in which monomer droplets are emulsified in a continuous phase of water forming an oil-in-water (O/W) heterogeneous system. For water-soluble monomers, such as acrylamide (AAm),[70] aniline[34,35] and pyrrole,[48] the monomers dissolved in water are emulsified in a continuous oil phase to form a water-in-oil (W/O) emulsion which is called an inverse emulsion. For inorganic nanoparticles that are hydrophilic, they can be dispersed in water with dissolved monomers and then polymerized by inverse emulsion polymerization. After polymerization, well-dispersed nanoparticles in polymer nanocomposites are obtained. In some cases, surface-modified nanoparticles can be used as stabilizers which are located at the surface of latex particles. When particles instead of surfactant molecules are used to stabilize the emulsion, the term 'Pickering emulsion' is commonly used.[71] Here, nanoparticles at the interface form a monolayer to reduce the interfacial energy of the biphasic system. The dispersed droplets are stabilized by the nanoparticles, which are later captured in the surface of resultant polymer beads to form nanocomposite microspheres. In Pickering emulsion or Pickering suspension polymerization, the most important factor is the stabilizing ability of nanoparticles. Several studies used silica nanoparticles and a toluene–water emulsion as a model system to show that the wettability of nanoparticles determined the stability of the emulsion or suspension.[72–74] When highly hydrophilic or hydrophobic nanoparticles were used in the system, the emulsion particles tended to have a large size ($> 100 \, \mu m$) and were unstable and easy to coalesce. If the surface of the silica nanoparticles was modified by chemisorbed silane, the emulsion size could be adjusted to less than $1 \, \mu m$ without coalescence.[72–74] In the following, we first introduce the preparation of polymer nanocomposites by inverse emulsion polymerization with nanoparticles dispersed in water, then introduce the inverse Pickering emulsion.

The simplest way to make a polymer nanocomposite *via* inverse emulsion polymerization is to mix the water-soluble monomer and hydrophilic nanoparticles together in the water as a dispersed phase. Kim *et al.*[48] prepared intercalated conducting polypyrrole (PPy)/Na$^+$-MMT nanocomposite by inverse emulsion polymerization. They followed a similar preparation method introduced by Sun and Ruckenstein[75] for preparing conductive PPy–rubber composites. In Kim *et al.* 's study, dodecylbenzenesulfonic acid (DBSA) was used as an emulsifier and a dopant for the polymerization of pyrrole. A water-soluble initiator, ammonium peroxydisulfate (APS), was initially dissolved in water then mixed with a continuous phase of isooctane containing dissolved DBSA for emulsification. When the inverse emulsion was formed, it was mixed with a dispersion of Na^+-MMA in water for 12 h at 0 °C. A predetermined amount of pyrrole in toluene was then dropped into the inverse emulsion to carry out the polymerization at 0 °C for 24 h. Nanocomposite particles were obtained after the polymerization. According to the X-ray diffraction (XRD) pattern of the PPy–Na^+-MMT nanocomposite, the average interlayer gallery spacing was estimated to be 1.6 nm compared with 1.1 nm for the Na^+-MMT. Thus, MMT was intercalated in the nanocomposite particles. Following a similar procedure, Kim *et al.*[34] prepared PANI–multi-walled carbon nanotubes (MWCNTs) nanocomposite *via* inverse emulsion polymerization. In this study, MWCNTs were first treated with a 3:1 mixture of concentrated sulfuric acid and nitric acid. This acid treatment is used to provide carboxylic groups at the defects of the CNT surface to improve its dispersion in water and organic solvent.

Surface treatment of nanoparticles has been widely used in the preparation of polymer nanocomposites. It can change the surface properties of nanoparticles to make them more compatible with the surrounding media. For example, natural MMT is very hydrophilic and has negative charges in layers caused by isomorphic substitution (for example, Al^{3+} replaced by Mg^{2+} or Fe^{2+}, or Mg^{2+} replaced by Li^+). These negative charges are counterbalanced by exchangeable cations residing between MMT layers.[4] Thus, natural MMT can be surface modified by a quaternary ammonium salt *via* an ion-exchange reaction, after which the organically modified MMT becomes hydrophobic and more compatible with the oil phase. Sun *et al.*[35] prepared PANI–MMT nanocomposites *via* inverse emulsion polymerization in supercritical carbon dioxide (scCO$_2$). The Na^+-MMT was surface modified with different surfactants by an ion-exchange reaction before use. The surfactant structure, weight percentage and interlayer gallery spacing of modified MMT are listed in Table 7.1. In a typical polymerization, a mixture of the surfactants bis(2-ethylhexyl) sulfosuccinate (AOT) and polyether-modified polysiloxane (PeSi) were dissolved in the ethanol. Aniline, hydrochloric acid and modified MMT were mixed in a predetermined amount of water with ultrasonication. The ethanol solution and water mixture were injected into an autoclave containing an APS solution in a sealed container. After the autoclave had been filled with scCO$_2$, the inverse emulsion was stabilized by AOT, PeSi and ethanol. Then, the inverse emulsion was mixed uniformly with modified MMT in scCO$_2$. After the polymerization had been carried out for 24 h by the exuded initiator, the scCO$_2$ was released to obtain a nanocomposite. XRD patterns and transmission electron microscope (TEM) images of nano-composites indicated that aminated MMT (NHMMT) and fluorinated MMT (FMMT) could be exfoliated in the nanocomposite with up to 12 wt% of

Table 7.1 Physical data of modified MMT clays. Reproduced from reference
35 by permission of John Wiley & Sons

MMT	Organic modifier	d spacing (nm)[a]	Modifier intercalated (wt%)[b]
NaMMT	None	1.38	–
Cetyl-MMT (CeMMT)	$CH_3(CH_2)_{15}N^+(CH_3)_3$	1.66	35.9
Benzyl-MMT (BeMMT)	$CH_3(CH_2)_{11}N^+(CH_3)_2CH_2C_6H_5$	2.08	33.2
Hydroxy-MMT (OHMMT)	$CH_3(CH_2)_{11}N^+CH_3(CH_2CH_2OH)_2$	2.62	45.1
Aminated MMT (NHMMT)	$NH_2(CH_2)_3N^+(CH_3)_2CH_2C_6H_5$	1.52	20.9
Fluorinated MMT (FMMT)	$CF_3(CF_2)_7SO_2NH(CH_2)_3N^+(CH_3)_3$	1.47	36.7

[a]Determined by XRD.
[b]Determined by TGA.

surface-modified MMT, whereas other modified MMTs could only form an
intercalated structure in the nanocomposite even at a low MMT loading. The
authors concluded that the interaction between chemical groups of surfactants
residing in the MMT layers with scCO$_2$ was the most important factor for
MMT dispersion.

In addition to water-soluble monomers, some hydrophobic monomers can
also be polymerized in an inverse emulsion system. Chen *et al.*[49] used styrene to
prepare magnetite (Fe_3O_4)–PS nanocomposite *via* inverse emulsion polymerization. Yang *et al.*[64] used a similar method to prepare Fe_3O_4–PS nanocomposite microspheres with a hollow structure. In their study, Fe_3O_4
nanoparticles were surface modified with oleic acid to make them amphipathic.
Water droplets stabilized by sorbitan monooleate (Span-80) were used as a soft
template for the formation of the hollow structure. In the inverse miniemulsion,
the amphipathic Fe_3O_4 particles tended to assemble at the W/O interface. The
polymerization was initiated by irradiation using ^{60}Co γ-rays, which could
produce free radicals from the water. According to Scheme 7.1, these free
radicals could further lead to the formation of free radicals on the surface of
Fe_3O_4 particles, on which the styrene diffused from oil phase was polymerized
to form a shell. The TEM and scanning electron microscope (SEM) images of
nanocomposite spheres with a hollow structure are shown in Figure 7.2. Later,
the same group used a reactive surfactant, 12-acryloxy-9-octadecenoic acid
(AOA), to copolymerize with styrene to control the structure of nanocomposite
hollow microspheres.[50]

The first study using inverse Pickering emulsion polymerization to prepare
polymer nanocomposites was reported by Voorn *et al.*[70] An organoclay,
Cloisite 20A (20A) from Southern Clay Company (Gonzalez, TX, USA), was

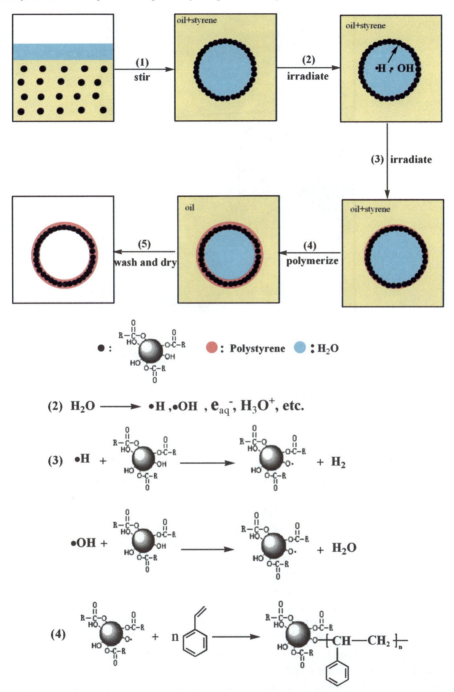

Scheme 7.1 Schematic illustration of the procedure for preparing hollow super-paramagnetic nanocomposite microspheres. Reproduced from reference 64 by permission of John Wiley & Sons.

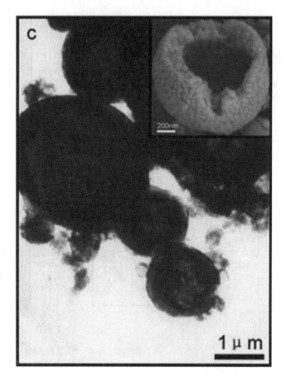

Figure 7.2   TEM images of hollow superparamagnetic nanocomposite microspheres (SEM image of a broken sphere shown in the inset; the scale bar of the inset is 200 nm). Reproduced from reference 64 by permission of John Wiley & Sons.

dispersed in cyclohexane by ultrasonication followed by centrifugation. This procedure was repeated twice to improve the dispersion of 20A in cyclohexane, which was characterized by dynamic light scattering (DLS). A water solution of monomer (AAm), initiator and crosslinker (not used in some experiments) were added to a mixture of 20A in cyclohexane to prepare an inverse emulsion stabilized by 20A. A possible stabilizing mechanism of 20A proposed by the authors was that the hydrophobic surfactant on the MMT surface could realign itself during the emulsification in order to maximize the hydrophilic surface which was in contact with the water droplets. This rearrangement would make the clay platelet a surfactant-like particle.[70] After emulsification, the temperature was increased to carry out the polymerization. The scheme of emulsification and polymerization is shown in Scheme 7.2. The SEM images in Figure 7.3 show the rugged surface morphology of polyacrylamide (PAAm) composite particles. indicating that the composite particles were covered by 20A. The cryo-TEM images in Figure 3 give a clearer picture in which nanoclay layers are located at the surface of the PAAm latex particles. The XRD results demonstrated that 20A was exfoliated in the polymer nanocomposites after polymerization.

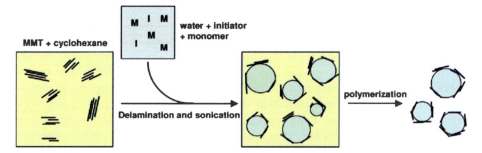

Scheme 7.2 Schematic representation of the inverse emulsion polymerization and the stabilizing function of the hydrophobic clay platelets. Reproduced from reference 70 by permission of the American Chemical Society.

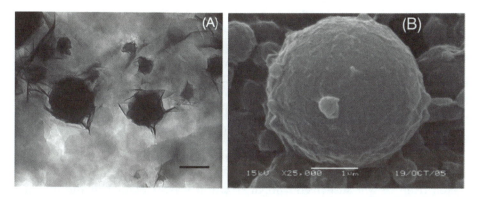

Figure 7.3 (A) Cryo-TEM images of PAAm latexes prepared with a water-soluble initiator; (B) SEM images of PAAm latex particles stabilized by Cloisite 20A clay platelets; the micrographs clearly show the rugged surface morphology of the PAAm composite particles (scale bar is 1 μm). Reproduced from reference 70 by permission of the American Chemical Society.

7.3 Preparation of Polymer Nanocomposites *via* Suspension Polymerization

Suspension polymerization is a polymerization process in which dispersed monomer droplets are stabilized by surfactant and mechanical agitation in a liquid phase such as water. As in emulsion polymerization, the monomers used for suspension polymerization are usually insoluble in water. Advantages of suspension polymerization compared with other polymerization methods include easy removal of reaction heat, effective temperature control, low viscosity of the reaction system, simple polymerization mechanism, high purity of the product, easy separation and purification of the product and shaped product in particle form. Polymers prepared by suspension polymerization include poly(glycidyl methacrylate) (PGMA),[76] phenolic resin, PMMA,[40–46] poly (urethane acrylate) (PUA),[77] PS,[51,53–56] PVAc[65,66] and PVC.[67] Unlike emulsion polymerization, in which the polymerization takes place in the micelles and the monomer molecules are transferred from the monomer droplets to micelles for further polymerization, suspension polymerization has a relatively simple mechanism. The polymerization is carried out in monomer droplets. It can be regarded as bulk polymerization in small reactors which are dispersed in a continuous liquid phase. If nanoparticles are dispersed in the monomer droplets which can be stabilized, it is much easier to make polymer nanocomposites *via in situ* suspension polymerization than emulsion polymerization.

The first study using suspension polymerization to prepare polymer nanocomposites was reported by Huang and Brittain,[40] who prepared a PMMA–layered silicate nanocomposite. In their study, two types of hydrophilic layered silicate were used, Gelwhite GP (MMT) and Laponite XLS (a synthetic layered silicate). In a typical suspension polymerization, layered silicate was exfoliated in water mixed with different organic modifiers such as *n*-decyltrimethylammonium chloride, 2,2′-azobis(isobutylamidine hydrochloride) and [2-(methacryloyloxy)ethyl]trimethylammonium chloride. The chemical structures of these three surfactants are shown in Scheme 7.3: *n*-decyltrimethylammonium chloride is an alkyl chain surfactant, 2,2′-azobis(isobutylamidine hydrochloride) is an initiator surfactant and [2-(methacryloyloxy)ethyl]trimethylammonium chloride is a co-monomer surfactant. A mixture of monomer (MMA) and initiator (2,2′-azobis(2-methylpropionitrile)) (AIBN) was added with vigorous stirring to form a suspension. Modified layered silicates were speculated to be absorbed on the surface of monomer droplets as a stabilizer. According to the XRD pattern, the MMT in nanocomposite pellets taken from *in situ* suspension polymerization was exfoliated in the polymer matrix up to 5 wt%. After melt pressing, however, the alkyl chain-modified MMT showed an aggregated structure. The exfoliation of MMT modified by other two surfactants was preserved. Compared with pristine PMMA, the

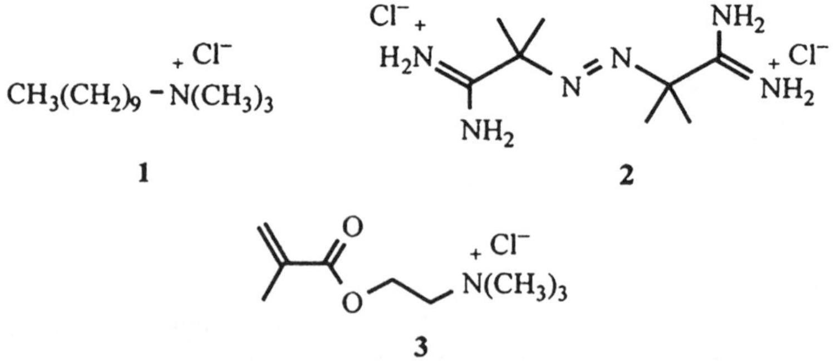

Scheme 7.3 Structure of onium ions used as modifiers for the layered silicate. Reproduced from reference 40 by the permission of the American Chemical Society.

nanocomposite showed an increase of 15 °C in glass transition temperature (T_g) and 60 °C in thermal degradation temperature. Wang *et al.*[41] reported a different method to prepare PMMA–MMT nanocomposites. The organically modified MMT was initially dispersed in MMA and then the mixture was mixed in water to carry out the suspension polymerization. They also concluded that a C=C double bond in the surfactant was beneficial for the exfoliation of MMT during polymerization. Kim *et al.*[42] used a similar method to prepare PMMA–MMT nanocomposites. PVA was used as a stabilizer for the suspension polymerization.

In a suspension system, nanoparticles can be dispersed in either the continuous phase (water), the dispersed phase (monomer droplets) or the interface as a stabilizer. For inorganic nanoparticles that are hydrophilic, it is relatively simple to disperse unmodified inorganic nanoparticles in the continuous phase such as water. Yeum *et al.*[65] prepared high molecular weight (HMW) PVAc–silver nanocomposite microspheres *via* suspension polymerization. Their reaction started with dissolving a suspending agent in water under a nitrogen atmosphere with constant stirring for 20 min. The monomer, vinyl acetate, the silver nanoparticles in aqueous solution and the initiator, 2,20-azobis(2,4-dimethylvaleronitrile) (ADMVN), were added together to the reactor. After polymerization for a period of time, the suspension was cooled for more than 24 h, allowing complete sedimentation of PVAc–silver nanocomposite microspheres at the bottom. From the SEM images in Figure 7.4, a golfball-shaped surface on the microspheres was observed, which might have resulted from aggregation of silver nanoparticles. Following a similar method, the same group prepared PMMA–silver nanocomposite microspheres.[43] Kwon *et al.*[44] prepared PMMA–MWCNTs *via* suspension polymerization in a continuous phase of methanol. Purified MWCNTs which had carboxyl groups on the surface were hydrophilic, so they were initially dispersed in methanol. The experimental results showed that MWCNTs were not only embedded in the microspheres but were also present on the sphere surface (Figure 7.5).

In most cases, *in situ* suspension polymerization is carried out with nanoparticles dispersed in monomers. The surface modification of nanoparticles is usually required in order to achieve a good dispersion. Since the mechanism of *in situ* suspension polymerization is similar to that of *in situ* bulk polymerization, the surface modification used in bulk polymerization, including ion-exchange reaction, surfactant and coupling agent, polymer grafting and acid treatment can also be applied here.

As mentioned before, MMT can be easily surface modified by ion-exchange reaction. Xie *et al.*[51] prepared PS–MMT nanocomposites *via* suspension polymerization. Four different quaternary ammonium salts were used to modify the surface of MMT which were later shown to be exfoliated in the PS matrix according to XRD results. Hwu *et al.*[52] prepared surface-modified MMT in different solvents and compared the dispersion of MMT in nanocomposites which were synthesized by suspension polymerization. Jung *et al.*[66] prepared PVAc/PVA–MMT nanocomposite microspheres by suspension

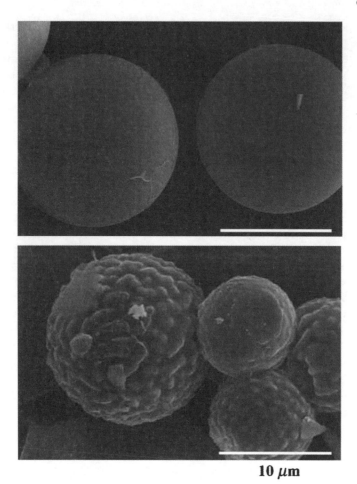

10 μm

Figure 7.4 SEM images of HMW PVAc and PVAc–silver nanocomposite micro-spheres prepared by suspension polymerization at a suspending agent concentration of 9.0 g dl^{-1} of water and [ADMVN] = 10^{-4} mol per mole of VAc. Reproduced from reference 65 by permission of John Wiley & Sons.

polymerization and saponification. In their study, MMT was mixed with a cationic surfactant, cetyltrimethylammonium bromide (CTAB), for surface modification in the VAc monomer. The mixture was then poured into water containing dissolved suspending agent. The suspension polymerization was carried out immediately when the initiator ADMVN was added at a fixed temperature. The PVAc/PVA–MMT nanocomposite microspheres were prepared by heterogeneous saponification. The SEM image showed that when more MMT was added to the monomer, the surface of the nanocomposite microsphere appeared rougher. Observation of the fracture surface of nano-composite microspheres indicated that MMT was embedded inside the PVAc matrix. Jun and Suh[77] prepared PUA–MMT nanocomposite particles by *in situ*

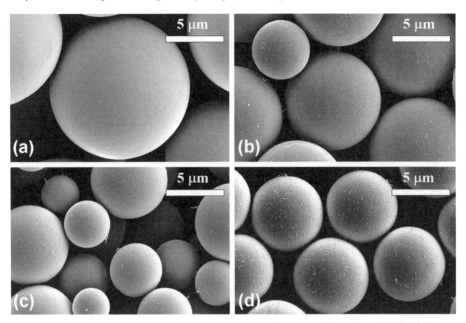

Figure 7.5 SEM images of PMMA and PMMA–MWCNT microspheres prepared by *in situ* dispersion polymerization: (a) PMMA; (b) PMMA–MWCNT (0.01 wt%); (c) PMMA–MWCNT (0.02 wt%); (d) PMMA–MWCNT (0.03 wt%). Reproduced from reference 44 by permission of John Wiley & Sons.

suspension polymerization using a commercial organoclay, Cloisite 25A, which showed an intercalated structure in the nanocomposite.

The surface modification of layered double hydroxides (LDHs), titanium dioxide (TiO$_2$) and SiO$_2$ nanoparticles can be carried out by mixing nanoparticles with surfactants or reaction with coupling agents. Ding and Qu[53] prepared PS–LDH nanocomposites *via* suspension polymerization with two different surfactants, *N*-lauroyl glutamate (LG) and sodium dodecyl sulfate (SDS), and a long-chain spacer, *n*-hexadecane. Molecular level dispersion of LDHs in PS could be achieved using LG and *n*-hexadecane as the surfactant and long-chain spacer, respectively. Bao *et al.*[67] prepared PVC–LDH nanocomposites *via* suspension polymerization in which the LDH was intercalated by dodecyl sulfate (DS) anions. Compared with melt blending of PVC with DS-intercalated LDHs (LDH-DS), suspension polymerization provided a much better dispersion of LDHs in the PVC matrix, leading to better mechanical properties. Horak *et al.*[76] synthesized PGMA–La$_{0.75}$Sr$_{0.25}$MnO$_3$ nanocomposite particles for magnetic separation *via* suspension polymerization. La$_{0.75}$Sr$_{0.25}$MnO$_3$ nanoparticles were surface modified using penta(methylethylene glycol) phosphate methacrylate (PMGPMA) to improve compatibility and then polymerized within PGMA. For SiO$_2$ nanoparticles, the coupling agent silane was usually used for surface modification. Zheng *et al.*[45] prepared

PMMA–SiO$_2$ nanocomposites *via* suspension polymerization. SiO$_2$ nano-particles were surface modified by two methods: one was modification by γ-methacyloxypropyltrimethoxysilane (KH570) and lauryl alcohol (12COH), and the other was grafting PMMA on to the surface of SiO$_2$ treated with KH570. The dispersion of SiO$_2$ with grafted PMMA was much better than that of SiO$_2$ treated with KH570. Yang *et al.*[46] prepared PMMA–Ag and PMMA–Cu nanocomposites by suspension polymerization and H$_2$-assisted reduction. An organometallic precursor complex was directly added into the MMA solution followed by a suspension polymerization.

Carbon nanotubes (CNTs) and carbon nanofibers (CNFs) have a chemically inert surface and it is difficult to carry out surface modification. Acid treatment is usually used to introduce carboxylic groups at the defects of the CNT and CNF surface, which can improve particle dispersion. However, the acid treatment may damage the nanoparticles causing a reduction of mechanical properties. Park *et al.*[54] prepared PMMA–MWCNT nanocomposites *via* suspension polymerization. Acid-treated MWCNTs were dispersed in MMA by ultrasonication, then an initiator (AIBN) was dissolved in the MMA–MWCNT mixture which was later poured into a solution with a stabilizer (PVA) to carry out polymerization at 95 °C for 3.5 h. After polymerization, the PS–MWCNT nanocomposite particles were collected by filtration.

Inorganic nanoparticles can also be dispersed at the interface between the continuous phase and the dispersed phase as a stabilizer. If the suspension is only stabilized by inorganic nanoparticles, this specific polymerization is called Pickering suspension polymerization. When using hydrophilic layered silicates as a stabilizer for Pickering emulsion or suspension polymerization, surface modification is usually required. The first Pickering suspension polymerization to synthesize polymer nanocomposites was reported by Yang *et al.*.[55] The MMT surface was modified with an ammonium free radical initiator *via* ion-exchange reaction. Styrene and 2-(dimethylamino)ethyl methacrylate (DMAEMA) were polymerized in two steps to form a mixed hydrophobic and hydrophilic polymer brush on the MMT surface. The modified MMT particles were used as a stabilizer for Pickering suspension polymerization. Later, the same group proposed a new strategy to prepare modified MMTs for Picking suspension polymerization.[56] In that study, the surface of MMT was not modified to maintain it hydrophilic, but the edges of the MMT were grafted with hydrophobic PS brushes by atom transfer radical polymerization (ATRP). The modified MMT was dispersed in tetrahydrofuran (THF) with ultra-sonication to form a uniform mixture which was later added to water. The solvent THF was removed under reduced pressure. After styrene monomer and initiator had been added, suspension polymerization was carried out at 60 °C with stirring. Since the surface of the MMT layers was still hydrophilic but the edges became hydrophobic because of the grafted PS brushes, the PS brushes would penetrate into the suspension particles while keeping the MMT layers in the aqueous phase. As a result, the modified MMT stayed on the surface of the PS particles like a stabilizer. TEM images of MMT–PS stabilized particles and a proposed particle structure

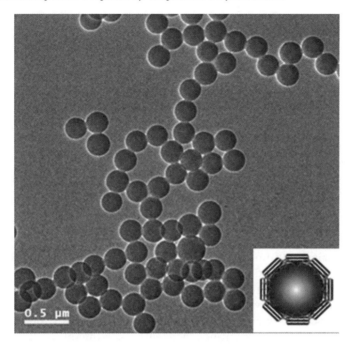

Figure 7.6 TEM image of PS colloidal particles stabilized by clay–PS (inset: schematic illustration of the PS colloidal particles stabilized by clay–PS).
Reproduced from reference 56 by permission of John Wiley & Sons.

are shown in Figure 7.6.[56] The TEM images also indicate that the particle size is
very uniform.

Gao *et al.*[78] prepared thermo-sensitive hybrid microcapsules by inverse
Pickering suspension polymerization. Hydrophilic SiO_2 nanoparticles were
surface modified by dimethyldichlorosilane to be hydrophobic and then dispersed in *n*-hexane. Monomer *N*-isopropylacrylamide (NIPAm) and initiators
were dissolved in water, which was later mixed with *n*-hexane to form a Pickering inverse suspension stabilized with the modified SiO_2 nanoparticles.
Polymerization was carried out at 60 or 0 °C using various initiators to form
different structures of microcapsules.

7.4 Polymer Synthesis *via* Suspension Polymerization of Inverse Emulsion

The concept of suspension polymerization of an inverse emulsion was first
proposed by Crevecoeur *et al.*[57] Basically, this complex polymerization was a
conventional suspension polymerization process in which the suspension particle was an 'inverse emulsion'. Inside the suspension particle, water was
emulsified by the surfactant in a mixture of styrene and PS forming a W/O

inverse emulsion. Particles were dispersed in water to form a water-in-oil-in-water (W/O/W) system for suspension polymerization.

Suspension polymerization of inverse emulsion was developed to prepare water-expandable PS (WEPS).[57] In the traditional method to prepare expandable PS (EPS) beads, styrene is suspended with a stabilizer in water as droplets and can then be polymerized. An organic blowing agent, usually a mixture of pentane isomers, is added to the reaction system, contained in a closed vessel in the final stage of polymerization. The temperature in the vessel is kept higher than the T_g of PS, which is beneficial for diffusion of pentanes in the PS beads. The blowing agent is entrapped in the PS beads after several hours for equilibrium before the temperature is reduced to a lower level. Since the organic blowing agent is unsafe and not environmentally benign, water is proposed as a replacement for foaming applications. To introduce water into the PS beads, the diffusion process cannot work because of the immiscibility between water and styrene–PS mixture. An alternative approach is to make a water-in-styrene emulsion in a continuous-phase polymerization. Here, water droplets are dispersed and wrapped in a solid PS matrix. Detailed experimental procedures in Crevecoeur *et al.*'s study are briefly as follows:[57] surfactant (AOT) and initiators (dibenzoyl peroxide and *tert*-butyl peroxybenzoate) were initially dissolved in styrene. The polymerization was carried out at 90 °C for 100–120 min under a nitrogen atmosphere with a stirring rate of 300 rpm, which is called prepolymerization. When the conversion of styrene was about 20–60%, a predetermined amount of water with sodium chloride was added and quickly emulsified within 5 min assisted by stirring at a rate of 800 rpm (Figure 7.7a). The prepolymerization is important to fix the water droplets in a viscous styrene–PS continuous phase which will later be suspended in the water with a suitable suspending agent to form styrene–PS beads (Figure 7.7b). The viscous mixture was added to a suspension reaction vessel with water and a suspension stabilizer to form a W/O/W suspension system. The suspension polymerization of inverse emulsion was carried out at 90 °C under a nitrogen atmosphere for 4 h with a constant stirring rate of 350 rpm. The reaction temperature was then increased to 125 °C under a nitrogen pressure of 4 bar for a further 3 h. Finally, the suspension was cooled to room temperature and the spherical beads were filtered and washed with clean water. Using the same procedure, WEPS beads with different water contents could be synthesized. When the water in the beads diffused out, empty holes in the PS beads could be observed by SEM as shown in Figure 7.8. If the water stays in the PS beads, the beads can be expanded by heating at a temperature higher that the T_g of PS.[58]

The same group later used a similar method to synthesize an amphiphilic copolymer, PS-*block*-poly(4-vinylbenzenesulfonic acid sodium salt) (PS-*block*-PSSS).[59] First, water, trioctylmethylammonium chloride (TOMAC, phase-transfer catalyst) and initiator were mixed in styrene then the mixture was heated to 90 °C to carry out the bulk polymerization. SSS monomer which was initially dissolved in the water phase, was transferred to the styrene phase to react with PS. The formed amphiphilic PS-*block*-PSSS acted as a surfactant which could emulsify the W/O inverse emulsion. Following a similar

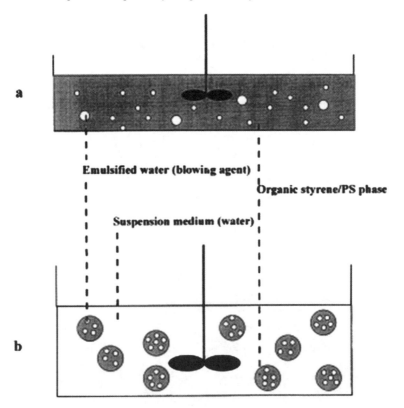

Figure 7.7 Schematic representation of the process for the preparation of WEPS: (a) emulsification of water (blowing agent) in a styrene–PS mixture and (b) suspension polymerization of styrene–PS droplets containing emulsified water. Reproduced from reference 57 by permission of Elsevier.

suspension polymerization they proposed,[57] the beads of water in PS were synthesized. SEM images indicated that the water droplets in the PS beads using PS-*block*-PSSS as a surfactant were smaller and more uniform than those using commercial surfactants.[59]

Pallay and Berghmans[60] used starch as the emulsifier and the water-swellable phase. Prepolymerization of styrene–starch mixture was carried out to a conversion of approximately 30%. The viscous mixture was subsequently transferred to a water medium containing suitable suspension agents. Suspension polymerization was carried out to achieve complete conversion of styrene and the water was directly absorbed into the starch inclusions.[60,61]

7.5 Preparation of Polymer Nanocomposites *via* Suspension Polymerization of Inverse Emulsion

The first attempt to prepare polymer nanocomposites *via* suspension polymerization of inverse emulsion was reported by Shen *et al.*[62] They improved

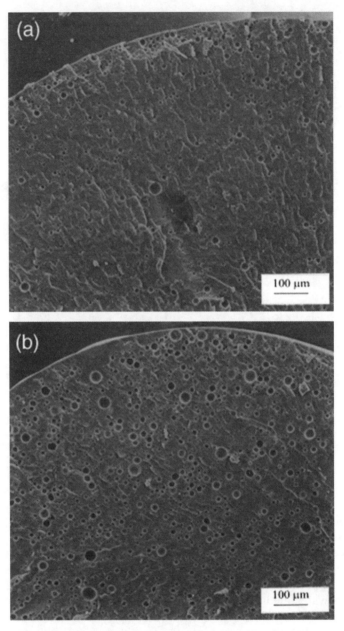

Figure 7.8   SEM images of samples: (a) 0.2 wt% AOT, 2.0 wt% emulsified water, 3.2
wt% incorporated water and (b) 0.8 wt% AOT, 8.0 wt% emulsified water,
9.3 wt% incorporated water prepared *via* suspension polymerization.
Reproduced from reference 57 by permission of Elsevier.

Crevecoeur *et al.*'s suspension polymerization method to prepare water-expandable PS–nanoclay nanocomposites (WEPSCN). MMT is usually surface modified with quaternary ammonium compounds to improve the dispersion of MMT in the polymer matrix. The presence of these low molecular weight hydrocarbons, however, results in a negative impact on fire resistance performance. Therefore, it is highly desirable to develop a method to incorporate MMT into polymers with uniform dispersion, but without the use of fire-hazardous surfactants. For this reason, a modified suspension polymerization of inverse emulsion method was applied to prepare surfactant-free expandable PS–MMT nanocomposites. Instead of emulsifying pure water, a mixture of water and a raw clay, sodium montmorillonite (Na^+-MMT), was emulsified in the organic styrene–PS phase. Due to the hydrophilicity of the Na^+-MMT surface, a uniform and stable dispersion of MMT in water was achieved. Therefore, using water as the carrier, MMT could be incorporated into the polymer system. The whole process consisted of three steps, the first being the preparation of MMT–water mixture. MMT was dispersed in emulsified water with the aid of ultrasonication. The ultrasonication time was controlled until the formation of a uniform and stable water–MMT mixture. Subsequently, 0.5 wt% NaCl (based on emulsified water) was added to the mixture. The salt could facilitate the emulsification of water droplets in a later stage of the process. However, the concentration and addition sequence of NaCl needed to be done with care since NaCl could cause the aggregation of MMT layers. While a low concentration of electrolyte can prevent particle aggregation due to osmotic repulsion, a high concentration will lead to the compression of double layers at both the planer and edge surfaces. As the concentration of electrolyte increases to the critical flocculation concentration (CFC), all three modes (face-to-face, face-to-edge, edge-to-edge) of clay aggregation will occur. For Na^+-MMT used in this study, the highest NaCl concentration which could still maintain a stable water–clay suspension was 0.5 wt%. The time to add NaCl was another factor that affected the stability of the water–clay suspension. If both nanoclay and NaCl were added to the emulsified water simultaneously, the exfoliation of clay became more difficult. However, once the clay had been exfoliated in water, the addition of NaCl merely reduced the stability of this suspension. The second step was the preparation of inverse emulsion. In this step, AIBN (0.25 wt% based on styrene), benzoyl peroxide (BPO) (0.25 wt% based on styrene) and AOT (10 wt% based on emulsified water) were dissolved in styrene. The mixture was heated to 90 °C under a nitrogen atmosphere at a stirring rate of 350 rpm. The reaction was performed in the bulk phase to a conversion of approximately 60% [determined by offline differential scanning calorimetry (DSC)], at which point the viscosity of the continuous phase was sufficiently high to fixate the water droplets. At a higher stirring rate (700 rpm), water–NaCl–MMT was added to the styrene–PS mixture to form a W/O reaction medium. Polymerization was continued for a further 10 min. A schematic of the emulsion system is shown in Figure 7.9a. After the inverse emulsion was prepared, in the

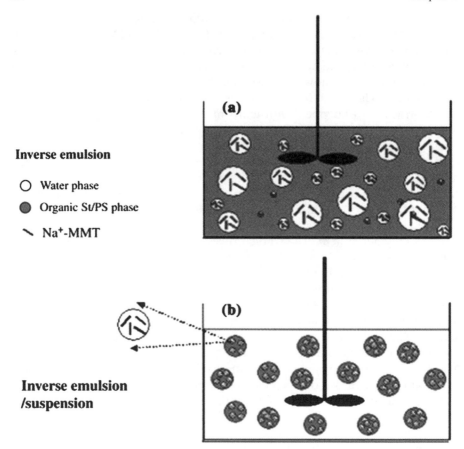

Inverse emulsion

○ Water phase

● Organic St/PS phase

↘ Na⁺-MMT

Figure 7.9 Schematic representation of the process for the preparation of WEPSCN:
(a) emulsification of water–clay mixture in prepolymerized styrene–PS and
(b) suspension polymerization of styrene–PS droplets containing emulsified water–clay droplets. Reproduced from reference 62 by permission of
Elsevier.

third step the viscous mixture was suspended in water with the aid of suspension stabilizers, hydroxyethyl (0.6 wt% based on suspension water) and
PVA (0.005 wt% based on suspension water). Polymerization was continued
under a nitrogen atmosphere. The stirring rate and temperature were kept at
350 rpm and 90 °C, respectively. Finally, the suspension was cooled to room
temperature and the spherical products were recovered by filtration.
A schematic of the suspension system is shown in Figure 7.9b. Water-
expandable PS–MMT nanocomposites were successfully synthesized *via* this
suspension polymerization of water-in-oil inverse emulsion. Using water as the
carrier, surfactant-free nanoclay can be incorporated into the polymer beads.

The addition of nanoclay can trap more water in the beads during synthesis and reduce the water loss during storage.

Yang *et al.*[63] used a similar procedure to prepare water-expandable PS–activated carbon composite (WEPSAC). The purpose of their study was to prepare water-containing PS composite beads without any emulsifiers. Unlike the WEPS and WEPSCN, this WEPSAC can be applied in both batch and extrusion foaming processes. Activated carbon (AC) has an exceptionally high surface area, which makes it an excellent water-absorbent material. Prior to this study, the same group had studied the effect of moisture content in the extrusion foaming process by adding activated carbon particles containing different amounts of water together with PS to a twin screw extruder. However, most of the water was evaporated by the heat generated from the extruder. Therefore, AC presaturated with water was introduced into a styrene–PS solution. Since water was fully absorbed in AC, there was no need to use any emulsifiers. The viscous mixture was subsequently transferred to a water medium containing suspension agents. PS beads containing AC–water droplets were obtained by suspension polymerization. Applying these beads in the extrusion foaming process with CO_2 as a blowing agent, more water could be trapped. The synthesis of WEPSAC was different from that of WEPSCN in the preparation of viscous PS–monomer mixture. In the former, the viscous PS–monomer mixture is prepared by mixing PS with styrene, whereas in the latter this is achieved by bulk polymerization. There are two advantages of dissolving PS in styrene: one is to simplify the process, since the mixture can be directly suspended in the water containing stabilizer when the PS/styrene ratio is approximately 50/50 (w/w); the other is that the dissolved PS increases the viscosity, which avoids the aggregation of AC particles. CO_2 was used as a co-blowing agent to make WEPSCN and WEPSAC foams. SEM images indicated that the cell size has a bimodal distribution, which was caused by two different foaming mechanisms.

Lee *et al.*[47] prepared multi-hollow structured PMMA–Ag nanocomposite microspheres by suspension polymerization of inverse emulsion. In their study, two different dispersion agents were used, PVA and poly(ethylene glycol) (30EO) dipolyhydroxystearate (Arlacel P135). PVA was used as a stabilizer for the suspension polymerization. Arlacel P135 is a lipophilic surfactant, used for the W/O inverse emulsion. Silver nanoparticles dispersed in a water suspension were added to MMA with the surfactant Arlacel P15. When the silver nanoparticle absorbed the surfactant Arlacel P135, the particle surface was converted from hydrophilic to hydrophobic and then transferred from the aqueous phase to the monomer phase. After the mixture had been ultrasonicated for several minutes at room temperature, a W/O inverse emulsion was formed and stabilized by Arlacel P135. Subsequently, the inverse emulsion was poured into a water solution of PVA to form a W/O/W suspension, which was polymerized under a nitrogen atmosphere. After a predetermined time of polymerization, the suspension was cooled completely and recovered by filtration. SEM images of the cross-section in Figure 7.10 show that various

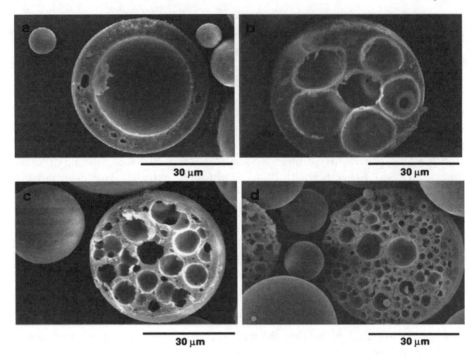

Figure 7.10   Various hollow structures of PMMA–silver nanocomposite microspheres with (a) 0.01, (b) 0.03, (c) 0.06 and (d) 0.09 wt% surfactant
concentration of MMA. (ADMVN concentration, 0.0001 mol per mole
of MMA; concentration of suspension agent PVA, 0.5 g l^{-1} water; silver
nanoparticles dispersion content, 10 wt% of MMA). Reproduced from
reference 47 by permission of Springer.

hollow structures of PMMA–silver nanocomposite microspheres were obtained
by adding different amounts of the surfactant Arlacel P135.

7.6   Conclusion

Compared with emulsion polymerization and suspension polymerization, the
manufacturing process of suspension polymerization of inverse emulsion is more
complicated. The advantage of using this technology is to make a capsular
structure which has a polymer shell and liquid inner phase. The polymer shell can
provide protection by preventing the inner phase from leaking out. When
nanoparticles are added to the polymer shell, they can inhibit mass transfer and
reinforce the polymer matrix to make the nanocomposites stronger or thermally
stable. More importantly, the presence of nanoparticles can partially replace or
completely eliminate the use of emulsifiers in the nanocomposite bead synthesis
process. One application, as mentioned before, is to make PS nanocomposite

particles containing water as a physical blowing agent for PS foam. Other applications include drug delivery and functional particle preparation.

References

1. S. Iijima and T. Ichihashi, *Nature*, 1993, **363**, 603.
2. A. Usuki, M. Kawasumi, Y. Kojima, A. Okada, T. Kurauchi and O. Kamigaito, *J. Mater. Res.*, 1993, **8**, 1174.
3. K. S. Novoselov, A. K. Geim, S. V. Morozov, D. Jiang, Y. Zhang, S. V. Dubonos, I. V. Grigorieva and A. A. Firsov, *Science*, 2004, **306**, 666.
4. S. S. Ray and M. Okamoto, *Prog. Polym. Sci.*, 2003, **28**, 539.
5. D. Tasis, N. Tagmatarchis, A. Bianco and M. Prato, *Chem. Rev.*, 2006, **106**, 1105.
6. D. W. Schaefer and R. S. Justice, *Macromolecules*, 2007, **40**, 8501.
7. C. M. Homenick, G. Lawson and A. Adronov, *Polym. Rev.*, 2007, **47**, 265.
8. L. J. Lee, C. C. Zeng, X. Cao, X. M. Han, J. Shen and G. J. Xu, *Compos. Sci. Technol.*, 2005, **65**, 2344.
9. S. M. Liff, N. Kumar and G. H. McKinley, *Nat. Mater.*, 2007, **6**, 76.
10. W. B. Zha, C. D. Han, S. H. Han, D. H. Lee, J. K. Kim, M. M. Guo and P. L. Rinaldi, *Polymer*, 2009, **50**, 2411.
11. N. Jouault, P. Vallat, F. Dalmas, S. Said, J. Jestin and F. Boue, *Macromolecules*, 2009, **42**, 2031.
12. T. Ramanathan, A. A. Abdala, S. Stankovich, D. A. Dikin, M. Herrera-Alonso, R. D. Piner, D. H Adamson, H. C. Schniepp, X. Chen, R. S. Ruoff, S. T. Nguyen, I. A. Aksay, R. K. Prud'homme and L. C. Brinson, *Nat. Nanotechnol.*, 2008, **3**, 327.
13. A. Bansal, H. C. Yang, C. Z. Li, K. W. Cho, B. C. Benicewicz, S. K. Kumar and L. S. Schadler, *Nat. Mater.*, 2005, **4**, 693.
14. J. Shen, C. C. Zeng and L. J. Lee, *Polymer*, 2005, **46**, 5218.
15. C. C. Zeng, X. M. Han, L. J. Lee, K. W. Koelling and D. L. Tomasko, *Adv. Mater.*, 2003, **15**, 1743.
16. S. Stankovich, D. A. Dikin, G. H. B. Dommett, K. M. Kohlhaas, E. J. Zimney, E. A. Stach, R. D. Piner, S. T. Nguyen and R. S. Ruoff, *Nature*, 2006, **442**, 282.
17. Y. L. Yang, M. C. Gupta, K. L. Dudley and R. W. Lawrence, *Adv. Mater.*, 2005, **17**, 1999.
18. Y. L. Yang and M. C. Gupta, *Nano Lett.*, 2005, **5**, 2131.
19. L. Liu and J. C. Grunlan, *Adv. Funct. Mater.*, 2007, **17**, 2343.
20. Y. Kojima, A. Usuki, M. Kawasumi, A. Okada, Y. Fukushima, T. Kurauchi and O. Kamigaito, *J. Mater. Res.*, 1993, **8**, 1185.
21. T. D. Fornes, P. J. Yoon, H. Keskkula and D. R. Paul, *Polymer*, 2001, **42**, 9929.
22. Z. Wang, T. Lan and T. J. Pinnavaia, *Chem. Mater.*, 1996, **8**, 2200.

23. P. Podsiadlo, A. K. Kaushik, E. M. Arruda, A. M. Waas, B. S. Shim, J. D. Xu, H. Nandivada, B. G. Pumplin, J. Lahann, A. Ramamoorthy and N. A. Kotov, *Science*, 2007, **318**, 80.

24. E. P. Giannelis, R. Krishnamoorti and E. Manias, *Adv. Polym. Sci.*, 1999, **138**, 107.

25. S. Pavlidou and C. D. Papaspyrides, *Prog. Polym. Sci.*, 2008, **33**, 1119.

26. S. S. Ray, K. Yamada, M. Okamoto and K. Ueda, *Polymer*, 2003, **44**, 857.

27. J. Zhu, A. B. Morgan, F. J. Lamelas and C. A. Wilkie, *Chem. Mater.*, 2001, **13**, 3774.

28. T. Ebina and F. Mizukami, *Adv. Mater.*, 2007, **19**, 2450.

29. G. Choudalakis and A. D. Gotsis, *Eur. Polym. J.*, 2009, **45**, 967.

30. J. W. Gilman, C. L. Jackson, A. B. Morgan, R. Harris, E. Manias, E. P. Giannelis, M. Wuthenow, D. Hilton and S. H. Phillips, *Chem. Mater.*, 2000, **12**, 1866.

31. M. S. Toprak, B. J. McKenna, J. H. Waite and G. D. Stucky, *Chem. Mater.*, 2007, **19**, 4263.

32. Y. Z. Bao, Z. M. Huang, S. X. Li and Z. X. Weng, *Polym. Degrad. Stabil.*, 2008, **93**, 448.

33. E. P. Giannelis, *Adv. Mater.*, 1996, **8**, 29.

34. D. K. Kim, K. W. Oh and S. H. Kim, *J. Polym. Sci., Part B: Polym. Phys.*, 2008, **46**, 2255.

35. F. Sun, Y. H. Pan, J. Wang, Z. Wang, C. P. Hu and Q. Z. Dong, *Polym. Compos.*, 2010, **31**, 163.

36. W. J. Bae, K. H. Kim, W. H. Jo and Y. H. Park, *Macromolecules*, 2004, **37**, 9850.

37. B. H. Kim, J. H. Jung, S. H. Hong, J. Joo, A. J. Epstein, K. Mizoguchi, J. W. Kim and H. J. Choi, *Macromolecules*, 2002, **35**, 1419.

38. B. Lepoittevin, N. Pantoustier, M. Devalckenaere, M. Alexandre, D. Kubies, C. Calberg, R. Jerome and P. Dubois, *Macromolecules*, 2002, **35**, 8385.

39. P. Werner, R. Verdejo, F. Woellecke, V. Altstaedt, J. K. W. Sandler and M. S. P. Shaffer, *Adv. Mater.*, 2005, **17**, 2864.

40. X. Y. Huang and W. J. Brittain, *Macromolecules*, 2001, **34**, 3255.

41. D. Y. Wang, J. Zhu, Q. Yao and C. A. Wilkie, *Chem. Mater.*, 2002, **14**, 3837.

42. S. S. Kim, T. S. Park, B. C. Shin and Y. B. Kim, *J. Appl. Polym. Sci.*, 2005, **97**, 2340.

43. J. H. Yeum and Y. L. Deng, *Colloid Polym. Sci.*, 2005, **283**, 1172.

44. S. M. Kwon, H. S. Kim, S. J. Myung and H. J. Jin, *J. Polym. Sci., Part B: Polym. Phys.*, 2008, **46**, 182.

45. J. P. Zheng, R. Zhu, Z. H. He, G. Cheng, H. Y. Wang and K. D. Yao, *J. Appl. Polym. Sci.*, 2010, **115**, 1975.

46. J. X. Yang, T. Hasell, W. X. Wang and S. M. Howdle, *Eur. Polym. J.*, 2008, **44**, 1331.

47. E. M. Lee, H. W. Lee, J. H. Park, Y. A. Han, B. C. Ji, W. T. Oh, Y. L. Deng and J. H. Yeum, *Colloid Polym. Sci.*, 2008, **286**, 1379.
48. J. W. Kim, F. Liu, H. J. Choi, S. H. Hong and J. Joo, *Polymer*, 2003, **44**, 289.
49. Y. Chen, Z. Qian and Z. C. Zhang, *Colloids Surf. A*, 2008, **312**, 209.
50. S. Yang, H. R. Liu, H. F. Huang and Z. C. Zhang, *J. Colloid Interface Sci.*, 2009, **338**, 584.
51. W. Xie, J. M. Hwu, G. J. Jiang, T. M. Buthelezi and W. P. Pan, *Polym. Eng. Sci.*, 2003, **43**, 214.
52. J. M. Hwu, T. H. Ko, W. T. Yang, J. C. Lin, G. J. Jiang, W. Xie and W. P. Pan, *J. Appl. Polym. Sci.*, 2004, **91**, 101.
53. P. Ding and B. J. Qu, *J. Appl. Polym. Sci.*, 2006, **101**, 3758.
54. S. J. Park, M. S. Cho, S. T. Lim, H. J. Choi and M. S. Jhon, *Macromol. Rapid Commun.*, 2005, **26**, 1563.
55. Y. F. Yang, J. Zhang, L. Liu, C. X. Li and H. Y. Zhao, *J. Polym. Sci., Part A: Polym. Chem.*, 2007, **45**, 5759.
56. Y. N. Wu, J. Zhang and H. Y. Zhao, *J. Polym. Sci., Part A: Polym. Chem.*, 2009, **47**, 1535.
57. J. J. Crevecoeur, L. Nelissen and P. J. Lemstra, *Polymer*, 1999, **40**, 3685.
58. J. J. Crevecoeur, J. F. Coolegem, L. Nelissen and P. J. Lemstra, *Polymer*, 1999, **40**, 3697.
59. J. J. Crevecoeur, L. Nelissen and P. J. Lemstra, *Polymer*, 1999, **40**, 3691.
60. J. Pallay and H. Berghmans, *Cell. Polym.*, 2002, **21**, 19.
61. J. Pallay, P. Kelemen, H. Berghmans and D. Van Dommelen, *Macromol. Mater. Eng.*, 2000, **275**, 18.
62. J. Shen, X. Cao and L. J. Lee, *Polymer*, 2006, **47**, 6303.
63. J. T. Yang, S. K. Yeh, N. R. Chiou, Z. H. Guo, T. Daniel and L. J. Lee, *Polymer*, 2009, **50**, 3169.
64. S. Yang, H. R. Liu and Z. C. Zhang, *J. Polym. Sci., Part A: Polym. Chem.*, 2008, **46**, 3900.
65. J. H. Yeum, Q. H. Sun and Y. L. Deng, *Macromol. Mater. Eng.*, 2005, **290**, 78.
66. H. M. Jung, E. M. Lee, B. C. Ji, Y. L. Deng, J. D. Yun and J. H. Yeum, *Colloid Polym. Sci.*, 2007, **285**, 705.
67. Y. Z. Bao, Z. M. Huang and Z. X. Weng, *J. Appl. Polym. Sci.*, 2006, **102**, 1471.
68. T. Lan, P. D. Kaviratna and T. J. Pinnavaia, *Chem. Mater.*, 1995, **7**, 2144.
69. C. Zilg, R. Thomann, R. Mulhaupt and J. Finter, *Adv. Mater.*, 1999, **11**, 49.
70. D. J. Voorn, W. Ming and A. M. van Herk, *Macromolecules*, 2006, **39**, 2137.
71. S. U. Pickering, *J. Chem. Soc.*, 1907, **91**, 2001.
72. B. P. Binks and S. O. Lumsdon, *Langmuir*, 2000, **16**, 8622.
73. B. P. Binks and J. H. Clint, *Langmuir*, 2002, **18**, 1270.

74. I. Akartuna, A. R. Studart, E. Tervoort, U. T. Gonzenbach and L. J. Gauckler, *Langmuir*, 2008, **24**, 7161.
75. Y. Sun and E. Ruckenstein, *Synth. Met.*, 1995, **72**, 261.
76. D. Horak, E. Pollert, M. Trchova and J. Kovarova, *Eur. Polym. J.*, 2009, **45**, 1009.
77. J. B. Jun and K. D. Suh, *J. Appl. Polym. Sci.*, 2003, **90**, 458.
78. Q. X. Gao, C. Y. Wang, H. X. Liu, C. H. Wang, X. X. Liu and Z. Tong, *Polymer*, 2009, **50**, 2587.

CHAPTER 8

Polymer Nanocomposites by Radiolytic Polymerization

SEONG-HO CHOI AND HAI-DOO KWEN

Department of Chemistry, Hannam University, Daejeon 305-811, Republic of Korea

8.1 Introduction

Polymeric nanocomposites are hybrid materials in which inorganic substances with nanometric dimensions are dispersed in a polymeric matrix to improve the performance properties of the polymer.[1,2] These systems have attracted great industrial and scientific interest due to the important improvements in their mechanical properties and chemical and thermal resistance. These improvements can be obtained when low concentrations ($\sim$1–5 vol.%) of an inorganic nanoscale filler are added to the polymeric matrix and its layers are well dispersed and exfoliated.[1,3]

Over the past several years, polymer–clay nanocomposites have been studied in many research fields due to their excellent properties.[4,5] For example, polymer–clay nanocomposites provide, but are not limited to, high stiffness,[6] improved modulus,[7] gas barrier properties[8] and flame retardant properties.[9,10] Historically, the preparation of polymer–clay nanocomposites using laminated silicate began in the 1960s.[11,12] Many methods have been developed for preparing polymer–clay nanocomposites such as direct melt blending, solution intercalation and *in situ* polymerization, but they are all physical or are initiated by a chemical initiator.[1,13–16] Research on polymer–clay nanocomposites expanded enormously after reports of the Toyota group's research on the development of the technology of polyamide–silicate nanocomposites prepared by intercalative polymerization of caprolactam between laminated silicate

RSC Nanoscience & Nanotechnology No. 16
Polymer Nanocomposites by Emulsion and Suspension Polymerization
Edited by Vikas Mittal
© Royal Society of Chemistry 2011
Published by the Royal Society of Chemistry, www.rsc.org

in 1987.[16–18] Furthermore, the Toyota group also successfully prepared exfoliated polypropylene–clay nanocomposites using the compounding method and successfully applied it in automobile components, and Toyota remain the leading researchers in the polymer–clay nanocomposites field. On the other hand, Wang and co-workers reported on the preparation of nanocomposites based on thermosetting plastics such as epoxy-type plastics and analyzed the mechanism of nanocomposite formation.[19–21] Giannelis and co-workers originally reported that the melted thermosetting plastic is directly inserted between laminated layers of silicate and presented the desired fundamental data, such as the distribution condition of the low molecular weight material intercalated *via* an intercalation and exfoliation mechanism.[22,23] However, little has been reported on how polymer–clay nanocomposites are prepared by γ-irradiation polymerization in spite of the ease of preparation at room temperature and ambient pressure.

High-energy ionizing radiation (gamma or electron beam) has attracted much interest, primarily due to its ability to produce crosslinked networks with various polymers.[24–26] Being energized by γ-ray irradiation, the polymer and active species, such as free radicals, absorb energy to initiate various chemical reactions. There are three fundamental processes that result from these reactions: (1) crosslinking, where polymer chains are connected to form a network, (2) degradation, where the molecular weight of polymer decreases through chain scissioning, and (3) grafting, where a new monomer is polymerized and grafted on to the base polymer chain.

Polymer–metal nanocomposites have been successfully prepared using γ-irradiation for applications in antimicrobial materials,[27] semiconductors,[28] *etc.* It was found that the γ-irradiation techniques have many advantages for preparation of polymer–metal nanocomposites. In particular, they need no radical initiators or reducing agents, which are needed in preparing metal-based nanoparticles. During γ-irradiation, free radicals are generated at low temperature, in the viscous or solid phase, and hydrated electrons are also formed in a solution state for reduction of metallic ions. Therefore, the polymer–metal nanocomposites can be simultaneously prepared using the free radicals and hydrated electrons generated during γ-irradiation, as shown in Figure 8.1. The generation of free radicals at room temperature and even in a solid phase is a very interesting technique for many research studies.

In addition to polymer–clay and polymer–metal nanocomposites, a third type of nanocomposite utilizes the attachment of functional groups or aliphatic carbon chains to the outer surface of carbon nanotubes (CNTs), thus dramatically increasing the solubility and enhancing the ability to control nanotube materials. The functionalization of CNTs has mainly been performed by chemical methods, and little has been reported on the functionalization of CNTs using γ-irradiation polymerization.

In this chapter, we describe the preparation of nanocomposites, polymer–clay nanocomposites, polymer–metal nanocomposites and polymer–CNT nanocomposites using γ-irradiation polymerization. Primarily, we used the free radicals generated during γ-ray irradiation as the initiator in the preparation of

$$H_2O \xrightarrow{\gamma\text{-ray}} e_{aq}^-, H^\cdot, OH^\cdot, \text{etc} \qquad (1)$$

$$CH_3\text{-}CH(OH)\text{-}CH_3 + {}^\cdot OH \longrightarrow CH_3\text{-}\dot{C}(OH) + H_2O \text{ (or } H_2) \qquad (2)$$

$$M^{n+} + 2e_{aq}^- \longrightarrow M^\circ \qquad (3)$$

$$nM^\circ \longrightarrow \text{nanoparticle} \qquad (4)$$

$$\text{Radicals} + MMA \xrightarrow{\gamma\text{-ray}} \text{polymer} \qquad (5)$$

Figure 8.1 Possible mechanism for polymer metal nanocomposites by γ-irradiation polymerization in aqueous solution with 2-propanal.

nanocomposites. We also discuss the characterization and application of the three nanocomposite types prepared by γ-irradiation polymerization.

8.2 Preparation of Polymer–Clay Nanocomposites by γ-Irradiation Polymerization

Three main methods have been developed to prepare polymer–clay nano-composites.[29–31] The first is exfoliation–adsorption: the layered clay is exfoliated into a single layer in solvent in which the polymer can dissolve. Owing to the weak forces between the clay layers, the polymer enters between the delaminated layers and, when the solvent is evaporated, the nanocomposites are formed. The second is melt intercalation: the polymer, in the molten state, is incorporated with the layered clay. The polymer then crawls into the interlayer space, causing the layers to separate and form nanocomposites. These two methods are physical and no chemical reaction occurs. The third method *in situ* intercalative polymerization: the modified layered clay absorbs the liquid monomer and polymerization can be carried out between the intercalated sheets as shown in Figure 8.2. So far, many polymer–clay nanocomposites have been prepared through *in situ* intercalative polymerization initiated by chemical agents but very few by irradiation.[32]

The γ-irradiation method can be widely used to initiate radical polymerization because it can be performed at room temperature and causes little harm to the environment. However, very little has been reported on the preparation of polymer–clay nanocomposites using γ-irradiation polymerization at room temperature and ambient pressure.

Zhang *et al.* reported that poly(methyl methacrylate)–clay nanocomposites could be prepared by *in situ* intercalative polymerization initiated by γ-irradiation.[33] The degree of dispersion and the intercalation spacing of these nanocomposites were investigated by X-ray diffraction (XRD) and high-resolution transmission electron microscopy (HRTEM), respectively. The

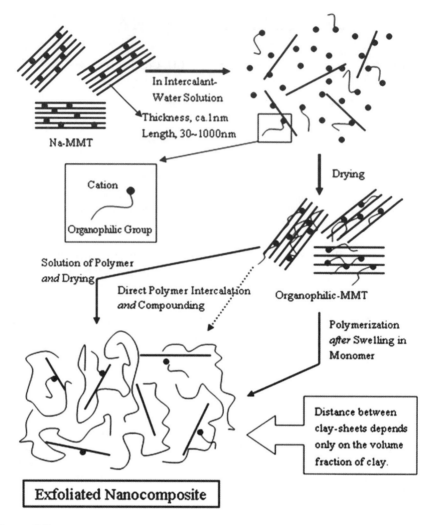

In Intercalant-
Water Solution

Thickness, ca.1nm
Length, 30~1000nm

Na-MMT

Cation

Organophilic Group

Drying

Solution of Polymer
and Drying

Direct Polymer Intercalation
and Compounding

Organophilic-MMT

Polymerization
after Swelling in
Monomer

Distance between
clay-sheets depends
only on the volume
fraction of clay.

Exfoliated Nanocomposite

Figure 8.2   Method of preparation of polymer–clay nanocomposites.

thermal stabilities of the nanocomposites were also studied by thermogravimetric analysis (TGA) and differential scanning calorimetry (DSC). Polystyrene (PS)–clay nanocomposites have also been successfully prepared by
γ-irradiation polymerization.[34] Four different types of organic compound-
modified clay were used: three of the four contained a reactive group, whereas
the other did not. Exfoliated PS–clay nanocomposites can be obtained by using
reactive organophilic clay. The intercalated PS–clay nanocomposites can be
formed by using non-reactive clay, which was confirmed by XRD and TEM.
In the formation of exfoliated PS–clay nanocomposites, the effect of double
bonds in the clay intercalated agents is much more important than the alkyl
chain length. The enhanced thermal properties of PS–clay nanocomposites

were characterized by TGA and DSC. In particular, the enhancement of the thermal properties of PS–clay nanocomposites prepared with reactive organophilic clay was much higher than that of the PS–clay nanocomposites incorporating non-reactive clay.

Yoon *et al.* also reported the preparation of poly(styrene-*co*-divinylbenzene)–clay (polymer–clay) nanocomposites by γ-irradiation copolymerization.[32] The intercalated structure of the polymer–clay nanocomposites was observed by XRD and energy-filtered transmission electron microscopy (EF-TEM). The enhancement of the thermal properties of the polymer–clay nanocomposites was confirmed via TGA. The mechanism of the intercalate structure formation in γ-irradiation polymerization was explained as shown in Figure 8.1.

The polymer–clay nanocomposites with core-poly(butyl acrylate) (PBA) intercalated organophilic montmorillonite (OMMT) and shell-PS were successfully prepared using γ-irradiation emulsion polymerization by Wang *et al.*,[35] as shown in Figure 8.3. The prepared core-PBA–OMMT and shell-PS structure nanocomposites can be incorporated in a pure PS matrix; the impact strength of the blend was dramatically improved and the tensile strength of PS also remained high.

Many researchers have studied the effects of γ-irradiation on the mechanical and thermal properties of polymer–clay nanocomposites.[36] In this case, the polymer–clay nanocomposites were prepared by the conventional method, not the γ-irradiation method. Ahmadi *et al.* studied the effects of γ-irradiation on the properties of ethylene–propylene diene methylene-linked rubber

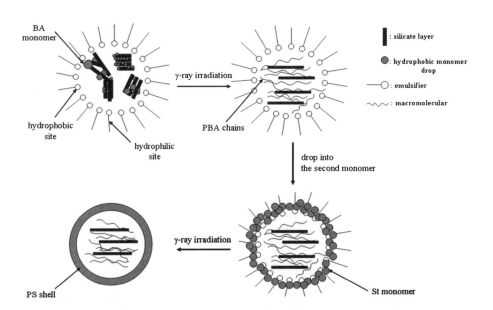

Figure 8.3   Schematic illustration of the preparation of polymer–clay nanocomposites with core (PBA–OMMT)–shell (PS) structure by γ-irradiation.

(EPDM)–clay nanocomposites and the conventional composites with pristine clay.[36] The dispersion of the silicate layers in the EPDM matrix was characterized by XRD and TEM; the results indicated that alkylamonium chains are intercalated between the nanolayer silicates after modification. The mechanical behavior of the nanocomposites was investigated using dynamic mechanical analysis and tensile test (DMTA) after γ-irradiation. The DMTA results demonstrated that the α-relaxation peaks were shifted to higher temperature and the storage modulus increased with increase in irradiation dose. The experimental data suggest that the EPDM hybrids improved the tensile strength of the nanocomposites with irradiation dose due to the crosslinking effect. The nanocomposites exhibit superior irradiation resistance properties to unfilled EPDM and conventional composites.

Lu *et al.* reported on the effects of γ-irradiation on the mechanical and morphological properties of high-density polyethylene–ethylene–vinyl acetate–organically modified montmorillonite (HDPE–EVA–OMT) nanocomposites.[37] The results showed that the nanocomposites with irradiation had superior mechanical properties to an HDPE–EVA blend. A TEM study verified that a face-to-face ordered nanostructure had been induced by γ-irradiation.

8.3 Preparation of Polymer–Metal Nanocomposites by γ-Irradiation Polymerization

In recent years, research on polymer–metal nanocomposites and their properties has attracted considerable attention due to their potential applications in catalysts, electronics and non-linear optics.[38–40] Polymer–metal composites consist of polymer and metal particles on the surface or interior (core) of the polymer matrix.[41] They exhibit various properties depending on the type of metal and polymer, especially the former. These composites not only combine the advantageous properties of polymers and metals but also exhibit many new characteristics that single-phase materials do not have. Precious metals such as silver and gold have been studied most extensively among polymer–metal nanocomposites.[42] Many polymer–metal nanocomposites have been prepared in one step with γ-irradiation in an aqueous solution. Choi *et al.* reported on polyester–silver and nylon–silver nanocomposites prepared by the dispersion of silver nanoparticles, which were prepared by γ-irradiation, on to polyester and nylon during condensation polymerization to obtain anti-bacterial fibers.[43] It was found that the silver nanoparticles aggregated in the polyester matrix, whereas the silver nanoparticles dramatically dispersed in the nylon matrix.

The emulsion polymerization of vinyl and acrylic monomers has received much attention for practical and academic reasons and with an eye toward the synthesis of polymers for biomedical, coating and adhesive applications. Wang and Pan reported on poly(styrene-*co*-acrylonitrile)–nickel nanocomposites, the post-emulsion polymerization of acrylonitrile (AN) and styrene, the subsequent coupling of poly(styrene-*co*-acrylonitrile) and a small amount of PdCl$_4$ as a reducing agent and nickel ion reduction with poly(acrylonitrile-*co*-styrene)–palladium.[44] However, less has been reported on the emulsion polymerization of vinyl and acrylic monomers in the presence of nanocomposites prepared by γ-irradiation.

Choi *et al.* reported the preparation of CdS and polyacrylonitrile (PAN)–CdS nanocomposites by γ-irradiation polymerization.[45] The prepared CdS and PAN–CdS nanocomposites were characterized by powder XRD, IR spectroscopy, Fourier transform (FT) Raman spectroscopy, TEM, X-ray photoelectron spectroscopy (XPS) and TGA (thermogravimetric analysis/dynamic thermal analysis). In photoluminescence (PL) spectroscopic analysis, the maximum peak of PAN–CdS nanocomposites prepared by γ-irradiation polymerization was at about 485 nm, whereas the maximum peak of CdS nanocomposites was at about 460 nm.

Conductive poly(*N*-vinylcarbazole) (PVK)–CdS, PVK–Ag, PVK–Pd$_{50}$/Ag$_{50}$ and PVK–Pt$_{50}$/Ru$_{50}$ nanocomposites were prepared by γ-irradiation polymerization in a tetrahydrofuran–water mixture (3:1, v/v) at room temperature and ambient pressure.[46] For comparison, polyvinylpyrrolidone (PVP)–CdS, PVP–Ag, PVP–Pd$_{50}$/Ag$_{50}$ and PVP–Pt$_{50}$/Ru$_{50}$ nanocomposites were also prepared by γ-irradiation. Ultraviolet–visible (UV/VIS) spectroscopy, TEM, XRD analysis and PL spectroscopy were used for the characterization of PVP–Pd$_{50}$/Ag$_{50}$ and PVP–Pt$_{50}$/Ru$_{50}$ nanocomposites and PVK–CdS, PVK–Ag, PVK–Pd$_{50}$/Ag$_{50}$ and PVK–Pt$_{50}$/Ru$_{50}$ nanocomposites. The absorption spectrum of the PVK–CdS nanocomposites revealed a quantum confinement effect. The emission spectrum of the PVK–CdS nanocomposites indicated the block effect of PVK for surface recombination.

Polyaniline (PANI)–Ag nanocomposites were prepared by two different methods using γ-irradiation.[47] The morphology of the Ag nanoparticles in the PANI–Ag nanocomposites was followed by TEM. In method I, PANI–Ag nanocomposites were prepared by the following sequential steps. PVP-stabilized Ag colloids were prepared by γ-irradiation, aniline was added and oxidative polymerization was performed. Method II involved the preparation of PANI–Ag nanocomposites by oxidative polymerization of aniline-stabilized Ag colloids prepared by γ-irradiation. The average size of the PVP-stabilized Ag sphere-type nanoparticles was 13 nm. The morphology of Ag nanoparticles in PANI–Ag nanocomposites prepared by method I was spherical. In contrast, the morphology of aniline-stabilized Ag nanoparticles prepared by γ-irradiation was hexagonal. The size of aniline-stabilized Ag nanoparticles was found to depend on the weight ratio of aniline to Ag ions used in the preparation. Choi *et al.* discussed the changes in the morphology of Ag nanoparticles in the PANI–Ag nanocomposites in connection with the method of preparation, source of protection and polymerization.

Polystyrene (PS)–Ag spherical nanocomposites were successfully prepared using two synthetic methodologies.[48] In the first, PS beads were prepared via emulsion polymerization, with Ag nanoparticles subsequently loaded on to the surface of the PS beads using γ-irradiation. The polymerization of styrene was induced radiolytically in an ethanol (EtOH)–water medium, generating PS beads. Subsequently, Ag nanoparticles were loaded on to the PS beads via the reduction of Ag ions. The results of morphological studies, using field emission transmission electron microscopy (FE-TEM), revealed that the PS particles were spherical and nanosized and the average size of the PS spherical particles decreased with increasing volume percentage of water in the polymerization medium. The size of the PS spherical particles increased with increase in the radiation dose for polymerization. Also, the Ag nanoparticle loading could be increased by increasing the irradiation dose for the reduction of Ag ions. In the second method, the polymerization of styrene and reduction of Ag ions were performed simultaneously by irradiating a solution containing styrene and Ag ions in an EtOH–water medium. Interestingly, the Ag nanoparticles were preferentially homogeneously distributed within the PS particles (not on the surface of the PS particles). Thus, Ag nanoparticles were distributed on to the surface of the PS particles using the first approach, but into the PS clusters of the particles *via* the second approach. The antimicrobial efficiency of a cloth coated with the PS–Ag composite nanoparticles was tested against bacteria, such as *Staphylococcus aureus* and *Klebsiella pneumoniae*, in 100 water washing cycles.

Poly (styrene-*co*-MMA-*co*-AA)–MoO_2 nanocomposite microspheres have been successfully synthesized in a microemulsion system by γ-irradiation at room temperature and ambient pressure.[13] The influence of the experimental conditions on the formation of nanocomposites was discussed, such as the oil/water ratio, monomer content and concentration of ammonium molybdate. The products were characterized by XRD, TEM, FTIR, TGA and XPS methods. The catalytic properties were investigated in the conversion of CO to CO_2, which showed ∼80% conversion at 300 °C.

8.4   Preparation of Polymer–CNT Nanocomposites by γ-Irradiation Polymerization

Graft polymerization can be initiated by various methods, *e.g.* high-energy ionizing irradiation (γ-ray, electron beam),[49,50] plasma treatment,[51] UV radiation,[52] chemical initiators[53] and oxidation of polymers.[54] Among these methods, the high-energy irradiation process is one of the most convenient and most effective techniques for industrial use due to the ease of creating active sites on many kinds of polymers that penetrate effectively into the polymer–CNT under moderate reaction conditions. There have been several reports on radiation-initiated polymerization of acrylic and methacrylic acid on to various substrates.[55–59] These include both the direct grafting method and the pre-irradiation method to synthesize ion-exchange membranes using an ordinary polymer matrix.

Polymer–multi walled carbon nanotube (MWNT) nanocomposites were first prepared by γ-irradiation polymerization of vinyl monomers with functional groups for application as electron transfer materials in aqueous solution at room temperature and ambient pressure by Yang *et al.*[60] The vinyl monomers

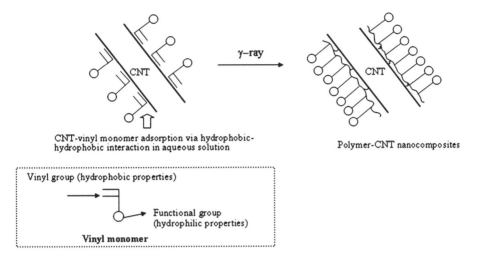

Figure 8.4 Possible mechanism for polymer–CNT nanocomposites by γ-irradiation polymerization.

used were acrylic acid (AAc), methacrylic acid (MAc), glycidyl methacrylate (GMA), maleic anhydride (MAn) and 4-vinylphenylboronic acid (VPBAc), as shown in Figure 8.4.

Polymer–MWNT nanocomposites have been obtained by γ-irradiation polymerization of various vinyl monomers. Among them, the poly(VPBAc)–MWNT nanocomposite was used as sensing sites in enzyme-free glucose sensors for the detection of glucose without enzymes. The prepared poly(VPBAc)–MWNT nanocomposite electrodes displayed an excellent linear response to glucose concentration in the range 1.0–10 mM.

Kim *et al.* reported on the preparation of a tyrosinase-immobilized biosensor using poly(AAc)–MWNT nanocomposites and poly(MAn)–MWNT nanocomposites prepared by γ-irradiation.[61] The tyrosinase-immobilized biosensor was prepared on an indium tin oxide (ITO) glass electrode using a coating of chitosan solution with tyrosinase-immobilized poly(AAc)–MWNT nanocomposites and poly(MAn)-MWNT nanocomposites. The sensing ranges of the tyrosinase-immobilized biosensor based on poly(AAc)–MWNT nanocomposites and poly(MAn)–MWNT nanocomposites were in the concentration range 0.2–0.9 mM and in the range 0.1–0.5 mM for phenol in phosphate buffer solution, respectively. The optimum pH and temperature conditions for sensing various phenolic compounds with the tyrosinase-immobilized biosensor were determined. The total phenolic content for three commercial red wines using the tyrosinase-immobilized biosensor was also determined. As a result, the polymer with carboxylic acid–MWNT nanocomposites was confirmed to be a good material for immobilization of biomolecules and an enhanced electron transfer material.

Polymer–MWNT nanocomposites in ionic liquids were prepared by the immobilization of 1-butylimidazole bromide on to an epoxy group on a poly(glycidyl methacrylate)–MWNT nanocomposite, which was synthesized by γ-irradiation polymerization of glycidyl methacrylate on to MWNT in an aqueous solution.[62] Subsequently, a MWNT ionic liquid electrode was fabricated by hand-casting MWNT ionic liquid, tyrosinase and chitosan solution as a binder on an ITO glass electrode. The sensing range of the MWNT ionic liquid electrode with immobilized tyrosinase was 0.01–0.08 mM in a phosphate buffer solution. The optimum conditions of pH and temperature and the effects of different phenolic compounds were determined. The total phenolics contents of three commercial red wines were also determined using the tyrosinase-immobilized biosensor.

An electrogenerated chemiluminescence (ECL) biosensor based on polymer–MWNT nanocomposites was prepared by γ-irradiation polymerization for ethanol sensing.[63] A higher sensing efficiency for ethanol using the ECL biosensor prepared with PAAc–MWCNT nanocomposites was measured compared with that of an ECL biosensor prepared with PMAc–MWCNT nanocomposites and purified MWCNTs. Experimental parameters affecting ethanol detection were also examined in terms of pH and the content of PAAc–MWCNT nanocomposites in Nafion. Little interference from other compounds was observed for the assay of ethanol. The results suggest that this ECL biosensor could be applied for ethanol detection in real samples.

Radiolytic deposition of Pt–Ru nanoparticles on polymer–MWNT nanocomposites was performed by γ-irradiation in aqueous solution at room temperature and ambient pressure.[64] The three polymers used were poly(AAc), poly(MAc) and poly(VPBAc). The Pt–Ru nanoparticles were then deposited on to polymer-MWNT nanocomposites by the reduction of metal ions using γ-irradiation to obtain polymer–MWNT with Pt–Ru nanoparticles. The catalysts obtained were then characterized by XRD, XPS, TEM and elemental analysis. The catalytic efficiency of the catalyst based on polymer–MWNT nanocomposites was examined for CO stripping and MeOH oxidation for use in a direct methanol fuel cell (DMFC). The catalyst based on polymer–MWNT nanocomposites shows enhanced activity for the electrooxidation of CO and MeOH oxidation over that of a commercial E-TEK catalyst.

Pt–M catalysts (M = Ru, Ni, Co, Sn and Au) based on polymer–MWNT nanocomposites were prepared using one-step γ-irradiation.[65] Two different types of functional polymers, poly(vinylphenylboronic acid) (PVPBAc) and polyvinylpyrrolidone (PVP), were used to prepare nanocomposites. The Pt–M catalysts obtained based on polymer–MWNT nanocomposites were then characterized by XRD, TEM and elemental analysis. The catalytic efficiency of the Pt–M catalysts based on polymer–MWNT nanocomposites was also examined for CO stripping and MeOH oxidation for use in a DMFC. The catalytic efficiency of the Pt–M catalyst based on polymer–MWNT nanocomposites for MeOH oxidation followed the order Pt–Sn > Pt–Co > Pt–Ru > Pt–Au > Pt–Ni catalysts. The CO adsorption capacity of the Pt–M catalyst based on polymer–MWNT nanocomposites for CO stripping decreased in the order Pt–Ru > Pt–Sn > Pt–Au > Pt–Co > Pt–Ni catalysts.

8.5 Conclusion

γ-Irradiation polymerization is an advantageous approach in the preparation of polymer–clay, polymer–metal and polymer–CNT nanocomposites that is environmentally friendly and energy efficient.

Two active species such as a free radical and a hydrated electron generated during γ-irradiation were used in preparing polymer nanocomposites. Various polymer–clay nanocomposites using γ-ray polymerization of the desired monomers can be prepared in a one-step process at room temperature and ambient pressure. The polymer–clay nanocomposites have enhanced moduli, decreased thermal expansion coefficients, reduced gas permeability and increased ionic conductivity. Precious metals have been studied most extensively among polymer–metal nanocomposites and used as catalysts in sensors, photochromic and electrochromic devices and recording materials. Various functional groups can be introduced on the CNT surface by γ-irradiation polymerization as a one-step process. The polymer–CNT nanocomposites can be used as supports to immobilize biomolecules in biosensors.

This radiolytic preparation of polymer nanocomposites is expected to have real applications in industrial fields because this method is very simple, inexpensive and can be performed at room temperature, with or without solvents.

References

1. M. Alexandre and P. Dubois, *Mater. Sci. Eng. R*, 2000, **28**, 1.
2. K. H. Wang, M. H. Choi, C. M. Koo, Y. S. Choi and I. J. Chung, *Polymer*, 2001, **42**, 9819.
3. S. C. Tjong, *Mater. Sci. Eng. R*, 2006, **53**, 73.
4. (a) T. J. Pinnavaia and G. W. Beall, *Polymer–Clay Nanocomposites*, Wiley, New York, 1997; (b) F. F. Fang and H. J. Choi, *J. Ind. Eng. Chem.*, 2006, **12**, 843.
5. J. Zhu, A. B. Morgan, F. J. Lamelas and C. A. Wilkie, *Chem. Mater.*, 2001, **13**, 3774.
6. Y. Kojima, A. Usuki, M. Kawasumi, A. Okada, A. Fukushima, T. Kurauchi and O. Kamigaito, *J. Mater. Res.*, 1993, **8**, 1185.
7. H. Li, Y. Yu and Y. Yang, *Eur. Polym. J.*, 2005, **41**, 2016.
8. Z. F. Wang, B. Wang, N. Qi, H. F. Zhang and L. Q. Zhang, *Polymer*, 2005, **46**, 719.
9. A. Laachachi, E. Leroy, M. Cochez, M. Ferriol and J. M. L. Cuesta, *Polym. Degrad. Stabil.*, 2005, **89**, 344.
10. L. Song, Y. Hu, Y. Tang, R. Zhang, Z. Chen and W. Fan, *Polym. Degrad. Stabil.*, 2005, **87**, 111.
11. B. K. G. Theng, *The Chemistry of Clay–Organic Reactions*, Wiley, New York, 1974.
12. M. Ogawa and K. Kuroda, *Bull. Chem. Soc. Jpn.*, 1997, **70**, 2593.
13. H. Liu and J. Du, *Solid State Sci.*, 2006, **8**, 526.

14. D. Y. Wang, J. Zhu, Q. Yao and C. A. Wilkie, *Chem. Mater.*, 2002, **14**, 3837.
15. X. Y. Huang and W. J. Brittain, *Macromolecules*, 2001, **34**, 3255.
16. M. Kawasumi, N. Hasegawa, M. Kato, A. Usuki and A. Okada, *Macromolecules*, 1997, **30**, 6333.
17. M. Kato, A. Usuki and A. Okata, *J. Appl. Polym. Sci.*, 1997, **66**, 1781.
18. N. Hasegawa, M. Kawasumi, M. Kato, A. Usuki and A. Okata, *J. Appl. Polym. Sci.*, 1998, **67**, 87.
19. M. S. Wang and T. J. Pinnavaia, *Chem. Mater.*, 1994, **6**, 468.
20. Z. Wang and T. J. Pinnavania, *Chem. Mater.*, 1998, **10**, 1820.
21. Z. Wang, T. Land and T. J. Pinnavaia, *Chem. Mater.*, 1996, **8**, 1584.
22. E. P. Giannelis, *Adv. Mat.*, 1996, **8**, 29.
23. R. A. Vaia, S. Vasudevan, W. Krawiec, L. G. Scanlon and E. P. Giannelis, *Adv. Mater.*, 1995, **7**, 154.
24. H. D. Lu, Y. Hu, J. F. Xiao, Q. H. Kong, Z. Y. Chen and W. C. Fan, *Mater. Lett.*, 2005, **59**, 648.
25. S. Chytiri, A. E. Goulas, K. A. Riganakos and M. G. Kontominas, *Radiat. Phys. Chem.*, 2006, **75**, 416.
26. W. A. Zhang, D. Z. Chen, H. Y. Xu, X. F. Shen and Y. E. Fang, *Eur. Polym. J.*, 2003, **39**, 2323.
27. S.-D. Oh, S. Lee, S.-H. Choi, I.-S. Lee, Y.-M. Lee, J.-H. Chun and H.-J. Park, *Colloids Surf. A.*, 2006, **275**, 228.
28. Y.-O. Kang, S.-H. Choi, A. Gopalan, K.-P. Lee, H. D. Kang and Y. S. Song, *J. Appl. Polym. Sci.*, 2006, **100**, 1809.
29. M. B. Ko and J. K. Kim, *Polym. Sci. Technol.*, 1999, **10**, 451.
30. S. S. Lee, M. Park, S. Lim and J. Kim, *Polym. Sci. Technol.*, 2007, **18**, 8.
31. M. Cho and Y. Lee, *Prospect. Ind. Chem.*, 2006, **9**, 22.
32. S. K. Yoon, B. S. Byun, S. Lee and S. H. Choi, *J. Ind. Eng. Chem.*, 2008, **14**, 417.
33. W. Zhang, Y. Li, L. Wei and Y. Fang, *Mater. Lett.*, 2003, **57**, 3366.
34. W. A. Zhang, D. Z. Chen, H. Y. Xu, X. F. Shen and Y. E. Fang, *Eur. Polym. J.*, 2003, **39**, 2323.
35. T. Wang, M. Wang, Z. Zhang, X. Gu and Y. Fang, *Mater. Lett.*, 2006, **60**, 2544.
36. S. J. Ahmadi, Y. D. Huang, N. Ren, A. Mohaddespour and S. Y. Ahmadi-Brooghani, *Compos. Sci. Technol.*, 2009, **69**, 997.
37. H. Lu, Y. Hu, Q. Kong, Z. Chen and W. Fan, *Polym. Adv. Technol.*, 2005, **16**, 688.
38. Z. Qiao, J. Xu, Y. Zhu and Y. Qian, *Mater. Res. Bull.*, 2000, **35**, 1355.
39. Z. Zhang, L. Zhang, S. Wang, W. Chen and Y. Lei, *Polymer*, 2001, **42**, 8315.
40. Z. Qiao, Y. Xie, M. Chen, J. Xu, Y. Zhu and Y. Qian, *Chem. Phys. Lett.*, 2000, **321**, 504.
41. P. K. Gupta, T. C. Hung, F. C. Lan and D. G. Perrier, *Int. J. Pharm.*, 1988, **43**, 167.

42. A. Biswas, D. K. Avasthi, D. Fink, J. Kanzow, U. Schurmann, S. J. Ding, O. C. Aktas and U. Saeed, *Nucl. Instrum. Methods Phys. Res. B.*, 2004, **217**, 39.

43. S.-H. Choi, K. P. Lee and S. B. Park, *Stud. Surf. Sci. Catal.*, 2003, **146**, 93.

44. P. H. Wang and C. Y. Pan, *Eur. Polym. J.*, 2000, **36**, 2297.

45. S.-H. Choi, M. S. Choi, K. P. Lee and H. D. Kang, *J. Appl. Polym. Sci.*, 2004, **91**, 2335.

46. Y.-O. Kang, S.-H. Choi, A. Gopalan, K.-P. Lee, H. D. Kang and Y. S. Song, *J. Appl. Polym. Sci.*, 2006, **100**, 1809.

47. Y.-O. Kang, S.-H. Choi, A. Gopalan, K.-P. Lee, H.-D. Kang and Y. S. Song, *J. Non-Cryst. Solids*, 2006, **352**, 463.

48. S.-D. Oh, B.-S. Byun, S. Lee and S.-H. Choi, *Macromol. Res.*, 2006, **14**, 194.

49. S.-H. Choi, G. T. Kim and Y. C. Nho, *J. Appl. Polym. Sci.*, 1999, **71**, 643.

50. S.-H. Choi and Y. C. Nho, *J. Appl. Polym. Sci.*, 1999, **71**, 999.

51. T. Seguchi and N. Tamura, *J. Polym. Sci. Polym. Chem. Ed.*, 1974, **12**, 1671.

52. Y. C. Nho, J. L. Garnett and P. A. Dworjanyn, *J. Polym. Sci. Polym. Chem. Ed.*, 1991, **31**, 1621.

53. S. Wu, *Polymer Interface and Adhesion*, Marcel Dekker, New York, 1982.

54. J. Wang, X. Liu, H. S. Choi and J. H. Kim, *J. Phys. Chem. B.*, 2008, **112**, 14829.

55. S.-H. Choi and Y. C. Nho, *J. Appl. Polym. Sci.*, 1999, **71**, 2227.

56. S.-H. Choi and Y. C. Nho, *Radiat. Phys. Chem.*, 2000, **57**, 187.

57. S.-H. Choi, S. Y. Park and Y. C. Nho, *Radiat. Phys. Chem.*, 2000, **57**, 179.

58. S.-H. Choi and Y. C. Nho, *Radiat. Phys. Chem.*, 2000, **58**, 157.

59. S.-H. Choi, K.-P. Lee, J.-G. Lee and Y. C. Nho, *J. Appl. Polym. Sci.*, 2000, **77**, 500.

60. D.-S. Yang, D.-J. Jung and S.-H. Choi, *Radiat. Phys. Chem.*, 2010, **79**, 434.

61. K.-I. Kim, J.-C. Lee, K. Robards and S.-H. Choi, *J. Nanosci. Nanotechnol.*, 2010, **10**, 3790.

62. K.-I. Kim, H.-Y. Kang, J.-C. Lee and S.-H. Choi, *Sensors*, 2009, **9**, 6701.

63. M.-H. Piao, D.-S. Yang, K.-R. Yoon, S.-H. Lee and S.-H. Choi, *Sensors*, 2009, **9**, 1662.

64. H.-B. Bae, S.-H. Oh, J.-C. Woo and S.-H. Choi, *J. Nanosci. Nanotechnol.*, 2010, **10**, 6901.

65. S. Yang, K.-S. Sim, H.-D. Kwen and S.-H. Choi, *J. Nanomater.*, in press.

CHAPTER 9

Polymer–Magnesium Hydroxide Nanocomposites by Emulsion Polymerization

XIAO-LIN XIE,[a] SHENG-PENG LIU,[a] FEI-PENG DU[a] AND YIU-WING MAI[b]

[a] State Key Laboratory of Materials Processing and Die and Mould Technology, School of Chemistry and Chemical Engineering, Huazhong University of Science and Technology, Wuhan 430074, China; [b] Centre for Advanced Materials Technology and School of Aerospace, Mechanical and Mechatronic Engineering, University of Sydney, Sydney, NSW 2006, Australia

9.1 Introduction

Polymer materials are widely used in many areas, such as automobiles, household appliances and electrical parts, owing to their light weight, excellent formability, high corrosion resistance, electrical insulating properties, balanced mechanical properties and low cost. However, a major problem is their high flammability, which has greatly limited their wider application.[1–3] Generally, halogen-based flame retardants (FRs) have been used as efficient elements to inhibit the flammability of a wide range of polymers without adversely affecting their mechanical performance. However, a large amount of smoke and noxious gases are emitted during combustion or high-temperature processing of these flame-retarded polymeric materials,[4,5] which cannot satisfy the increasing stringent requirements of environmental legislation and sustainability. Hence alternative, halogen-free FRs, such as metallic hydroxides, clays, carbon nanotubes, expandable graphite and intumescent flame phosphorus-containing

RSC Nanoscience & Nanotechnology No. 16
Polymer Nanocomposites by Emulsion and Suspension Polymerization
Edited by Vikas Mittal
Published by the Royal Society of Chemistry, www.rsc.org

compounds, have attracted considerable interest.[6–12] In particular, micron-sized magnesium hydroxide (MH) has been widely used due to its abundant sources, high endothermic decomposition temperature and suppression of smoke.[13,14] However, a high loading of over 60 wt% MH is required for adequate flame retardancy, which leads to a serious deterioration in processability and mechanical properties of the composites.[15,16] Interestingly, polymer composites filled with MH nanoparticles (nano-MHs) can achieve better performance than those with micron-sized fillers.[17,18] However, the major challenge is to achieve homogeneous dispersion of nano-MHs in a polymer owing to the strong tendency for aggregation and the high viscosity of the matrix. Some efforts have therefore been focused on modifying the surface of nano-MHs. For instance, organic small-molecule compounds, such as silane and titanate coupling agents,[19–21] are often used as surface modifiers to increase the hydrophobicity of inorganic surfaces. However, the mechanical properties of the composites are not as high as expected due to the weak physical entanglements between the polymer chains and these small molecules. Macromolecular compatibilizers, for example maleic anhydride-grafted polymers such as poly-propylene (PP-*graft*-MA), polyethylene (PE-*graft*-MA) and styrene–ethylene/butylene–styrene triblock copolymer (SEBS-*graft*-MA), have also been used to enhance the dispersion of nano-MHs and improve the interfacial adhesion.[22–25] Unfortunately, the existence of these compatibilizers may decrease the strength, stiffness and heat distortion temperature of the polymer–MH nanocomposites and increase their flammability. Also, *in situ* polymerization is used to prepare nanoparticle-filled composites with good dispersion, where the nanoparticles are first dispersed in a monomer and then the mixture is polymerized by a technique similar to bulk polymerization, such as free radical polymerization or surface-initiated atom transfer radical polymerization.[26–29] Especially emulsion polymerization is widely used in the preparation of polymer–inorganic particle nanocomposites[30,31] due to efficient encapsulation, environmental friendliness, low cost and simple processing. However, the method is unsuitable for poly-olefin-matrix nanocomposites because of the low catalytic activity of the Ziegler–Natta catalyst supported on the surface of a nanofiller and the low conversion of the monomer.

We modified micron-sized glass beads and nano-sized antimony trioxide with polymerization of a particular monomer, so-called *in situ* monomer–inorganic particle polymerization.[32,33] Since the polymer of the selected monomer is miscible thermodynamically or is compatible with the composite matrix, the polymer shell on the particles increases the interfacial adhesion of these inorganic particles with, and also promotes particle dispersion in, the composite matrix. Park *et al.* prepared core–shell MH–PMMA nanoparticles by emulsion polymerization using 50 nm MH modified by 3-(trimethoxysilyl)propyl methacrylate (γ-MPS) as the coupling agent and methyl methacrylate (MMA) as the shell monomer.[34] However, the hybrid particles are larger than 500 nm in dilute suspension and have multiple aggregated cores. Nonetheless, we have successfully prepared uniformly dispersed core–shell MH–PS hybrid nano-particles by ultrasonic wave-assisted *in situ* emulsion copolymerization.[35]

We believe that these hybrid nanoparticles can be homogeneously dispersed in polymer–MH nanocomposites.

In this chapter, we first introduce the synthesis of MH nanostructures, then review the preparation of polymer–MH nanocomposites by using different *in situ* emulsion polymerization methods and finally discuss their thermal, flammability and rheological properties.

9.2 Structure, Thermal Decomposition of MHs and Synthesis of MH Nanostructures

9.2.1 Crystal Structure of MHs

Figure 9.1 is a typical wide-angle X-ray diffraction (XRD) pattern of nano-MHs[36] showing that MH is a pure hexagonal phase (space group P$\bar{3}$m1) with calculated lattice constants $a = 3.15$ Å and $c = 4.78$ Å.

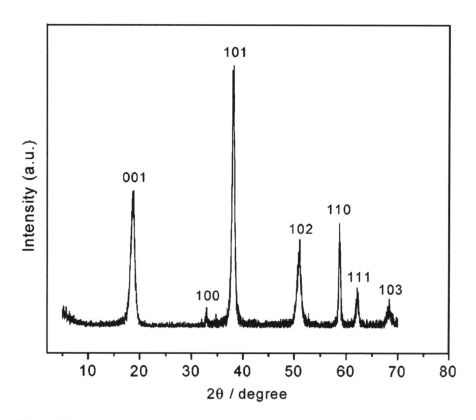

Figure 9.1 Wide-angle XRD pattern of nano-MHs. Reproduced from reference 36 with permission from the American Chemical Society.

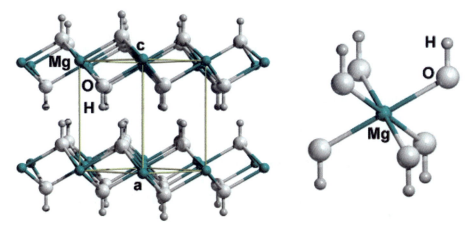

Figure 9.2 Portion of MH crystal viewed normal to the (110) plane and local coordination of Mg^{2+}. Reproduced from reference 38 with permission from Wiley.

Based on *ab initio* study, MH has a highly symmetric layered structure, where the atomic positions of Mg, O and H, respectively, are l(α) (0, 0, 0), 2(*d*) $\left(\frac{1}{3}, \frac{2}{3}, z\right)$ and 2(*d*) $\left(\frac{1}{3}, \frac{2}{3}, z\right)$, as shown in Figure 9.2.[37,38] Each Mg^{2+} ion is sixfold coordinated and the OH groups are vertical, with the shortest H . . . H interlayer and intralayer distances at 1.97 and 3.15 Å, respectively. Based on the IR spectrum of MH, a strong and remarkably sharp band exists at 3698 cm^{-1}, which is assigned to the stretching of OH.[39] Therefore, MH is a crystal with OH groups free from any H-bonding interactions.[37]

9.2.2 Thermal Decomposition of MHs

MH has been used as a flame retardant due to its ability to undergo endothermic dehydration in fire conditions,[40–44] which can be described as follows:

$$Mg(OH)_2(s) \rightarrow MgO + H_2O\,(g)$$

The decomposition of MH starts at about 300 °C, the associated endotherm is about 1450 J g^{-1} and 30.8% H_2O is released during the decomposition of MH.[45,46] As expected, the water release decreases the temperature and dilutes oxygen and flammable gases near the flame. Also, the product, MgO, acts as a barrier to minimize any further combustion and decomposition of the flammable polymer.

Chen *et al.*[47] investigated the thermal decomposition kinetics and mechanism of MH at high temperature (973–1123 K). They found that the apparent activation energy and reaction order of the thermal decomposed reaction are 50.9 kJ mol^{-1} and 0.55, respectively.

9.2.3   Synthesis of MH Nanostructures

Versatile morphological structures of nano-MH, such as needle, lamellar, wire, rod and flower-like, have been synthesized by the wet precipitation process,[48] the hydrothermal route,[49] solvothermal techniques,[50] the templating route[51] and the electrodeposition method.[52] Especially the precipitation and hydrothermal techniques are often used due to their ease of operation and economical costs.

For wet precipitation, magnesium chloride or magnesium nitrate hexahydrate is employed as precursor and sodium hydroxide or ammonia solution as precipitator. The reaction proceeds as follows:

$$MgCl_2 \cdot 6H_2O + 2NH_3 \rightarrow Mg(OH)_2 \downarrow + 2NH_4Cl + 6H_2O$$
$$MgCl_2 \cdot 6H_2O + 2NaOH \rightarrow Mg(OH)_2 \downarrow + 2NaCl + 6H_2O$$

For example, Qu and co-workers[53] synthesized different nano-MHs with needle, lamellar and rod-like morphologies by controlling the alkali solution concentration and reaction temperature. At low temperature (2 °C), needle- or rod-shaped nano-MHs are formed at low initial aqueous ammonia concentration (5 wt%), whereas platelet-shaped nano-MHs are fabricated at high initial aqueous ammonia concentration (25 wt%). They found that 20 °C is critical to form the lamellar structure at low initial aqueous ammonia concentration. When the reaction temperature is raised over the threshold temperature, *e.g.* to 30 °C, all nano-MHs are lamellar; also, their mean diameter increases with increase in reaction temperature, but their thickness is insensitive to temperature. At 20–50 °C, the thickness is about 30–40 nm and increases to 70 nm at 60 °C. Interestingly, the rough lamellar nano-MHs turn into well-defined hexagonal platelets with a perfectly smooth surface and monodispersed size of about 350 nm in diameter and 40 nm in thickness after 12 h of hydrothermal treatment at 180 °C.

Further, the surface hydrophobicity of nano-MHs can be adjusted by adding an organic compound as a surface modifier during coprecipitation.[54–56] Therefore, Lv *et al.*[54] prepared hydrophobic lamellar nano-MHs with an average width of 60 nm and a thickness of 19 nm using oleic acid (OA) as surface modifier during the precipitation reaction of magnesium chloride aqueous solution and sodium hydroxide aqueous solution in the pH range 10.5–11.0. Compared with the unmodified nano-MHs, the modified MH nanoplates obtained showed excellent hydrophobicity with a high water contact angle of 121°.

The hydrothermal route, which is one of the most extensively used methods to synthesize metal oxide nanostructures, has also been employed to fabricate nano-MHs. Under the same reaction conditions, the nano-MHs derived are highly homogeneous, crystalline and pure with a narrow particle size distribution.

Kumari *et al.*[57] prepared hexagonal MH nanoplates by a single-step hydrothermal reaction of magnesium nitrate hexahydrate solution and sodium hydroxide solution in an ultrasonic bath. They found that 200 °C is the optimum hydrothermal reaction temperature. When the reaction time is 12 h, the hexagonal MH nanoplates are about 60–140 nm wide and around 45 nm thick. More interestingly, Yu *et al.*[36] directly fabricated worm-like porous MH nanoplates with a pore size of ~3.8 nm, thickness 50–110 nm and lateral dimensions of several micrometers by using commercial bulk magnesium oxide powders *via* hydrothermal treatment at 160 °C for 24 h.

Moreover, nano-MHs have also been synthesized by a solvothermal process.[58] One-dimensional MH nanotubes and nanorods were prepared by Fan and co-workers[50,59] in different solvents at various temperatures with $Mg_{10}(OH)_{18}Cl_2 \cdot 5H_2O$ nanowires as precursors, which were synthesized from the reaction of magnesium chloride and magnesium oxide. When using a bidentate ligand such as ethylenediamine or 1,6-diaminohexane as the reaction solvent, MH nanotubes with outer diameter 80–300 nm, wall thickness 30–80 nm and several tens of micrometers in length were obtained. However, MH nanorods were prepared by choosing the monodentate ligand pyridine as the reaction solvent. Zhuo *et al.*[60] also prepared MH nanotubes with diameters of about 20 nm and lengths up to several micrometers using $MgCl_2$ as precursor and $NH_3 \cdot H_2O$ as precipitator at 250 °C for 36 h. They suggested that the formation of MH nanotubes involves the self-reorganization of the amorphous structures converted from the primary crystallized nanosheets. However, solvothermal methods often require long synthesis times or complex conditions, which make them impractical for mass production applications.

It should be noted that magnesium aluminum layered double hydroxides (MgAl-LDHs) have also been widely used as flame retardants for polymer materials.[16,61] Similarly, MgAl-LDHs can be synthesized by reaction of aluminum and magnesium salts as precursor and sodium hydroxide as precipitator *via* the coprecipitation method[62–67] and other preparation routes.[68]

9.3 Polymer–MH Nanocomposites by Emulsion Polymerization

Emulsion polymerization is one of the most common techniques used to prepare polymer nanocomposites and hybrid nanoparticles with a core–shell structure as it is environmentally friendly and economically feasible on an industrial scale. There are three general well-studied processes to prepare polymer–MH nanocomposites, *i.e. in situ* monomer–nano-MH emulsion polymerization, an *in situ* combined process of precipitation and emulsion polymerization and surface-initiated *in situ* emulsion polymerization. Each of these methods is described below.

9.3.1 *In Situ* Monomer–Nano-MH Emulsion Polymerization

There have been many reports about polymer–SiO_2[69,70] and polymer–TiO_2[71,72] nanocomposites *via in situ* emulsion polymerization, but only a few about polymer–MH nanocomposites. Park *et al.*[34] reported the synthesis of core–shell MH–PMMA nanoparticles by *in situ* emulsion polymerization. First, 50 nm MH was modified by 3-(trimethoxysilyl)propyl methacrylate(γ-MPS), then the silanol groups generated by hydrolysis of alkyloxy group of γ-MPS reacted with the hydroxyl groups on the MH surface through dehydration and condensation to form Si–O–Si bonds. Thus, the vinyl groups were covalently attached on the MH surface. Then, MMA was added to this mixture emulsion and copolymerized with the vinylated MHs. Subsequently, the PMMA macromolecular chains formed were covered on MH to produce core–shell MH–PMMA hybrid particles. However, there are multiple aggregated cores in the hybrid particles which are grape-like and have diameters larger than 500 nm in dilute suspension.

Ultrasonic irradiation is an effective way to disperse, emulsify and activate nanoparticle during chemical reaction.[73–76] In particular, ultrasonic waves were applied to enhance the dispersion of nano-silica, carbon nanotubes and nano-silver particles in polymer matrix synthesized by *in situ* polymerization. We also used ultrasonic wave-assisted *in situ* monomer–nano-MH emulsion polymerization to prepare the core–shell MH–polystyrene (PS) hybrid nanoparticles with vinylated MH in order to improve the dispersion of MH and enhance the interaction between the polymer shell and MH core.[35] First, 80 nm MHs, γ-MPS, aniline (100:10:0.2 by weight) and a mixture of toluene and deionized water (2:1 by weight) were poured into a reaction vessel immersed in a 40 kHz ultrasonic bath and reacted at toluene reflux temperature with stirring for 2 h under nitrogen. As shown in Figure 9.3, the hydroxyl groups on the MH surface react with the silanol groups generated by hydrolysis of the alkyloxy group of γ-MPS through dehydration and condensation to form Si–O–Si bonds. Therefore, the vinyl groups were covalently attached on the MH surface.

Upon completion, the product was purified by centrifugation (at 5000 rpm for 1 h) and washed with ethanol for three cycles to remove excess γ MPS. The vinylated MH was further modified by *in situ* copolymerization of styrene. Then, the vinylated MH, deionized water and sodium dodecyl sulfate (SDS) were all charged into the reaction vessel fitted with a mechanical stirrer and a thermometer and the condenser was used as a reactor. The reaction was carried out under nitrogen in a 40 kHz ultrasonic irradiation water-bath, with a stirring rate of 150 rpm. When the mixture was heated to 80 °C, a portion of potassium

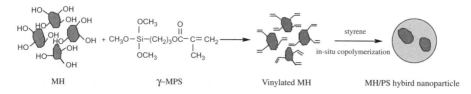

Figure 9.3 Schematic showing the preparation of core (MH)–shell (PS) nanocomposite. Reproduced from reference 35 with permission from Elsevier.

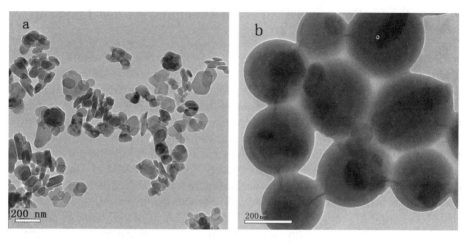

Figure 9.4 TEM images of (a) pristine MH and (b) MH–PS hybrid emulsion particles. Reproduced from reference 35 with permission from Elsevier.

persulfate (KPS) was divided into four equal parts by weight and added to the reactor at 0.75 h intervals. Styrene (St) was added dropwise. After reaction for 8 h, the emulsion was broken and centrifuged; and the MH–PS hybrid nanoparticles were dried at 60 °C in vacuum. FTIR spectra confirmed that the copolymerized PS chains were grafted on to the surface of the nano-MHs.

Figure 9.4 shows TEM images of pristine MH and MH–PS hybrid emulsion particles. It can be seen that the pristine nano-MHs are irregular thin sheets of nanometer size, significantly agglomerated, and there are empty spaces between the MH nanosheets. As illustrated in Figure 9.2, the interlayer force between MH nanosheets is weak, so they can be easily separated in a solvent. During the copolymerization reaction, St monomer enters the empty space between the primary nanosheets under ultrasonic irradiation. Subsequently, copolymerization of St and vinylated MH is initiated by diffusion of an initiator and the reaction heat released inside the gaps. This leads to breakdown of the agglomerated MH nanosheets and formation of uniformly dispersed MH–PS hybrid emulsion particles with MH as a core segment and PS chains as shell, as shown in Figure 9.4b. After being modified by *in situ* copolymerization of St, these MH nanosheets are uniformly dispersed in MH–PS hybrid latex particles with sizes of about 250–300 nm. In contrast to the agglomerated-core MH–PMMA hybrid nanoparticles prepared by Park *et al.*,[34] the uniformly dispersed core–shell MH–PS hybrid nanoparticles are attributed to ultrasonic cavitation and dispersion.

Samal *et al.*[77] prepared a poly(butyl methacrylate) (PBMA)–sodium silicate (SS)–MH ternary nanocomposite by *in situ* emulsion polymerization of BMA in the presence of layered silicate and MH with sorbitol as surfactant and benzoyl peroxide as initiator. The TEM images show that the silicate layers and nano-MHs were uniformly dispersed in the PBMA matrix at the nanoscale.

9.3.2   *In situ* Combined Process of Precipitation and Emulsion Polymerization

Recently, an *in situ* combined process of precipitation and emulsion polymerization has been used to enhance further the dispersion of nano-MHs in polymer–MH nanocomposites. In this process, MH nanoparticles are first synthesized *via in situ* precipitation reaction of the emulsion consisting of monomer, initiator and magnesium chloride or magnesium nitrate in aqueous solution containing sodium hydroxide, then the emulsion mixture containing nano-MHs undergoes *in situ* thermal initiation polymerization to prepare the polymer–MH nanocomposites. As reported by Qu and co-workers,[78] PMMA–MgAl layered double hydroxide nanocomposite was prepared by the following procedure. First, NaOH aqueous solution was added to an emulsion consisting of 0.225 mol of $Mg(NO_3)_2$, 0.075 mol of $Al(NO_3)_3$, 75 g of MMA, 0.75 g of benzoyl peroxide and 21.6 g of SDS with stirring under N_2 at room temperature. Then, thermal initiation polymerization took place at 80 °C for 5 h under N_2. By thermal compression molding, a transparent nanocomposite sheet was obtained. The elemental analysis results indicated that the filler loading of the MgAl layered double hydroxide in PMMA is about 33.90 wt%. XRD and TEM results showed that the intercalated MgAl layered double hydroxide layers are hexagonal sheets with thickness 25–40 nm and width 60–120 nm and are homogeneously dispersed in the PMMA matrix. The SEM image of the failure surface of the transparent PMMA–MgAl nanocomposite sheet indicates that there are no voids between the intercalated layers and PMMA matrix due to the strong interfacial interaction. It is noteworthy that although the loading is more than 30 wt%, the nanocomposite still has good transparency since the intercalated inorganic layers are very small and dispersed homogeneously in the matrix.

To enhance further the polymer–nano-MH interfacial interaction and improve the dispersion of nano-MHs in the polymer matrix, Pankow and Schmidt-Naake[79] prepared PS–MH and poly(butyl methacrylate) (PBuMA)–MH nanocomposites using 3-(methacryloxy)propyltrimethoxysilane (MPTMS) as a coupling agent and $MgCl_2$ as an inorganic precursor by an *in situ* combined process of precipitation and free-radical emulsion copolymerization. First, an emulsion was prepared with monomer, SDS, doubly distilled water and potassium peroxodisulfate (PPS) as water-soluble initiator. The monomer:MPTMS and MPTMS:$MgCl_2$ molar ratios were 90:10 and 1:2, respectively. After hydrolysis and condensation of MPTMS and $MgCl_2$ initiated by NaOH ($MgCl_2$:NaOH molar ratio 1:2), the emulsion was purged for 0.5 h with nitrogen followed by polymerization at 70 °C for 4 h. As shown in Figure 9.5, the addition of MPTMS produced vinyl groups on the MH surface, improved the stability of the emulsion containing monomer and provided a bridge between polymer and nano-MHs. In particular, the polymer nanocomposite was kept stable by the emulsifier due to the small particle size and homogeneous dispersion of the MHs.

Soap- or surfactant-free emulsion polymerization has attracted much attention for the manufacture of polymeric materials since no surfactants are

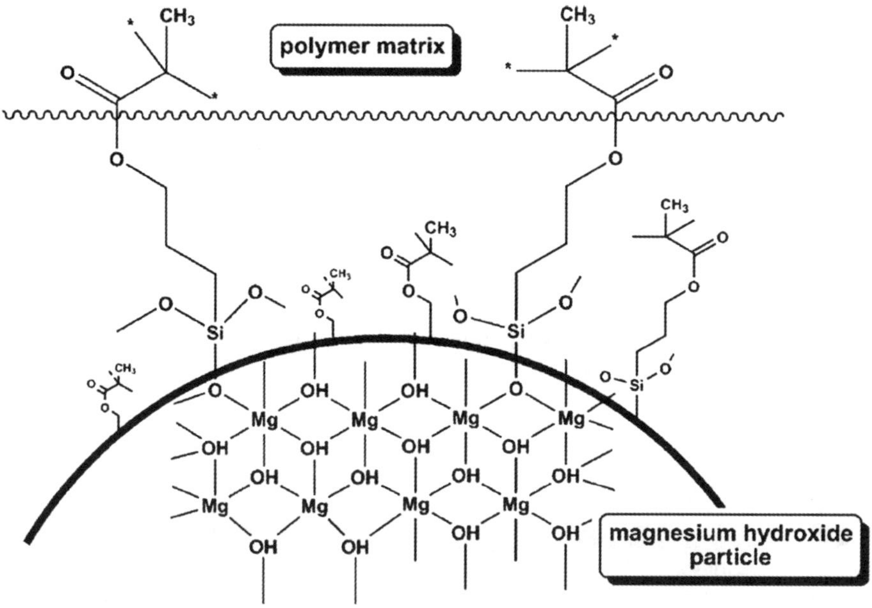

Figure 9.5 Schematic of polymer–MH nanocomposites using MPTMS as a coupling agent. Reproduced from reference 79 with permission from Wiley.

needed. In this polymerization system, the emulsion is stabilized by ionized monomers or initiators. Recently, soap-free emulsion polymerization has been used to fabricate polymer nanocomposites. Since there are no surfactants, these nanocomposites will have better performance than those prepared by the above-mentioned emulsion polymerization techniques.[80] According to Su and co-workers,[81] NaOH solution was injected into a magnesium chloride solution in the presence of oleic acid (OA), and MH nanoneedles modified by oleic acid (OA–MH) were obtained. As shown in Figure 9.6, the vinyl groups in OA are attached on the surface of the MH nanoneedles, which can participate in the copolymerization to form PS–MH nanocomposites. Conversely, the OA segments on the surface of the MH nanoneedles serve as an emulsifier to form a stable emulsion for the ensuing soap-free emulsion polymerization of St. The TEM image shows that the MH nanoneedles are dispersed homogeneously in the PS matrix, which may be attributed to the improved interaction between the PS matrix and inorganic MH nanoneedles by the grafted PS chains.

9.3.3 Surface-initiated *In Situ* Polymerization

Surface-initiated *in situ* polymerization, as a recent technology, has become increasingly used to prepare polymer–inorganic nanocomposite particles.[82–84] Liu and Yi[85] prepared PS–MH nanocomposite particles *via* surface-initiated *in situ* emulsion polymerization. As shown in Figure 9.7, ammonium persulfate

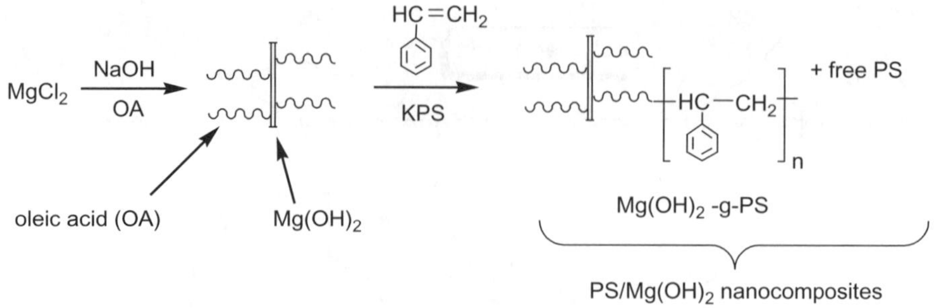

Figure 9.6 Procedure for the preparation of PS–MH nanocomposite. Adapted from reference 81.

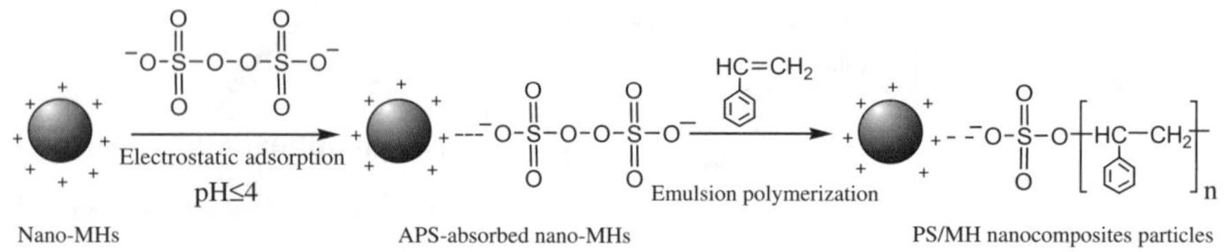

Figure 9.7 Procedure for the preparation of PS–MH nanocomposite particles. Adapted from reference 85.

(APS) as a water-soluble anionic initiator was first adsorbed on the surface of positively charged nano-MHs (~ 80 nm) at low pH (≤ 4) *via* electrostatic adsorption. Hence the active polymeric points are transferred to the surface of nano-MHs to initiate polymerization as a 'grafting from' model. Through *in situ* emulsion polymerization of styrene on APS-adsorbed nano-MHs, PS chains were covered on the nano-MHs and PS–MH nanocomposite particles were prepared. Compared with the traditional *in situ* polymerization, a higher grafting ratio of 15.2% was obtained in this surface-initiated *in situ* emulsion polymerization as the amount of APS was greater than 0.08 mmol g^{-1} nano-MHs. It is believed that the modified nano-MHs are easily dispersed in PS or other polymer matrices and the composite performance will be enhanced more efficiently.

9.4 Properties of Polymer–MH Nanocomposites by Emulsion Polymerization

9.4.1 Thermal Stability

Thermogravimetric analysis (TGA) is a common method for characterizing thermal stability. The weight loss due to the emitting volatile products after decomposition or degradation at high temperature is monitored as a function of temperature. Thermal stability plays an important role in the processing and performance of polymers. As discussed, MHs start to decompose to magnesium oxide and release water at relatively high temperature (300–320 °C), allowing it to be processed in plastics. Generally, adding only a small amount of nano-MHs to a polymer matrix can improve its thermal stability.

Samal *et al.*[77] found that the PBMA–SS–MH nanocomposite prepared by *in situ* emulsion polymerization has a higher thermal stability than neat PBMA. The TGA curves show that the thermal decomposition occurred at ~200, ~230 and ~250 °C for neat PBMA, PBMA–SS and PBMA–SS–MH nano-composites, respectively. These results clearly show that the decomposition temperature increases with the addition of nano-MHs.

Similar thermal behavior was observed with PS–MH nanocomposites prepared by *in situ* soap-free emulsion polymerization.[81] Compared with pure PS, the PS–MH nanocomposite demonstrates a higher degradation temperature and a lower weight loss. In addition, the glass transition temperature of the PS–MH nanocomposite is increased from 99 to 107 °C, which correlates with the improved dispersion of nanoparticles and better matrix/filler interface adhesion. All these results indicate that the thermal stability of the nanocomposite is also enhanced by the introduction of MH nanoneedles.

Liu and Yi[85] investigated the decomposition of PS–MH nanocomposite prepared by surface-initiated *in situ* polymerization by TGA. The weight loss of MH under heating can be divided into three stages. The first stage, from room temperature to 315 °C, is attributed to desorption of the physically adsorbed water. The second stage, in the range 315–400 °C, is due to the decomposition of MHs with a notable endothermic peak observed. The final stage is the result of the decomposition of residual MHs. Different from the decomposition of neat MH, the TGA curve of PS–MH nanocomposite is made up of four stages. The first and final stages are identical with those for neat MH. The second stage, from 285 to 410 °C, is mainly caused by the decomposition of MHs and some PS. The third stage, from 410 to 535 °C, is entirely due to the decomposition of residual PS.

The thermal stability of PMMA–MgAl nanocomposites synthesized *via in situ* polymerization from an emulsion consisting of Mg and Al ions and MMA monomer has also been studied using TGA.[78] Compared with the PMMA–MgAl microcomposites prepared by mixing $Mg_3Al(OH)_8(C_{12}H_{25}SO_4)$ powder in an acetone solution of PMMA, PMMA–MgAl nanocomposites show a relatively higher thermal decomposition temperature and charred residues. It is suggested that the dehydration of MgAl hydroxide layers is retarded by the intercalation of PMMA chains into the galleries of the MgAl hydroxide layers, which enhances remarkably the thermal stability of the nanocomposites.

9.4.2 Flammability

Although limiting oxygen index (LOI) and UL 94 are popular measurements for evaluating fire extinction, the cone calorimeter is one of the most effective approaches for studying fire retardants. There are fire-relevant properties such

Table 9.1   Cone calorimeter and LOI data for PBMA and its nanocomposites.
Adapted from reference 77

		Cone calorimeter data				
Sample	*TI (s)*	*PHRR* $(kW\ m^{-2})$	*THR* $(MJ\ m^{-2})$	*MLR* $(g\ s^{-1})$	*Residue* (%)	*LOI*
PBMA	70	756	100.2	20	1.2	18
PBMA–SS	62	562	94.5	15	4.5	22
PBMA–SS–MH	72	453	75.4	13	12.3	30

as heat release rate (HRR), total heat release (THR), mass loss rate (MLR), time-to-ignition (TI), smoke production and CO_2 and CO yield to evaluate the fire safety of materials. Among these, the HRR, especially peak HRR (PHRR), is the most important parameter for evaluating the fire retardants. Samal *et al.*[77] reported cone calorimeter results and LOI data (Table 9.1) for PBMA–SS–MH nanocomposite prepared by *in situ* emulsion polymerization and for neat PBMA. Compared with the PBMA–SS nanocomposite prepared under the same conditions, the TI value of the PBMA–SS–MH nanocomposite is increased due to the addition of MH. In fact, adding MH not only decreases the PHRR, THR and MLR, but also reduces smoke production, which may be caused by the water release and the formation of a fine flame retardant material (MgO) holding back the flammable polymer to prevent further combustion. MHs are used for chemical control through catalytic effects and concurrently for the formation of a stable protective layer, both of which increase the effectiveness of the flame retardant additives. In addition, the presence of a small amount of SS can decrease the MH loading needed for better flame retardancy by the nanocomposite. The LOI values revealed that there is no remarkable distinction between neat PBMA and the PBMA–SS nanocomposite. However, there is obvious improvement for the PBMA–SS–MH nanocomposite, possibly due to the addition of MH. Compared with the neat polymer, the thermal stability and the flame retardant properties of PBMA–SS–MH nanocomposite show significant improvements of higher decomposition temperature and lower heat release rates associated with the MH nanoparticles.

Chang *et al.*[26] also reported on the effects of the shell thickness of PS-encapsulated MH on the flame retardancy of PS–MH composites filled with modified MH particles *via in situ* polymerization by using cone calorimetry and the horizontal burning rate. It was found that the fire retardancy of these composites is significantly improved compared with those composites containing untreated MHs. Also, there seems to be a critical PS to MH ratio, ∼6.0 wt%, for optimum flame retarding properties.

9.4.3   Rheological Properties

Rheological measurements have been conducted to evaluate the dispersion and adhesion qualities between the inorganic particles and polymer matrix and also the processing properties. Chang *et al.*[26] studied the effects of micron-sized MH particles encapsulated with PS by *in situ* polymerization and the shell thickness on the rheological properties of HIPS–MH composites. Compared with HIPS–MH composites filled with untreated MH, composites containing encapsulated MH have a stronger solid-like response at low frequency. This result demonstrates that PS shell-encapsulated MHs contribute to the enhancement of the interfacial interaction between MHs and HIPS. However, regarding PS shell thickness, *i.e.* PS to MH ratio, at 6.0 wt% and above the dynamic viscosity, loss modulus and storage modulus of HIPS–MH composites all decrease. The optimum PS to MH ratio is 4.5 wt%, which was determined using a new 'cross-over point' rheological method.

We recently investigated the rheological properties of polypropylene (PP)–MH nanocomposites prepared by melt compounding of PP with modified MH *via in situ* emulsion polymerization of styrene.[86] It was found that the 'ball-bearing' effect of the modified MH particles decreases the apparent viscosity of the PP–MH composites. Thus, as shown in Figure 9.8, PP–MH composites containing polymer-encapsulated MH *via in situ* emulsion polymerization exhibit better melt flowability than those composites filled with pristine MH, especially at low shear rates and high MH loadings.

9.5 Conclusions and Future Trends

With ever increasing awareness of environmental sustainability and tighter legislation, the *in situ* emulsion polymerization process has been widely adopted

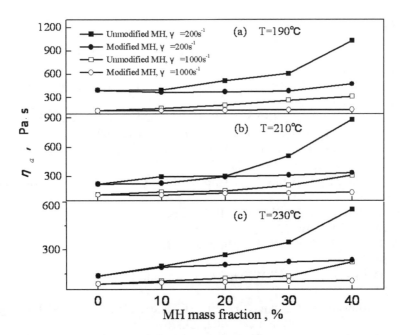

Figure 9.8 Effect of MH loading on apparent viscosity of PP–MH composites. Adapted from reference 86.

to prepare halogen-free flame retardant polymer–MH nanocomposites with homogenous dispersion. Such composites possess excellent thermal stability, low flammability, good rheological properties and superb mechanical properties. They can be used beneficially as flame retardant adhesives and coatings, elastomers and plastics.

We believe that *in situ* emulsion polymerization is an efficient method for modifying the surface of nano-MHs. We can adjust the thickness of the polymer shell covered on the nano-MH surface by controlling the ratio of monomer to nano-MHs; we can easily adjust the solubility parameter of polymers or copolymers covered on the nano-MHs and disperse the modified nano-MHs in any polymer matrix uniformly and hence enhance the interfacial interaction between nano-MHs and polymers. Hence we have the tools to fabricate high-performance polymer–MH nanocomposites. Finally, to overcome the formation of homopolymer during *in situ* monomer–nano-MH emulsion polymerization or copolymerization, surface-initiated *in situ* emulsion polymerization seems to be a promising method that has not yet been well explored.

Acknowledgment

The authors acknowledge financial support from the Outstanding Youth Fund of the National Natural Science Foundation of China (50825301).

References

1. H. L. Qin, S. M. Zhang, C. G. Zhao, G. J. Hu and M. S. Yang, *Polymer*, 2005, **46**, 8386.
2. B. Li, H. Jia, L. M. Guan, B. C. Bing and J. F. Dai, *J. Appl. Polym. Sci.*, 2009, **114**, 3626.
3. S. Zhang and A. R. Horrocks, *Prog. Polym. Sci.*, 2003, **28**, 1517.
4. C. M. Tai and R. K. Y. Li, *J. Appl. Polym. Sci.*, 2001, **80**, 2718.
5. M. A. Barnes, P. J. Briggs, M. M. Hirschler, A. F. Matheson and T. J. O'Neill, *Fire Mater.*, 1996, **20**, 17.
6. C. M. Jiao, Z. Z. Wang, X. L. Chen and H. Ya, *J. Appl. Polym. Sci.*, 2008, **107**, 2626.
7. H. A. Stretz, M. W. Wootan, P. E. Cassidy and J. H. Koo, *Polym. Adv. Technol.*, 2005, **16**, 239.
8. A. Dasari, Z. Z. Yu, Y.-W. Mai and S. L. Liu, *Nanotechnology*, 2007, **18**, 445602.
9. A. Dasari, Z. Z. Yu, Y.-W. Mai, G. P. Cai and H. Song, *Polymer*, 2009, **50**, 1577.
10. T. Kashiwagi, E. Grulke, J. Hilding, R. Harris, W. Awad and J. Douglas, *Macromol. Rapid. Commun.*, 2002, **23**, 761.
11. Y. F. Shih, Y. T. Wang, R. J. Jeng and K. M. Wei, *Polym. Degrad. Stabil.*, 2004, **86**, 339.

12. Z. L. Ma, X. Y. Pang, J. Zhang and H. T. Ding, *J. Appl. Polym. Sci.*, 2002, **84**, 522.
13. U. Hippi, J. Mattila, M. Korhonen and J. Seppälä, *Polymer*, 2003, **44**, 1193.
14. C. H. Hong, Y. B. Lee, J. W. Bae, J. Y. Jho, B. U. Nam, D. H. Chang, S. H. Yoon and K. J. Lee, *J. Appl. Polym. Sci.*, 2005, **97**, 2311.
15. S. Q. Chang, T. X. Xie and G. S. Yang, *J. Appl. Polym. Sci.*, 2006, **102**, 5184.
16. G. B. Zhang, P. Ding, M. Zhang and B. J. Qu, *Polym. Degrad. Stabil.*, 2007, **92**, 1715.
17. S. Mishra, S. H. Sonawane, R. P. Singh, A. Bendale and K. Patil, *J. Appl. Polym. Sci.*, 2004, **94**, 116.
18. Q. Zhang, M. Tian, Y. P. Wu, G. Lin and L. Q. Zhang, *J. Appl. Polym. Sci.*, 2004, **94**, 2341.
19. S. Bose, H. Raghu and P. A. Mahanwar, *J. Appl. Polym. Sci.*, 2006, **100**, 4074.
20. Z. L. Ma, J. H. Wang and X. Y. Zhang, *J. Appl. Polym. Sci.*, 2008, **107**, 1000.
21. S. Zhu, Y. Zhang and Y. X. Zhang, *J. Appl. Polym. Sci.*, 2003, **89**, 3248.
22. Z. D. Lin, Y. X. Qiu and K. C. Mai, *J. Appl. Polym. Sci.*, 2004, **91**, 3899.
23. W. Chen and B. J. Qu, *Chem. Mater.*, 2003, **15**, 3208.
24. C. Manzi-Nshuti, P. Songtipya, E. Manias, M. del Mar Jimenez-Gasco, J. M. Hassenlopp and C. A. Wilkie, *Polym. Degrad. Stabil.*, 2009, **94**, 2042.
25. S. P. Liu, J. R. Ying, X. P. Zhou, X. L. Xie and Y.-W. Mai, *Compos. Sci. Technol.*, 2009, **69**, 1873.
26. S. Q. Chang, T. X. Xie and G. S. Yang, *Polym. Int.*, 2007, **56**, 1135.
27. X. H. Gong, C. Y. Tang, H. C. Hu, X. P. Zhou and X. L. Xie, *J. Mater. Sci. Mater. Med.*, 2004, **15**, 1141.
28. H. Yan, X. H. Zhang, L. Q. Wei, X. G. Liu and B. S. Xu, *Powder Technol.*, 2009, **193**, 125.
29. M. J. Chang, J. Y. Tsai, C. W. Chang, H. M. Chang and G. J. Jiang, *J. Appl. Polym. Sci.*, 2007, **103**, 3680.
30. J. Zhou, S. W. Zhang, X. G. Qiao, X. Q. Li and L. M. Wu, *J. Polym. Sci., Part. A: Polym. Chem.*, 2006, **44**, 3202.
31. P. Ding and B. J. Qu, *J. Appl. Polym. Sci.*, 2006, **101**, 3758.
32. X. L. Xie, C. Y. Tang, X. P. Zhou, R. K. Y. Li, Z. Z. Yu, Q. X. Zhang and Y.-W. Mai, *Chem. Mater.*, 2004, **16**, 133.
33. X. L. Xie, R. K. Y. Li, Q. X. Liu and Y.-W. Mai, *Polymer*, 2004, **45**, 2793–2802.
34. E. J. Park, J. H. Kim, M. J. Moon, C. Park and K. T. Lim, in *New Developments and Application in Chemical Reaction Engineering*, ed. H. K. Rhee, I. S. Nam and J. M. Park, Elsevier, Amsterdam, 2006, p. 777.
35. S. P. Liu, J. R. Ying, X. P. Zhou and X. L. Xie, *Mater. Lett.*, 2009, **63**, 911.
36. J. C. Yu, A. W. Xu, L. Z. Zhang, R. Q. Song and L. Wu, *J. Phys. Chem. B*, 2004, **108**, 64–70.

37. L. Desgranges, G. Calvarin and G. Chevrier, *Acta. Crystallogr. Sect. B*, 1996, **52**, 82.
38. P. Ugliengo, F. Pascale, M. Mérawa, P. Labéguerie, S. Tosoni and R. Dovesi, *J. Phys. Chem. B*, 2004, **108**, 13632.
39. E. F. de Oliveira and Y. Hase, *Vib. Spectrosc.*, 2001, **25**, 53.
40. E. Kay and N. W. Gregory, *J. Phys. Chem.*, 1958, **62**, 1079.
41. R. S. Gordon and W. D. Kingery, *J. Am. Ceram. Soc.*, 1967, **50**, 8.
42. D. Beruto, P. F. Rossi and A. W. Searcy, *J. Phys. Chem.*, 1985, **89**, 1695.
43. D. R. Vollet and J. A. Varela, *J. Am. Ceram. Soc.*, 1991, **74**, 2683.
44. A. V. G. Chizmeshya, M. J. McKelvy, R. Sharma, R. W. Carpenter and H. Bearat, *Mater. Chem. Phys.*, 2002, **77**, 416.
45. R. N. Rothon and P. R. Horsby, *Polym. Degrad. Stabil.*, 1996, **54**, 383.
46. P. A. Larcey, J. P. Redfern and G. M. Bell, *Fire Mater.*, 1995, **19**, 283.
47. I. Chen, S. K. Hwang and S. Chen, *Ind. Eng. Chem. Res.*, 1989, **28**, 738.
48. W. J. Jiang, X. Hua, Q. F. Han, X. J. Yang, L. D. Lu and X. Wang, *Powder Technol.*, 2009, **191**, 227.
49. Y. Ding, G. T. Zhang, H. Wu, B. Hai, L. B. Wang and Y. T. Qian, *Chem. Mater.*, 2001, **13**, 435.
50. W. L. Fan, S. X. Sun, L. P. You, G. X. Cao, X. Y. Song, W. M. Zhang and H. Y. Yu, *J. Mater. Chem.*, 2003, **13**, 3062.
51. K. T. Ranjit and K. J. Klabunde, *Langmuir*, 2005, **21**, 12386.
52. G. L. Zou, W. X. Chen, R. Liu and Z. D. Xu, *Mater. Chem. Phys.*, 2008, **107**, 85.
53. J. P. Lv, L. Z. Qiu and B. J. Qu, *J. Cryst. Growth*, 2004, **267**, 676.
54. X. T. Lv, Hari-Bala, M. G. Li, X. K. Ma, S. S. Ma, Y. Gao, L. Q. Tang, J. Z. Zhao, Y. P. Guo, X. Zhao and Z. C. Wang, *Colloids Surf. A*, 2007, **296**, 97.
55. T. Kameda, H. Takeuchi and T. Yoshioka, *J. Phys. Chem. Solids*, 2009, **70**, 1104.
56. B. J. Li, Y. D. Zhang, Y. B. Zhao, Z. S. Wu and Z. J. Zhang, *Mater. Sci. Eng. A*, 2007, **452–453**, 302.
57. L. Kumari, W. Z. Li, C. H. Vannoy, R. M. Leblanc and D. Z. Wang, *Ceram. Int.*, 2009, **35**, 3355.
58. P. Jeevanandam, R. S. Mulukutla, Z. Yang, H. Kwen and K. J. Klabunde, *Chem. Mater.*, 2007, **19**, 5395.
59. W. L. Fan, S. X. Sun, X. Y. Song, W. M. Zhang, H. Y. Yu, X. J. Tan and G. X. Cao, *J. Solid State Chem.*, 2004, **177**, 2329.
60. L. H. Zhuo, J. C. Ge, L. H. Cao and B. Tang, *Cryst. Growth Des.*, 2009, **9**, 1.
61. M. Zammarano, M. Franceschi, S. Bellayer, J. W. Gilman and S. Meriani, *Polymer*, 2005, **46**, 9314.
62. F. Z. Zhang, M. Sun, S. L. Xu, L. L. Zhao and B. W. Zhang, *Chem. Eng. J.*, 2008, **141**, 362.
63. J. W. Boclair, P. S. Braterman, B. D. Brister and F. Yarberry, *Chem. Mater.*, 1999, **11**, 2199.

64. Q. Yuan, M. Wei, D. G. Evans and X. Duan, *J. Phys. Chem. B*, 2004, **108**, 12381.
65. M. C. Richardson and P. S. Braterman, *J. Phys. Chem. C*, 2007, **111**, 4209.
66. M. Wei, J. Guo, Z. Y. Shi and Q. Yuan, *J. Mater. Sci.*, 2007, **42**, 2684.
67. L. Y. Wang and G. Q. Wu, *Mater. Chem. Phys.*, 2007, **104**, 133.
68. K. Putyera, T. J. Bandosz, J. Jagieo and J. A. Schwarz, *Carbon*, 1996, **34**, 1559.
69. X. F. Ding, J. Z. Zhao, Y. H. Liu, H. B. Zhang and Z. C. Wang, *Mater. Lett.*, 2004, **58**, 3126.
70. C. P. L. Rubinger, L. C. Costa, A. C. C. Esteves, A. Barros-Timmons and J. A. Martins, *J. Mater. Sci.*, 2008, **43**, 3333.
71. M. J. Yang and Y. Dan, *J. Appl. Polym. Sci.*, 2006, **101**, 4056.
72. Z. Q. Ai, G. L. Sun, Q. L. Zhou and C. S. Xie, *J. Appl. Polym. Sci.*, 2006, **102**, 1466.
73. H. S. Xia, Q. Wang and G. H. Qiu, *Chem. Mater.*, 2003, **15**, 3879.
74. H. S. Xia, C. H. Zhang and Q. Wang, *J. Appl. Polym. Sci.*, 2001, **80**, 1130.
75. K. S. Suslick, *Science*, 1990, **247**, 1439.
76. K. Zhang, Q. Fu, J. H. Fan and D. H. Zhou, *Mater. Lett.*, 2005, **59**, 3682.
77. R. Samal, P. K. Rana, G. P. Mishra and P. K. Sahoo, *Polym. Compos.*, 2008, **29**, 173.
78. W. Chen, L. Feng and B. J. Qu, *Solid State Commun.*, 2004, **130**, 259.
79. O. Pankow and G. Schmidt-Naake, *Macromol. Mater. Eng.*, 2006, **291**, 1348.
80. P. Liu, *Colloids Surf. A*, 2006, **291**, 155.
81. F. Z. Zhang, H. Zhang and Z. X. Su, *Polym. Bull.*, 2008, **60**, 251.
82. X. W. Fan, C. J. Xia and R. C. Advincula, *Langmuir*, 2003, **19**, 4381.
83. G. Ni, W. Yang, L. L. Bo, H. Guo, W. H. Zhang and J. Z. Gao, *Chin. Sci. Bull.*, 2006, **51**, 1644.
84. H. Liu, H. Q. Ye and Y. C. Zhang, *Appl. Surf. Sci.*, 2007, **253**, 7219.
85. H. Liu and J. H. Yi, *Appl. Surf. Sci.*, 2009, **255**, 5714.
86. S. P. Liu, J. R. Ying, X. P. Zhou and X. L. Xie, *Acta Mater. Compos. Sin.*, 2009, **26**, 33.

CHAPTER 10

Polymer–Clay Nanocomposites by Miniemulsion Polymerization

MATEJ MIČUŠÍK*, YURI REYES, MARÍA PAULIS AND
JOSE RAMON LEIZA

Institute for Polymer Materials, POLYMAT, Kimika Aplikatutako Dptua.,
Kimika Zientzien Fakultatea, University of the Basque Country, Joxe Mari
Korta Zentroa, Tolosa Etorbidea 72, 20018 Donostia-San Sebastián, Spain

10.1 Introduction

There are several different routes to prepare polymer–clay nanocomposites
(PCNs), the two most common being melt intercalation and *in situ* intercalative
polymerization. One of the approaches to achieve exfoliation of clay platelets,
by means of *in situ* polymerization, is miniemulsion polymerization. Miniemulsion polymerization is now widely studied in the academic field and it is
also attracting interest in industry.[1] The main difference between miniemulsion
and conventional emulsion polymerization is that mass transport of the
monomer through the aqueous phase and thus heteronucleation are prevented,
having ideally 1:1 copy of droplet:particle number. Therefore, if the clay is
initially located in all monomer droplets, all final polymer particles will contain
clay. This will occur when the surface area of the monomer droplets is large
compared with that of micelles, which requires submicron sized monomer
droplets.[1] To achieve a droplet size of the level of several hundred nanometers,
high shear forces are needed, which are usually provided by homogenizer
devices[2] or ultrasound fingers with high amplitudes.[3,4] To prevent Ostwald

---

*On leave from Slovak Academy of Sciences, Polymer Institute, Department of Composite
 Materials, Dúbravská cesta 9, 842 36 Bratislava, Slovakia

RSC Nanoscience & Nanotechnology No. 16
Polymer Nanocomposites by Emulsion and Suspension Polymerization
Edited by Vikas Mittal
© Royal Society of Chemistry 2011
Published by the Royal Society of Chemistry, www.rsc.org

ripening, a hydrophobe is added to the formulations. The advantage of mini-emulsion polymerization over the classical emulsion polymerization approach is the possibility of direct dispersion of hydrophobic inclusions in the monomer droplets and encapsulation of these particles upon polymerization of the miniemulsion droplets. As inorganic particles (in our case the clay platelets) are mostly hydrophilic in nature, it is necessary to hydrophobize their surface to disperse them throughout the organic phase.

The interest in expandable clay (smectite type) comes from their anisometric particle shape and unique layered silicate structure, where the individual layer has a thickness on the nanometric scale (~ 1 nm). Therefore, if delamination of clay is achieved in a polymeric matrix, a new material, a so-called nano-composite, is produced.[5] Clays carry two kinds of electrical charges: variable pH-dependent charge resulting from proton adsorption–desorption reactions on surface-edge hydroxyl groups and a structural negative charge coming from isomorphous substitutions within the clay structure.[6] The consequence of the negative potential is that clays have very important cation-adsorption and cation-exchange properties.[6,7] Since there is need to disperse the clay in the organic phase, this exchange of inorganic cations compensating the structural charge with organic cations is essential. Among the natural smectite clays, the most commonly used are the montmorillonite and hectorite types.[8] Amongst the synthetic prepared clays, the most common is synthetic hectorite (Lapo-nite), but also other types with various characteristics are used in nano-composite preparation.[9,10] The advantages of using synthetic clays rather than natural clays are the smaller size (usually less than 50 nm in length) and also the narrower distribution in size,[11] which facilitates easier encapsulation and better control over the miniemulsion process. On the other hand, advantages of natural clay are its ready availability and low cost.

One of the main goals of academic research is the encapsulation of organi-cally modified clay in miniemulsion droplets and, upon polymerization, in latex particles. In order to achieve this, it is necessary to have only droplet nucleation and to avoid secondary nucleation. There are examples in the literature of encapsulation of inorganic particles via miniemulsion polymerization,[12,13] but there is no clear evidence of successful encapsulation of clay or only under limited conditions (low solids content, high surfactant loading) in papers devoted to miniemulsion polymerization of waterborne polymer–clay nano-composites.[14-22]

In this chapter, we discuss the work devoted to miniemulsion polymerization in the presence of clay and try to shed more light on the encapsulation phe-nomenon, not only reviewing the state of the art, but also confronting it with our experience and recent modeling results in this field.

10.2 Organomodification of Clay

One of the most important issues regarding miniemulsion stability and clay location is the compatibility between the clay and organic phase employed.

Therefore, understanding of the organomodification process of the clay and thus possible control of the degree of organophilization are of paramount importance.

Since the most studied clay regarding adsorption of organic substances is montmorillonite (MMT), this section will be mostly devoted to this type of clay. As mentioned above, MMT carries two kinds of electrical charges: variable pH-dependent charge resulting from proton adsorption–desorption reactions on surface hydroxyl groups and a structural negative charge coming from isomorphous substitutions within the clay structure.[6] In MMT (and other phyllosilicates), this structural charge is compensated by inorganic cations (Na^+, K^+, Ca^{2+}), which can be in the interlayer space or on the surface. Surface hydroxyl groups are on the platelet edges and originate from broken and hydrolyzed Al–O and Si–O bonds,[23] whereas the structural charges come from substitutions in the octahedral alumina layer separated from the aqueous solution by two tetrahedral silicate layers carrying almost no reactive groups.[23,24] Avena and De Pauli[25] concluded that the MMT edges become positively charged at pH lower than ~ 8.5, so basic pH is needed to avoid the positive edge to negative face interactions. The consequence of the negative potential is that MMT has very important cation-adsorption and cation-exchange properties.[6,26] Teppen and Aggarwal[27] described in detail this cation-exchange mechanism regarding the thermodynamics of the process and they concluded that the swelling selectivity of clay minerals for larger organic cations over smaller ones is driven by the stronger aqueous phase hydration of the unselected cation. The cation-exchange capacity (CEC) of MMT varies from 0.7 to 1.2 mequiv. g^{-1} clay.[28]

Two types of bonding of surfactant molecules in the organoclays have been proposed, namely bonding to the clay surface by electrostatic interaction and to other surfactant molecules by hydrophobic interaction.[29,30] It was found that it is possible to remove the excess of organomodifier (which binds to other surfactant molecules) by washing with ethanol.[31,32] However, even if the amount of organomodifier is close to the CEC of clay, there is still the possibility that the organomodifier may not compensate the structural charge of clay, but interact hydrophobically with other surfactant molecules. This was observed by Cervantes-Uc *et al.*,[33] who found the presence of chlorine anion from methyl-bis-2-hydroxyethyl-tallow($\sim 65\%$ C_{18}, $\sim 30\%$ C_{16}, $\sim 5\%$ C_{14})-ammonium chloride (30B) modifier in the commercial MMT Cloisite 30B (C30B) with the amount of organomodifier corresponding to CEC of original Cloisite-Na. This also means that not all of the hydrophilic inorganic cation was exchanged, which brings additional hydrophilicity to the clay.

In most studies, the amount of organomodifier exchanged was calculated via thermogravimetric analysis (TGA), but TGA might induce errors as it also takes into account the adsorbed surfactant ion pairs, *i.e.* the total amount of surfactant. Hence it is necessary to know their CEC and the amount of organic surfactant that can enter between the layers, thus ensuring complete modification of the MMT. This method was suggested by Mermut.[34] Vazquez *et al.*[35] proposed an ammonium acetate method to quantify the ion-exchanged surfactant, which seemed to be more adequate than TGA. In this method, inorganic cations still present in the organomodified clay (Na^+, Ca^{2+}) were exchanged with small ammonium cations (NH_4^+) that have more affinity to the MMT. The ammonium cations were then removed from the MMT by 1 M potassium chloride solution and then determined in a Kjeldahl distillation unit. Although the TGA results suggested total modification of the MMT; the ammonium acetate method revealed that modification was not complete. As mentioned above, weight loss measurements might cause errors since the TGA technique measures the total amount of adsorbed surfactants, whereas the ammonium acetate method only measures Na^+ cations still exchangeable at the MMT.

Table 10.1 shows the common organomodifiers used for clay modification or for studying the adsorption behavior in clay dispersions.

Table 10.1 Summary of common organomodifiers used for clay modification or for studying adsorption behavior in clay dispersions.

Organomodifier[a]	Clay	CEC[b] (%)	d_{001} (nm)	Ref.
Non-reactive				
TMA-Cl	MMT	40, 80, 160	1.47, 1.41, 1.41	36
HDTMA-Br	MMT	40, 80, 160	1.59, 2.23, 1.86	36
TMA-Cl	MMT	103	1.36	37
ODTMA-Cl	MMT	103	1.80	37
Bis(OD)DMA-Cl	MMT	103	2.42	37
HDTMA-Cl	MMT	30, 100, 300	1.32, 1.74, 2.39	30
HDTMA-Br	MMT	140	1.91	38
TDTMA-Br	MMT	103	1.80	32
DDTMA-Br	MMT	91	1.62	32
HDTMA-Br	MMT	150, 500	1.98, 2.26	39
TBA-Cl	Laponite	38	1.52	40
Jeffamine D2000	MMT	100	4.55	41
POSS	MMT	47	1.61	42
Reactive				
MEHDMA-Br	MMT	95, 237	2.05, 3.87	32
AMPS	MMT	30, 100	1.50, 1.82	43
AIBA-Cl	Laponite	57, 200	1.40, 1.55	44
MA16-Br	MMT	130	1.96	45
VB16-Cl	MMT	125	2.87	46

[a]Abbreviations: AIBA-Cl, 2,2-azobis(2-methylpropionamidine) hydrochloride; AMPS, 2-acryla-mido-2-methyl-1-propanesulfonic acid; bis(OD)DMA-Cl, bis(octadecyl)dimethylammonium chloride; DDTMA-Br, dodecyltrimethylammonium bromide; HDTMA-Br, hexadecyltrimethyl-ammonium bromide; HDTMA-Cl, hexadecyltrimethylammonium chloride; Jeffamine D2000, polyetheramine with M_w ~2000 g mol^{-1}; MA16-Br, 2-methacryloyloxyethylhexadecyldimethyl-ammonium bromide; MEHDMA-Br, 2-(methacryloyloxy)ethylhexadecyldimethylammonium bromide; ODTMA-Cl, octadecyltrimethylammonium chloride; POSS, aminopropylisobutyl polyhedral oligomeric silsesquioxane; TBA-Cl, tri-*n*-butylammonium chloride; TDTMA-Br, tet-radecyltrimethylammonium bromide; TMA-Cl, tetramethylammonium chloride; VB16-Cl, styr-yldimethylhexadecylammonium chloride.
[b]The amount of organomodifier initially used for cation exchange expressed as a percentage of the cation-exchange capacity (CEC) of clay employed.

An excess of organomodifier is not necessarily undesirable for the dispersion of organoclays in the organic phase. Sobisch and Lerche[47] noted that excess modifier is used as an 'activator' to enhance organoclay dispersion; they found that an organoclay containing excess surfactant produced larger gel volumes in cyclohexane. Also, Slade and Gates[48] found that excess organic salt in organoclay increased the absorption of aromatic solvents. Of course, there are other studies showing the opposite.[49-51] Additionally, it depends strongly on the type of organoclays and solvents used in the various studies and also differences in the amount, and possible location, of excess modifier in the organoclays.[32] Moreover, the dispersion of organoclay in organic solvents is a complex process, governed by the balance of cohesive and solvation energies in the organoclay and also the electrostatic forces, which are particularly influenced by excess organic salt and the entropic effects in the system.[49]

In a monomer miniemulsion system, the monomer/water interface plays an additional important role and the studies mentioned above on the dispersion of clay in the organic phase should be extended to consider this factor. Systematic studies of the influence of an excess of organomodifier on the clay–monomer–water system in the miniemulsion–latex and consequently on the clay location and stability are still lacking. Only Moraes *et al.*[16] have compared the different methods of modification of MMT, keeping the same amount of cetyltrimethylammonium chloride, on the stability and degree of exfoliation in the final films. The clay prepared by the method that produced the highest interlayer space (42.5 Å) gave the final film with the highest degree of exfoliation confirmed by wide-angle X-ray diffraction (WAXD) and small amplitude oscillatory shear. They concluded that efficient cation exchange is desirable to obtain an exfoliated structure by miniemulsion polymerization.

From our experience, and as will be shown later in the simulation model for the morphology of hybrid monomer–clay miniemulsion droplets, it is advantageous to avoid any excess of organomodifier (because of the possible interaction with water due to its cationic character) and the organoclay has to be as compatible as possible with the monomer system employed.

Another method of organomodification of clay platelets is by modification of the edges. In this case, the incorporation of the organomodifier is typically achieved by reaction with the OH groups. Typical reagents employed are silanes and titanates.[52-54] Song and Sand[55] observed no change in the *d*-spacing after modifying the edges of MMT using octadecyltrichlorosilane or octadecyltrimethoxysilane, whereas Negrete-Herrera *et al.*,[56] on modifying Laponite edges using methacryloxypropyldimethylmethoxysilane or methacryloxypropyltrimethoxysilane detected a small increase in basal spacing.

Wheeler *et al.*[57] performed dual modification of Laponite platelets, *i.e.* cation exchange of both structural charge and OH groups on the edges, to increase the degree of exfoliation of organoclay in a poly(methyl methacrylate) (PMMA) matrix. First they modified the edges with methacrylate-terminated ethoxysilane followed by cation exchange of Na$^+$ by cetyltrimethylammonium bromide (CTAB). They achieved only 40% CEC of the organic content of CTAB, although they used initially 160% CEC of CTAB for the modification. This indicates that the silane has either already blocked the surface (they observed a small increase in *d*-spacing after silanization as reported by Negrete-Herrera *et al.*[56]) or simply hindered the ionic interaction for full cation exchange. The opposite approach was employed by Chen and co-workers,[58,59] who modified already cation-exchanged commercial organo-Cloisites by reacting (glycidoxypropyl)trimethoxysilane with the OH groups. They did not observe any additional change in the *d*-spacing of organo-Cloisites, indicating that the silanization was running solely on the edges, which is consistent with the observations of Song and Sand[55] in the case of MMT. The dual functionalized clays were mixed with a polymeric matrix by the melt compounding[58–60] or by solvent blending[57] method and recently Faucheu *et al.*[61] reported the use of the dual functionalized clay in miniemulsion polymerization. They first modified Laponite with γ-methacryloxypropyl-trimethoxysilane (γ-MPTMS) on the edges and consequently cation-exchanged Na$^+$ for didodecyldimethylammonium bromide (DDAB). They observed encapsulation of the Laponite in the methyl methacrylate-butyll acrylate (MMA–BA) polymer particles.

Generally, by the proper choice of the modification method and the chemical nature of the organomodifiers, the compatibility between the reinforcing clay and polymer matrix can be optimized, yielding superior mechanical properties by maximizing the interaction between them.

10.3 Morphology of Hybrid Monomer–Clay Miniemulsion Droplets/Particles

The most important issue for producing waterborne polymer–clay nano-composites by miniemulsion polymerization of hybrid monomer–clay systems is to know the location of clay platelets in the dispersion; in other words, to understand the morphologies of the hybrid monomer–clay miniemulsions.

The possible morphologies that can be created during the sonication process are depicted in Figure 10.1. There are two thermodynamically stable cases regarding the clay location, namely the preferential location at the droplet surface (Figure 10.1e) and full encapsulation (Figure 10.1a). The first case is desirable when solid particles instead of surfactant molecules are used to stabilize a (mini)emulsion and one speaks of a Pickering (mini)emulsion.[62,63] The latter case of encapsulation of clay platelets into the latex particles is preferred when surfactant molecules are used to stabilize the (mini)emulsion.

We have recently developed an equilibrium morphology map of monomer–clay hybrid miniemulsion droplets. The complete description of the simulation is given elsewhere.[64] Briefly, the simulation was carried in two dimensions for better visualization of the results and because of computational efficiency. The

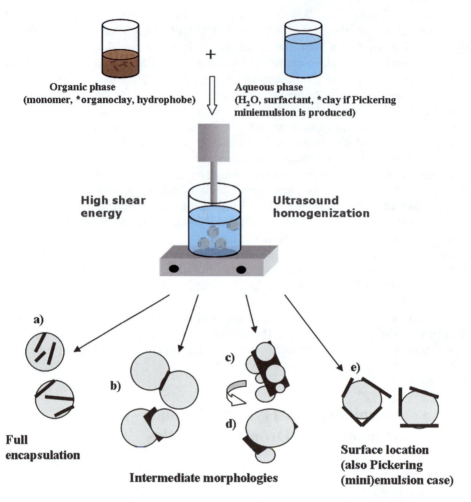

Figure 10.1 Schematic illustration of the miniemulsion process and the possible morphologies of polymer–clay droplets/particles during the mini-emulsification step and miniemulsion polymerization: (a) full encapsulated morphology; (b) dumbbell-like or snowman-like morphology; (c), (d) dispersed silicate platelets with miniemulsion droplets or polymer adsorbed to its surface; (e) clay platelets acting at the surfaces of droplets/particles.

three components in the system were water, monomer and clay. The interaction potential is defined by

$$U_{ij}(r) = \begin{cases} \infty & \text{if} \quad r \leq \sigma_{ij} \\ 0 & \text{if} \quad r > \sigma_{ij} \end{cases} \qquad (10.1)$$

$$j \neq i$$

where σ_{ij} is the cross interaction diameter, which is calculated by

$$\sigma_{ij} = 0.5(\sigma_i + \sigma_j)(1 + \Delta_{ij}) \qquad (10.2)$$

where $\Delta_{ij} \geq 0$ is a non-additive parameter. To calculate the interaction diameter of particles of the same type, $\Delta_{ii} = 0$, *i.e.* only the excluded volume is taken into account. This simple model potential was used due to computer efficiency and mainly because it is the simplest model that is able to show phase separation. Moreover, the larger Δ_{ij} is, the higher is the interfacial tension (or repulsion) between the components i and j.[65] Therefore, one can obtain the equilibrium location of the clay in miniemulsion monomer droplets as a function of the clay hydrophobicity and/or hydrophilicity. In the simulations the $\Delta_{\text{monomer–water}} = 0.9$ was maintained constant to reflect the insolubility of the monomer in water. The unit of length in the simulation is $\sigma_1 = 1$. The composition of the cell was constant using 1050 monomer and 2250 water disks. A large number of water disks is necessary to avoid the influence of the hard wall of the simulation cell on the structure of nanodroplets. The clay platelets were simulated by rigid chains with different numbers of beads (this mimics the high length:thickness ratio of the clay), to study the influence of the clay size on the equilibrium structure of the clay–monomer nanodroplets, keeping the clay:monomer ratio constant. The circular simulation cell had a diameter $D = 100$ in terms of the unit of length of the simulation. The above-described mixture initially inserted at random was left to equilibrate for 1×10^6 Monte Carlo steps according to the Metropolis algorithm, with an acceptance ratio around 45%.[66] The two-dimensional density profiles were acquired in an additional 1×10^6 Monte Carlo steps. The density profiles are average measurements which show the probability of finding a specific component of the mixture (clay, monomer or water) in a given region of the simulation cell. In all cases, two independent simulations were carried out, obtaining similar results.

The simulation results (Figure 10.2) predicted clay-encapsulated morphologies only if the clay, in addition to compatibility with monomer system (low $\Delta_{\text{clay–monomer}}$), would not provide favorable interactions with water (high $\Delta_{\text{clay–water}}$), *i.e.* the picture in the bottom right corner of Figure 10.2. This is not true for almost all organomodified clay, because even if the clay surface is ideally modified, the OH groups on the edges still provide some favorable interaction with water. Nevertheless, the dual modification by cation exchange and edge OH functionalization can bring some advantages in the full hydrophobization of the clay.[61]

As observed in the map, in most cases the clay is preferentially located at the water/monomer interface, regardless of the good compatibility of the clay and the organic phase. This phenomenon is related to the entropic increment when the anisotropic species, the clay, are located at the interface. Also, it should be noted that it is not easy to obtain individual clay platelets, but rather stacks in the organic or aqueous phase, since only the excluded volume is considered in the simulation and the clay aggregates to diminish the clay/monomer or clay/water interfacial area.

Figure 10.3 shows a density profile of the bottom left picture in Figure 10.2, in a narrow slide in the y direction. In this graph, we observe that the water is accumulated close to the wall of the simulation cell; however, there is a region where the density of water is constant. In addition, in the center of the

$\Delta_{\text{clay-monomer}}$

0.9

0.6

0.3

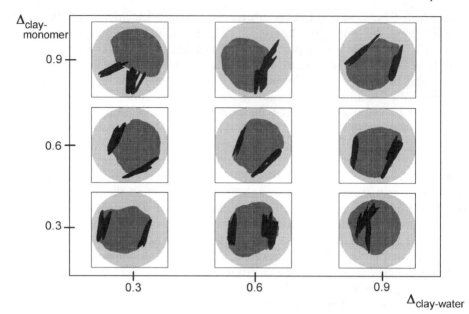

0.3   0.6   0.9

$\Delta_{\text{clay-water}}$

Figure 10.2 Equilibrium morphology map of a clay–monomer hybrid miniemulsion droplet. The clay is simulated by five rigid chains composed of 30 beads (simulating MMT clay). The density distribution, $\rho_i(x,y)$, is perpendicular to the xy plane. The water is depicted in light gray, monomer in dark gray and clay platelets in black. The monomer/water interface is the darker gray region Reprinted from reference 64 with permission from Wiley.

simulation cell one can observe that the density of the monomer droplet is also constant. In the region $60 < x < 80$, it is possible to observe clearly that the clay is located in the middle of the monomer/water interface. However, on the other side of the monomer droplet, this interface is disturbed by the presence of the clay, but it is still located between the aqueous and organic phases. It should be noted that the mobility of the clay is rather low and it remains preferentially at the interface, as deduced by the narrow peaks. A different behavior is observed when the clay is compatible with the monomer and incompatible with water, $\Delta_{\text{clay-water}} = 0.9$, $\Delta_{\text{clay-monomer}} = 0.3$ (bottom right corner of Figure 10.3b). In this case, one can observe that although the clay is located at the monomer/water interface, it is preferentially placed within the organic phase, but rather close to the interface, avoiding the unfavorable interaction with water.

In Figure 10.4, smaller clay was considered in the simulation in order to take into account smaller clay types (*e.g.* Laponite) and to discard just entropic effects on the clay location. The structure of the clay–monomer nanodroplets as a function of the clay hydrophobicity or hydrophilicity is fairly similar to that discussed above, *i.e.* in most cases the clay is located at the water/monomer interface (unless a very hydrophilic, unmodified, clay is used as in the case in the

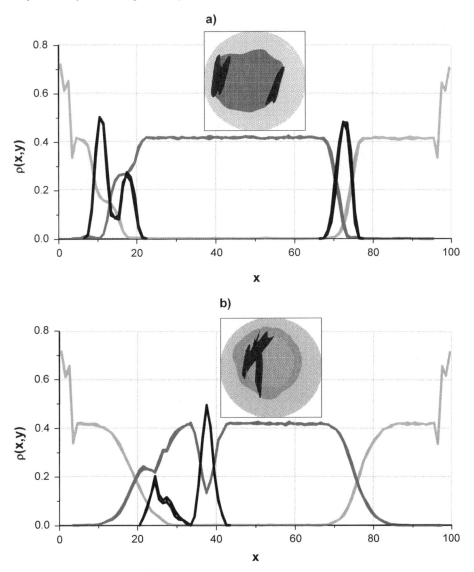

Figure 10.3 Density profiles of the point (a) $\Delta_{\text{clay–water}} = 0.3$, $\Delta_{\text{clay–monomer}} = 0.3$, from Figure 10.2 at ($48 < y < 50$), and (b) $\Delta_{\text{clay–water}} = 0.9$, $\Delta_{\text{clay–monomer}} = 0.3$ at ($50 < y < 52$). Water is depicted in gray, monomer in darker gray and clay in black.

upper left corner), and it can be encapsulated only if the clay is more compatible with the monomer than with the water. However, because of the reduction in the clay length, it can move easily, and therefore it is possible to locate the clay inside the organic phase even if the hydrophobicity is not so high ($\Delta_{\text{clay–water}} = 0.6$, $\Delta_{\text{clay–monomer}} = 0.3$).

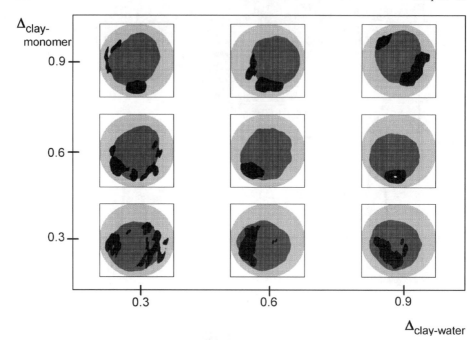

Figure 10.4 Equilibrium morphology map of a clay–monomer hybrid miniemulsion droplet. The clay is simulated by 15 rigid chains composed of 10 beads (simulating clay smaller than montmorillonite). The density distribution, $\rho_t(x,y)$, is perpendicular to the xy plane. The water is depicted in light gray, monomer in dark gray and clay platelets in black. The monomer/water interface is the darker gray region.

10.3.1 Pickering Miniemulsion

First we will discuss the preferential surface location of the clay platelets, *i.e.* the Pickering (mini)emulsion case (Figure 10.1e).

Cauvin *et al.*[67] synthesized polystyrene latexes with 5 wt% solids content, stabilized by Laponite RD clay. In order to adsorb the clay building blocks at the monomer/water interface and to provide Pickering stabilization, they have to possess surface characteristics which allow favorable partitioning on the interface in competition with the two bulk phases. This is illustrated by Figure 10.3a, where the clay is clearly fixed in the interface of these two phases. The average value of the zeta potential of the Laponite platelets dispersion at low electrolyte concentration was used to control the partitioning of the clay in the particle/water interface. They added sodium chloride, inducing slight colloidal instability, which led to flocculation of the clay platelets, but more importantly greatly enhanced the capacity of the clay to allow Pickering stabilization of oils in water.[68] In other words, the addition of salt compressed the double layer[69] and lowered the zeta potential, thereby reducing electrostatic repulsion, inducing possible clay flocculation and thus increasing its partitioning to the oil/water interface.

Ianchis *et al.*[70] reported surfactant-free miniemulsion polymerization of styrene by using five types of MMT modified by grafting monofunctional alkoxysilanes on to the OH groups on the edges of the MMT. The modified clay was dispersed in the water phase and then mixed with styrene to make miniemulsions. The zeta potential values of polystyrene (PSt)–unmodified MMT and PSt latexes were very close, indicating that the unmodified MMT platelets did not affect the particle charge and most probably remained dispersed in the water phase and not on the surface of the PSt particles. On the other hand, the zeta potentials of PSt-modified MMT were more negative (around -75 mV) than those of PSt–unmodified MMT (-55.0 mV) or of pure latex (-55.8 mV), indicating that the modified silicate lamellas acted as stabilizer for the PSt. The increase in the numerical value of the zeta potential followed the increase in hydrophobicity (increase in the chain length of the alkoxysilane) and thus better interaction between the clay and PSt particles improved the partitioning on the particle/water interface. TEM images (Figure 10.5) confirmed this trend and the more hydrophobic PSt–C8SiMMT exhibited particles with clear evidence of the clay on its surface (Figure 10.5e, f) than is the case with the less hydrophobic PSt–VSiMMT (Figure 10.5c, d).

One of the main disadvantages of the MMT Pickering miniemulsions approach is the low stability of the dispersion for long storage times and the low solids content achieved so far.[67,70]

10.3.2 Encapsulation of Clay

The encapsulation of the clay in the miniemulsion droplets/particles has attracted considerable interest in the academic field.

One of the first to report the encapsulation of clay platelets by miniemulsion polymerization were Sun *et al.*,[71] who polymerized styrene (18% solids content) by miniemulsion polymerization in the presence of 4.2% by weight based on the monomer (wbm) of CTAB-modified Laponite. Figure 10.6 shows the TEM images of PSt latexes with or without the modified Laponite. They emphasized the encapsulation of Laponite in the PSt particles, but from Figure 10.6b one can clearly see the Laponite platelet also on the surface of the latex particles, which is obvious also from the slightly non-spherical shape of the particles. Figure 10.6c confirmed the encapsulation, but a single particle cannot be taken as representative of the whole sample. Additionally, the authors speculated about the concerted manner of stabilization of the non-ionic surfactant TX-405 and CTAB organomodifier of organo-Laponite, which would be possible only if the clay were on the surface and not inside the particle, hence far from influencing the stability of the miniemulsion droplets/particles. Moreover, to obtain stable latexes a large amount of surfactant was employed (8.4 wbm%). The particle size before or after polymerization was basically within the same range (180–200 nm), indicating a true miniemulsion polymerization. Films

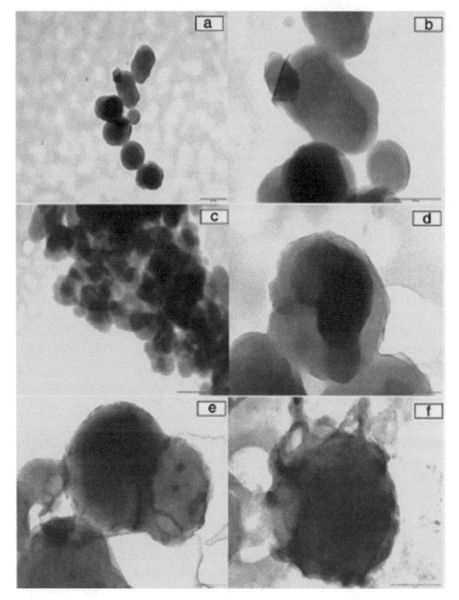

Figure 10.5 TEM images of: PSt–MMT (a) (scale bar = 200 nm) and (b) (scale bar = 100 nm)]; PSt–VSiMMT (vinyl dimethylethoxysilane modified MMT) (c) (scale bar = 500 nm) and (d) (scale bar = 100 nm)]; and PSt–C8SiMMT (octyl dimethylmethoxysilane modified MMT) (e) and (f) (scale bar = 50 nm). Reprinted from reference 70 with permission from Elsevier.

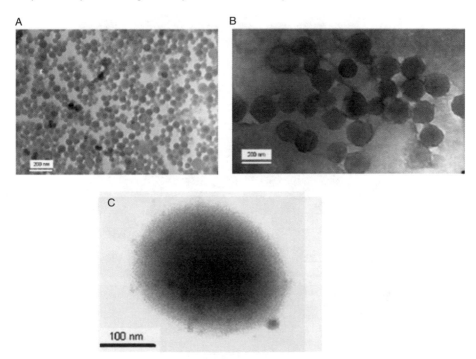

Figure 10.6 TEM images of (A) latex particles without clay and (B) latex particles with CTAB-modified Laponite; (C) TEM observation of the morphology of one latex particle containing CTAB-modified Laponite. Reprinted from reference 71 with permission from Wiley.

showed an exfoliated structure according to the WAXD pattern at a relatively high loading of Laponite (7.8 wbm%).

Tong and Deng[18] claimed the encapsulation of organo-synthetic saponite in polystyrene polymer particles synthesized by miniemulsion polymerization (see Figure 10.7) using azobisisobutyronitrile (AIBN) as initiator. They modified the saponite (50 nm) with reactive vinylbenzyltrimethylammonium chloride (VBTAC). The TEM images (Figure 10.7) show particles with smooth surfaces, which the authors attributed to encapsulated clay morphology. However, the TEM image of the melted polystyrene-containing clay presents evidence of honeycomb morphology, which can evidence the surface location of the clay, rather than its encapsulation.

As shown in the simulation of the morphology of hybrid monomer–clay miniemulsion droplets (Figures 10.2 and 10.4), the encapsulation of clay platelets is possible provided that the clay/water interfacial tension is very high (superhydrophobic clay) and that the clay/monomer interfacial tension is low (high compatibility between clay and monomer). Another aspect that the simulations show is that the size of the clay platelets might also play a role in the encapsulation of the clay in the monomer droplets (polymer particles).

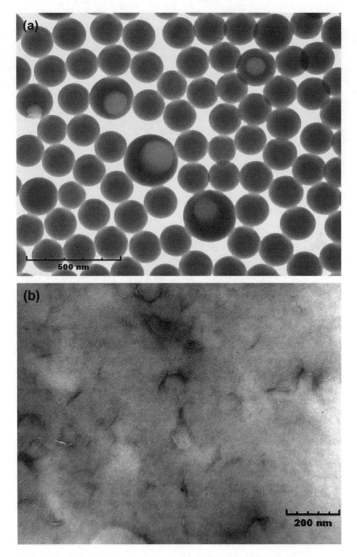

Figure 10.7 TEM images of polystyrene nanosaponite composite latex (a) before melting and (b) after melting. Reprinted from reference 18 with permission from Elsevier.

Thus, the smaller the size (Figure 10.4), the better is the encapsulation of the clay platelets. In practice, small-sized clays (< 100 nm) are only available synthetically. Natural clays, on the other hand, are larger on average and typically present a broader size distribution (ranging from 20 to 300 nm, with an average value of 118 nm[72] or 166 nm[73] depending on the evaluation method).

However, some groups have claimed the successful encapsulation of MMT platelets inside polymer particles.[14,20,74] In all three cases cited it is more than

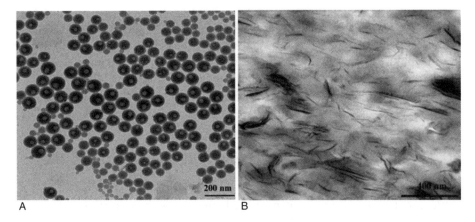

A   B

Figure 10.8 TEM images of (left) the miniemulsion latex of PS–PCDBAB–3.6% MMT nanocomposite with 11% hydrophobe and (right) PS–PCDBAB–3.6% MMT clay nanocomposite film. Reprinted from reference 74 with permission from Wiley.

questionable that full encapsulation was achieved. Samakande *et al.*[74] showed an encapsulated MMT using reversible addition–fragmentation chain transfer (RAFT)-mediated miniemulsion polymerization of styrene (Figure 10.8, left). The particle size was measured by dynamic light scattering (DLS) and, as can be seen from the TEM image (Figure 10.8, left), it was < 130 nm, whereas the TEM image of nanocomposite film prepared from the same latex also revealed clay platelets with a size of ~200 nm (which actually corresponds to the real size distribution of MMT), so the question is where these platelets were in the latex, when the particle size is much smaller?

Mirzataheri *et al.*[20] claimed that MMT platelets were encapsulated during the miniemulsion polymerization of St–BA in the presence of the commercial organoclay C30B using a mixture of anionic sodium lauryl sulfate (SLS) and non-ionic (Span 80) surfactants. Bouanani *et al.*[14] claimed partial encapsulation of MMT platelets in the miniemulsion polymerization of 1,3,5-trimethyl-1,3,5-bis(3,3,3-trifluoropropyl)cyclotrisiloxane monomer using a mixture of cationic and non-ionic surfactants and KOH as initiator. In both cases the proof of encapsulation was not sound. The TEM images had low resolution and platelet location could hardly be recognized. Mirzataheri *et al.*[20] argued that the increase in zeta potential for the clay-containing latex (−40.1 mV *versus* −60.2 mV for the blank latex) was further proof of encapsulation. However, this is not necessarily so because, as was shown by Ianchis *et al.*,[70] the zeta potential of the latex particles is strongly affected by the clay located on the surface. Therefore, it is more likely to find an increase in zeta potential (less negative), because the organomodified clay is influencing the zeta potential in a positive way[49,75] when the organomodified clay is located at the surface rather than when it is encapsulated in the polymer particles.

Generally, the conclusions drawn from TEM images are partly speculation, particularly in the case of clays, where the micrograph always depends on the orientation of the clay, since the clay having a thickness of $\sim 1\,nm$ is always invisible in the direction of incident beam perpendicular to the platelet surface. One way to improve the mapping of the clay location in miniemulsion droplets/particles is to obtain micrographs at different tilt angles and thus to detect also the clays which were invisible at certain angle.[53]

10.4 Kinetics of Miniemulsion Polymerization in the Presence of Clay

Little attention has been devoted to the kinetics of polymerization in the presence of clays, and this is also true for polymerizations carried out under miniemulsion conditions. The literature has shown contradictory results on the effect of clays in the free-radical polymerization of common monomers. Some authors have shown a catalytic effect of the clay in initiating the polymerization,[76] whereas others mentioned that the clay platelets can act as radical scavengers.[77] Tong and Deng[78] studied the miniemulsion polymerization of styrene with different amounts of nanosaponite and showed that on increasing the amount of saponite the polymerization rate decreased, as shown in Figure 10.9. This figure also shows that the polymerization rate is higher for the polymerization carried out without clay. The reason for the decrease cannot be attributed solely to the effect of the clay in the polymerization kinetics, but also to the effect of the clay on the droplet formation and hence the particle nucleation mechanism. Table 10.2 reproduces the data from the original work in terms of monomer droplet size and the ratio of the number of particles to the number of droplets (N_p/N_d) and also the final conversion achieved during the polymerization and the weight-average molecular weight (M_w). It can be concluded that increasing the amount of clay made both the droplet size and the particle size larger and hence the total number of polymer particles lower, consequently decreasing the polymerization rate, as shown in Figure 10.9. The decrease in the number of particles as the clay content increased was likely compensated by an increase in the average number of radicals per particle that made the effect less strong for higher clay contents (8 and 12%).

Bon and Colver[77] showed that the surface location of clay influenced the overall polymerization kinetics and Pickering miniemulsion polymerizations of styrene in the presence of clay showed compartmentalization. Moreover, retardation effects up to intermediate monomer conversions were observed; they were more prominent for the smaller particles and were ascribed to the Laponite clay.

Diaconu and co-workers[15] studied the effect of different organo-MMT (Cloisite 30B, MA16–MMT and living oligomer–MMT) loadings on the miniemulsion polymerization rate. They found that for Cloisite 30B loadings

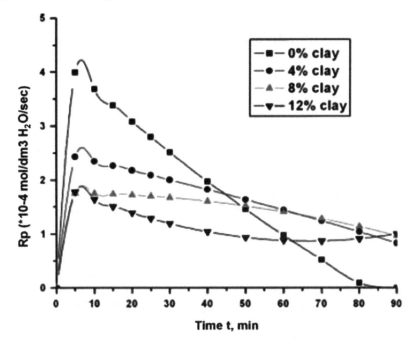

Figure 10.9 Rate of polymerization (R_P) as a function of polymerization time in the presence of organosaponite at different loadings. 10 wbm% hexadecane was used in the miniemulsion process; $T = 70\,°C$ Reprinted from reference 78 with permission from Wiley.

Table 10.2 Parameters of styrene miniemulsion polymerization in the presence of different contents of organosaponite (SSC).[78]

	Clay content (%)			
Parameter	*0*	*4*	*8*	*12*
$D_{m,i}$ (nm)	209	227	371	762
$D_{p,f}$ (nm)	161	209	292	481
$N_{p,f}/N_{m,i}$	1.6	1.0	1.4	2.5
Final conversion (%)	94.0	92.1	78.3	72.2
M_w (Da)	122 800	119 400	105 800	96 100

up to 4%, the polymerization rate was almost independent of the amount of clay. This was also true when MMT modified with an organocation bearing a double bond (MA16) was used in the miniemulsion polymerization. However, when the MMT was modified with a living oligomer bearing a nitroxide group, the polymerization rate decayed, probably due to the nitroxide group and not to the clay itself.

10.5 Final Properties of Polymer–Clay Nanocomposites Prepared by Miniemulsion Polymerization

Improvements in the application properties of films prepared from nano-composite latexes are sought. The exfoliation state of the clay (more exfoliated means a larger interfacial area) and the chemistry between the surface of the clay platelets and the polymer matrix are the key parameters to achieve a significant enhancement of the final properties. Samakande *et al.*[74] stated that the use of simple organic cations in the modification of clay layers and the subsequent use of the modified clays in the preparation of nanocomposites results in polymeric nanocomposites with only moderately enhanced properties (*e.g.* thermomechanical). They concluded that for exceptional property enhancement, functional or reactive surfactants are required, in order to compatibilize both phases further.

In their study, they used the RAFT agents *N,N*-dimethyl-*N*-(4-{[(phenyl-carbonothioyl)thio]methyl}benzyl)ethanammonium bromide (PCDBAB) and *N*-[4-({[(dodecylthio)carbonothioyl]thio}methyl)benzyl]-*N,N*-dimethylethan-ammonium bromide (DCTBAB) as organomodifiers for MMT and ran the styrene miniemulsion polymerization in the controlled manner of RAFT radical polymerization. It is true that they grew the polymeric chain out of the clay platelet surface, but the process being in a controlled manner the final molecular weights were very low (~ 100 kDa) and hence poor mechanical properties were achieved. For St–BA miniemulsion polymerization under identical conditions using similar RAFT-modified MMTs, the Young's modulus of the copolymers at room temperature with 5 wbm% of MMT–DCTAB or MMT–PCDBAB was still less than 1 MPa,[19] which is much lower than the value of 82 MPa achieved by Diaconu *et al.*[79] for the system BA–MMA (50:50, w/w) prepared by miniemulsion polymerization in the presence of 3 wbm% of non-reactive commercial C30B. The molecular weights obtained by Diaconu *et al.* were significantly larger as expected from a conventional free-radical polymerization mechanism. Therefore, the reinforcing effect that Samakande *et al.* expected from the better compatibilization between polymer and clay by using clay-attached RAFT agents was suppressed by the low molecular weights achieved.

Diaconu and co-workers[15,21] also studied the thermal and barrier properties of MMA–BA polymers prepared with different types of organo-MMT in miniemulsion polymerization. They observed that the thermal and barrier properties of the coating type of nanocomposite polymer were enhanced in the presence of clay. Nevertheless, the enhancement of the properties expected when a chemical link was provided between the reactive modified clay and the polymer was not found in this case (see Table 10.3).

However, this further enhancement was observed in a recent study by Mičušík *et al.*,[64] where the commercial non-reactive clay C30B and reactive organoclay CMA16 were compared. It was demonstrated that the reactive

Table 10.3 Thermal stability, water vapor permeability and oxygen permeability for MMA–BA copolymer prepared by miniemulsion polymerization and its nanocomposites with 3 wbm% of Cloisite 30B, 4 wbm% of living oligomer-MMT and 1.8 wbm% of CMA16-MMT.

Run	T_g (°C), DMTA	Decomposition temperature, $T_{d\,max}{}^a$ (°C)	WVTR ($g\ mm\ cm^{-2}\ day^{-1}$) As prepared	Rinsed film	Oxygen permeation (bar) (as prepared)
Blank	37	387	24.6	19.0	4.34
3% C30B	48.5	410	24.7	12.2	3.40
4% living-MMT	45.3	417	19.2	16.4	–
1.8% CMA16	40.6	419	17.4	15.7	–

^{a}Calculated from the maximum derivative weight loss versus temperature curve. WVTR: Water vapor transmission rate.

organoclay CMA16 is more compatible with the MMA–BA (10:90 ratio) monomers than C30B, which, combined with its reactivity, leads to the full exfoliation state and better dispersion of the clay platelets throughout the polymeric matrix. These nanocomposites were prepared with MMA–BA in the proportions of an adhesive formulation. The final adhesive properties of both types of films (one with non-reactive C30B and the other with reactive CMA16), were measured and better shear resistances compared with the blank film were obtained. However, what needs to be noted is that the shear enhancement for the film having the fully exfoliated CMA16 clay was much higher, with practically no loss of tack, than that for the film having C30B. Generally, the higher degree of exfoliation produces better properties of the final film.[72,80,81]

10.6 Towards Real Applications: High Solids Content and Reproducible Latexes

For real applications and thus being attractive to industry, the latexes have to be free of coagulum and with a high solids content (in industry, latexes with a solids content above 45% are used, because this saves process and transportation costs). There can be some exceptions regarding the solids content, but only if the final product has very specific properties so that very small amounts of the product have market potential.

It is worth mentioning here that the latex industry requires film-forming latexes. Therefore, all the waterborne polymer–clay nanocomposites synthesized with model styrene are not very useful for industrial applications.

Another important point is the reproducibility of the process. In the case of miniemulsion polymerization, the most problematic step is the miniemulsification. All of the studies mentioned in this chapter used laboratory ultrasound to create the desirable shear, but none of the groups properly and sufficiently described the miniemulsion process. The droplet size distribution is influenced by, apart from the properties of the dispersion (such as viscosity), the stirrer speed (if used during sonication), the system geometry and the ultrasound intensity.[82] In most of the papers the power output is just stated, but it is a meaningless number unless information about the amplitude of the oscillations, which depends also on the type of tip used, is provided. When applied to dispersions, higher amplitudes show a higher destructiveness towards solid particles,[3] which can be very important in the case of polymer–clay miniemulsions.

In the case of polymer–clay latexes prepared by miniemulsion polymerization, it is very problematic to obtain stable latexes at a reasonable solids content. Most of the studies presented up to now had very low ($<5\%$[14,67]), low ($\sim 10\%$[18,78,83]) or relatively low ($\sim 20\%$[17,20]) solids contents.

Only three papers have reported higher solids contents and all are by Diaconu and co-workers,[15,21,79] where 30 wt% solids content latexes were successfully prepared by batch miniemulsion polymerization and 40–42 wt% solids content latexes by semibatch miniemulsion polymerization of BA–MMA (50:50, w/w).[15,79] The latexes with higher solids contents were synthesized by two different routes. The latex with 40 wt% solids content was polymerized in batch at 70 °C with *tert*-butyl hydroperoxide (TBHP)–ascorbic acid (AsAc), from a miniemulsion with 40 wt% solids content and 3 wbm% of C30B. On the other hand, the latex with 42 wt% solids content was synthesized by seeded semibatch polymerization. The seed used in this case was the 30 wt% solids content latex with 3 wbm% C30B previously synthesized batchwise. Neat monomers and an oil-soluble initiator (AIBN) were fed to this seed at 70 °C, until a final nanocomposite latex with 42 wt% solids content and 1.8 wbm% C30B based on monomer was obtained. The selection of the different initiators for this semibatch step was done on the basis of not decreasing the solids content and pushing the polymerization loci to the already formed polymer particles. Stable and coagulum-free latexes were obtained. However, the relatively high surfactant loading (4 wbm% SLS) caused a secondary nucleation and therefore true miniemulsion polymerization was not achieved. Mičušík *et al.*[64] recently showed that it is possible to reduce the surfactant loading to ~ 2 wbm% while producing coagulum-free latexes at 30 wt% solids content, with a closer approximation to the ideal case of miniemulsion polymerization (1:1 copy droplet:particle number).

10.7   Conclusion

The preparation of PCNs by means of *in situ* miniemulsion polymerization is not an easy task, since clay can interact with all components in the system (monomer, surfactant, hydrophobe), thus influencing the droplet size distribution after the miniemulsification step and also the polymerization loci. As a result, the mechanisms of miniemulsion polymerization in the presence of the clay are not fully understood and stable latexes with reasonable solids content are difficult to achieve.

From the two approaches presented in this chapter, Pickering miniemulsion (clay dispersed in the water phase) and miniemulsions with surfactants as stabilizers (organoclay dispersed in the organic phase), the latter case seems to be more a industry-friendly approach. The ideal case of miniemulsions stabilized by surfactants is when the organoclay platelets are encapsulated inside the polymer particles. Even though clear evidence of fully encapsulated clay platelets inside the polymer particles has not been achieved up to now, the idea of having the clay inside the droplets/particles, not influencing the stability, at high solids content with low surfactant loading and generally not bringing another complexity to the already complex miniemulsion polymerization, is very attractive.

The simulation model to predict the morphology of monomer droplet–clay miniemulsions showed that the encapsulation of the clay platelets would only be possible if the clay, besides the compatibility with the monomer system (low $\Delta_{clay-monomer}$), would not provide favorable interactions with water (high $\Delta_{clay-water}$). This case seems to be achievable by dual functionalization of the clay (the cation exchange of both structural charge and OH groups on the edges) with modifiers having similar chemistry to the monomer system. Additionally, these modifiers can be reactive and the thermodynamic aspect, which drives the clay platelets to the surface of droplets/particles, can be overcome. As was shown, the reactivity of the clay is also beneficial for the degree of exfoliation of clay platelets and the resulting larger interfacial area leads to superior mechanical properties. Furthermore, it is considered that the clay encapsulated inside the droplets/particles would influence to a lower extent the kinetics of miniemulsion polymerization and thus better control of the process would be achieved.

The goal for the future is to optimize the dual functionalization of the clays so as to be able to tune the surface chemistry of the clay for a particular monomer system employed. This seems to be much easier for synthetic than for natural clays, not only because of the narrow size distribution of the synthetic clays, but also due to the chemical and structural diversity of the natural clays.

Acknowledgments

Financial support by the European Union (Napoleon project IP 011844-2), Basque Government (GV 07/16-IT-303-07) and Ministerio de Ciencia y Tecnologia (CTQ 2006-03412 and Programa Consolider-Ingenio 2010 'CIC nanoGUNE Consolider' contract CSD2006-00053) is gratefully acknowledged. The SGI/IZO-SGIker UPV/EHU (supported by the National Program for the Promotion of Human Resources within the National Plan of Scientific Research, Development and Innovation – Fondo Social Europeo, Gobierno Vasco and MCyT) is also gratefully acknowledged.

References

1. J. M. Asua, *Prog. Polym. Sci.*, 2002, **27**, 1283.
2. A. López, A. Chemtob, J. L. Milton, M. Manea, M. Paulis, M. J. Barandiaran, S. Theisinger, K. Landfester, W. D. Hergeth, R. Udagama, T. McKenna, F. Simal and J. M. Asua, *Ind. Eng. Chem. Res.*, 2008, **47**, 6289.
3. T. Hielscher, paper presented at ENS'05, Paris, 14–16 December 2005.
4. G. Kermabon-Avon, C. Bressy and A. Margaillan, *Eur. Polym. J.*, 2009, **45**, 1208.
5. M. Alexandre and P. Dubois, *Mater. Sci. Eng. Rep.*, 2000, **28**, 1.
6. G. Borchardt, in *Minerals in Soil Environments*, ed. J. B. Dixon and S. B. Weeds, Soil Science Society of America, Madison, WI, 1989, p. 675.
7. S. E. Miller and P. F. Low, *Langmuir*, 1990, **6**, 572.
8. B. Chen, J. R. G. Evans, H. C. Greenwell, P. Boulet, P. V. Coveney, A. A. Bowden and A. Whiting, *Chem. Soc. Rev.*, 2008, **37**, 568.
9. J. Tudor, L. Willington, D. O'Hare and B. Royan, *Chem. Commun.*, 1996, **17**, 2031.
10. P. Reichert, J. Kressler, R. Thomann, R. Mülhaupt and G. Stöppelmann, *Acta Polym.*, 1998, **49**, 116.
11. D. Tchoubar and N. Cohaut, in *Handbook of Clay Science 1*, ed. F. Bergaya, B. K. G. Theng and G. Lagaly, Elsevier, Amsterdam, 2006, p. 886.
12. B. Erdem, E. D. Sudol, V. L. Dimonie and M. S. El-Aasser, *J. Polym. Sci. Part A Polym. Chem.*, 2000, **38**, 4419.
13. K. Landfester, *Angew. Chem. Int. Ed.*, 2009, **48**, 4488.
14. F. Bouanani, D. Bendedouch, P. Hemery and B. Bounaceur, *Colloids Surf. A*, 2008, **317**, 751.
15. G. Diaconu, M. Paulis and J. R. Leiza, *Macromol. React. Eng.*, 2008, **2**, 80.
16. R. P. Moraes, A. M. Santos, P. C. Oliveira, F. C. T. Souza, M. do Amaral, T. S. Valera and N. R. Demarquette, *Macromol. Symp.*, 2006, **245**, 106.
17. Q. Sun, Y. Deng and Z. L. Wang, *Macromol. Mater. Eng.*, 2004, **289**, 288.
18. Z. Tong and Y. Deng, *Polymer*, 2007, **48**, 4337.
19. A. Samakande, R. D. Sanderson and P. C. Hartmann, *Polymer*, 2009, **50**, 42.
20. M. Mirzataheri, A. R. Mahdavian and M. Atai, *Colloid Polym. Sci.*, 2009, **287**, 725.
21. G. Diaconu, M. Mičušík, A. Bonnefond, M. Paulis and J. R. Leiza, *Macromolecules*, 2009, **42**, 3316.
22. R. P. Moraes, T. S. Valera, N. R. Demarquette, P. C. Oliveira, M. L. C. P. da Silva and A. M. Santos, *J. Appl. Polym. Sci.*, 2009, **112**, 1949.
23. W. F. Bleam, G. J. Welhouse and M. A. Janowiak, *Clays Clay Miner.*, 1993, **41**, 305.
24. W. F. Bleam, *Clays Clay Miner.*, 1990, **38**, 527.
25. M. J. Avena and C. P. De Pauli, *J. Colloid Interface Sci.*, 1998, **202**, 195.
26. S. E. Miller and P. F. Low, *Langmuir*, 1990, **6**, 572.
27. B. J. Teppen and V. Aggarwal, *Clays Clay Miner.*, 2007, **55**, 119.

28. K. Czurda, in *Handbook of Clay Science 1*, ed. F. Bergaya, B. K. G. Theng and G. Lagaly, Elsevier, Amsterdam, 2006, p. 697.
29. Y. Xi, W. Martens, H. He and R. L. Frost, *J. Thermal Anal. Calorim.*, 2005, **81**, 91.
30. S. I. Marras, A. Tsimpliaraki, I. Zuburtikudis and C. J. Panayiotou, *J. Colloid Interface Sci.*, 2007, **315**, 520.
31. Z. Klapyta, T. Fujita and N. Iyi, *Appl. Clay Sci.*, 2001, **19**, 5.
32. K. R. Ratinac, R. G. Gilbert, L. Ye, A. S. Jones and S. P. Ringer, *Polymer*, 2006, **47**, 6337.
33. J. M. Cervantes-Uc, J. V. Cauich-Rodríguez, H. Vázquez-Torres, L. F. Garfias-Mesías and D. R. Paul, *Thermochim. Acta*, 2007, **457**, 92.
34. A. R. Mermut, *Clay Miner. Soc. CMS Workshop*, 1994, **6**, 106.
35. A. Vazquez, M. López, G. Kortaberria, L. Martín and I. Mondragon, *Appl. Clay Sci.*, 2008, **41**, 24.
36. C.-C. Wang, L.-C. Juang, C.-K. Lee, T.-C. Hsu, J.-F. Lee and H.-P. Chao, *J. Colloid Interface Sci.*, 2004, **280**, 27.
37. T. D. Fornes, D. L. Hunter and D. R. Paul, *Macromolecules*, 2004, **37**, 1793.
38. S. M. L. Silva, P. E. R. Araújo, K. M. Ferreira, E. L. Canedo, L. H. Carvalho and C. M. O. Raposo, *Polym. Eng. Sci.*, 2009, **49**, 1696.
39. H. He, R. L. Frost, T. Bostrom, P. Yuan, L. Duong, D. Yang, Y. Xi and J. T. Kloprogge, *Appl. Clay Sci.*, 2006, **31**, 262.
40. D. Ovadyahu, I. Lapides and S. Yariv, *J. Thermal Anal. Calorim.*, 2007, **87**, 125.
41. C. S. Triantafillidis, P. C. LeBaron and T. J. Pinnavaia, *J. Solid State Chem.*, 2002, **167**, 354.
42. D.-R. Yei, S.-W. Kuo, Y.-C. Su and F.-C. Chang, *Polymer*, 2004, **45**, 2633.
43. N. Greesh, P. C. Hartmann, V. Cloete and R. D. Sanderson, *J. Colloid Interface Sci.*, 2008, **319**, 2.
44. N. Negrete-Herrera, J.-L. Putaux, L. David and E. Bourgeat-Lami, *Macromolecules*, 2006, **39**, 9177.
45. N. Salem and D. A. Shipp, *Polymer*, 2005, **46**, 8573.
46. J. Zhu, A. B. Morgan, F. J. Lamelas and C. A. Wilkie, *Chem. Mater.*, 2001, **13**, 3774.
47. T. Sobisch and D. Lerche, *Colloid Polym. Sci.*, 2000, **278**, 369.
48. P. G. Slade and W. P. Gates, *Appl. Clay Sci.*, 2004, **25**, 93.
49. V. N. Moraru, *Appl. Clay Sci.*, 2001, **19**, 11.
50. D. L. Ho, R. M. Briber and C. J. Glinka, *Chem. Mater.*, 2001, **13**, 1923.
51. D. L. Ho and C. J. Glinka, *Chem. Mater.*, 2003, **15**, 1309.
52. L. M. Daniel, R. L. Frost and H. Y. Zhu, *J. Colloid Interface Sci.*, 2008, **321**, 302.
53. D. J. Voorn, W. Ming and A. M. van Herk, *Macromolecules*, 2006, **39**, 4654.
54. N. Negrete-Herrera, J.-L. Putaux and E. Bourgeat-Lami, *Prog. Solid State Chem.*, 2006, **34**, 121.
55. K. Song and G. Sand, *Clays Clay Miner.*, 2001, **49**, 119.

56. N. Negrete-Herrera, J.-M. Letoffe, J.-P. Reymond and E. Bourgeat-Lami, *J. Mater. Chem.*, 2005, **15**, 863.
57. P. A. Wheeler, J. Wang and L. J. Mathias, *Chem. Mater.*, 2006, **18**, 3937.
58. G.-X. Chen, J. B. Choi and J.-S. Yoon, *Macromol. Rapid Commun.*, 2005, **26**, 183.
59. G.-X. Chen and J.-S. Yoon, *Macromol. Rapid Commun.*, 2005, **26**, 899.
60. G.-X. Chen, H.-S. Kim, J.-H. Shim and J.-S. Yoon, *Macromolecules*, 2005, **38**, 3738.
61. J. Faucheu, C. Gauthier, L. Chazeau, Y. Cavaille, V. Mellon and E. Bourgeat-Lami, *Polymer*, 2010, **51**, 6.
62. W. Ramsden, *Proc. R. Soc. London*, 1903, **72**, 156.
63. S. U. Pickering, *J. Chem. Soc.*, 1907, **91**, 2001.
64. M. Mičušík, A. Bonnefond, Y. Reyes, M. Paulis and J. R. Leiza, *Macromol. React. Eng.*, DOI: 10.1002/mren200900084.
65. Y. Duda, E. Vakarin and J. Alejandre, *J. Colloid Interface Sci.*, 2003, **258**, 10.
66. D. P. Landau and K. Binder, *A Guide to Monte-Carlo Simulation in Statistical Physics*, Cambridge University Press, Cambridge, 2005.
67. S. Cauvin, P. J. Colver and S. A. F. Bon, *Macromolecules*, 2005, **38**, 7887.
68. N. P. Ashby and B. P. Binks, *Phys. Chem. Chem. Phys.*, 2000, **2**, 5640.
69. S. L. Tawari, D. L. Koch and C. Cohen, *J. Colloid Interface Sci.*, 2001, **240**, 54.
70. R. Ianchis, D. Donescu, C. Petcu, M. Ghiurea, D. F. Anghel, G. Stanga and A. Marcu, *Appl. Clay Sci.*, 2009, **45**, 164.
71. Q. Sun, Y. Deng and Z. L. Wang, *Macromol. Mater. Eng.*, 2004, **289**, 288.
72. T. D. Fornes and D. R. Paul, *Polymer*, 2003, **44**, 4993.
73. H. J. Ploehn and C. Liu, *Ind. Eng. Chem. Res.*, 2006, **45**, 7025.
74. A. Samakande, R. D. Sanderson and P. C. Hartmann, *J. Polym. Sci. Polym. Chem.*, 2008, **46**, 7114.
75. S. H. Xu and S. A. Boyd, *Langmuir*, 1995, **1**, 2508.
76. S. Talapatra, S. C. Guhaniyogi and S. K. Chakravarti, *J. Macromol. Sci. Chem.*, 1985, **A22**, 1611.
77. S. A. F. Bon and P. J. Colver, *Langmuir*, 2007, **23**, 8316.
78. Z. Tong and Y. Deng, *Macromol. Mater. Eng.*, 2008, **293**, 529.
79. G. Diaconu, J. M. Asua, M. Paulis and J. R. Leiza, *Macromol. Symp.*, 2007, **259**, 305.
80. M. A. Osman, J. E. P. Rupp and U. W. Suter, *Polymer*, 2005, **46**, 1653.
81. S. C. Tjong, *Mater. Sci. Eng. R*, 2006, **53**, 73.
82. M. do Amaral, A. Arevalillo, J. L. Santos and J. M. Asua, *Prog. Colloid Polym. Sci.*, 2004, **124**, 103.
83. Z. Tong and Y. Deng, *Ind. Eng. Chem. Res.*, 2006, **45**, 2641.

CHAPTER 11

PAN–Silica–Clay Nanocomposites by Emulsion Polymerization

CHUNHUA CAI, JIAPING LIN AND LAN WEI

Shanghai Key Laboratory of Advanced Polymeric Materials, State Key Laboratory of Bioreactor Engineering, School of Materials Science and Engineering, East China University of Science and Technology, Shanghai 200237, China

11.1 Introduction

The past two decades have witnessed significant developments in polymer nanocomposites, although they have been known for about a century.[1] Polymer nanocomposites are a new class of materials composed of an organic polymer matrix and a dispersed nanometer-scale inorganic component (nanofiller).[2–10] Because of the large interfacial area between the polymer and the nanofiller, nanocomposites possess unique properties that are not shared by conventional materials and offer new technological and economic opportunities. Usually, a significant improvement in the mechanical, thermal, optical, electrical and barrier properties of the polymer can be obtained with very low loadings of nanofiller.

Polymer–clay nanocomposites are the most commonly studied and applied nanocomposites.[11–23] Natural clay is composed of oxide layers with cations between the layers. The thickness of the layers is about 1 nm. In practice, there exist three types of polymer–clay composite materials, as shown in

RSC Nanoscience & Nanotechnology No. 16
Polymer Nanocomposites by Emulsion and Suspension Polymerization
Edited by Vikas Mittal
© Royal Society of Chemistry 2011
Published by the Royal Society of Chemistry, www.rsc.org

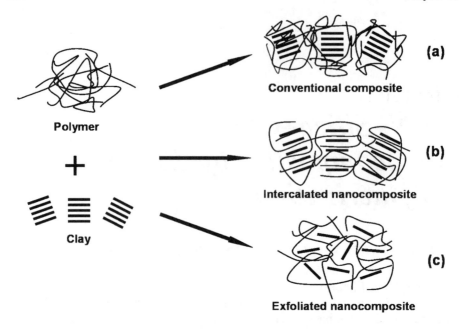

Figure 11.1   Schematic representation of various types of composites: conventional
polymer–clay composite (a); intercalated polymer–clay nanocomposite
(b); and exfoliated polymer–clay nanocomposite (c).

Figure 11.1.[24-26] The first type is conventional composites, where non-swollen
clay tactoids with layers are dispersed simply as a segregated phase in a polymer
matrix, which results in poor mechanical properties of the composite material.
The second type is intercalated polymer–clay nanocomposites, which are well-
ordered multilayered structures formed by the insertion of polymer chains into
the gallery space between parallel individual clay layers. Although the layer
spacing of clay increases, there are still attractive forces between the silicate
layers to stack the layers with uniform spacing. The last type is exfoliated
polymer–clay nanocomposites, in which the clays are well dispersed in the
polymer matrix. Compared with the intercalated polymer–clay nanocomposites, the layer spacing in the exfoliated polymer–clay nanocomposites increases
to the point that there are no longer sufficient attractions between the silicate
layers to maintain a uniform layer spacing. The last two hybrid structures are
nanocomposites and the exfoliated polymer–clay nanocomposites are especially desirable for improved properties because of the homogeneous dispersion
of clay and huge interfacial area between the polymer and clay.

   The preparation methods for polymer–clay nanocomposites can be classified
into two approaches. One is blending of polymers with clay either in solution or
in melt.[27-38] Because of the large scale and low mobility of polymer chains,
usually intercalated polymer–clay nanocomposites are obtained with this
blending strategy. The other method is *in situ* polymerization.[39-48] Clay is first

dispersed in the monomer or monomer solution and then polymerized. By this method, polymers directly attach to the clay layer and usually exfoliated polymer–clay nanocomposites are formed. The key of *in situ* polymerization is to increase the compatibility of clay in the monomer. Most monomers and polymers are hydrophobic, whereas clays are hydrophilic. Therefore, organic modification of the clay (organoclay) is common.[49–54] Since the clay dissolves well in water, if the *in situ* polymerization can be performed in aqueous solution the formation of exfoliated polymer–clay nanocomposites can be expected and also the clay can be used without premodification. With this aim, *in situ* emulsion polymerization was introduced to prepare exfoliated polymer–clay nanocomposites.[55–66] This approach displays several significant advantages for the preparation of polymer–clay nanocomposites: (1) clays are fully exfoliated and stably suspended in water owing to the highly absorbent properties of water; therefore, clay can be directly used without surface modification; (2) the emulsifier can further improve the compatibility of clay and polymer, which makes it easy for the monomers and initiators to enter the interlayers of clay; (3) the reaction medium is water, which eliminates toxicological and environmental problems during industrial production.

In addition to clay, silica (SiO_2) is another important and widely used nanofiller. The thermal and mechanical properties of the polymer matrix can be considerably enhanced when SiO_2 nanoparticles are introduced. Polymer–silica nanocomposites can be prepared by blending,[68–73] sol–gel reaction[74–78] and *in situ* polymerization.[79–83] Blending, including mainly solution and melt, is a simplest method, but the silica particles usually tend to agglomerate. Sol–gel reaction of tetraethoxysilane (TEOS) in the polymer matrix is a commonly used method for the preparation of polymer–silica nanocomposites. The major advantage of this process is the mild conditions, such as relatively low temperature and pressure. The sol–gel process can be viewed as a two-step network-forming process, the first step being the hydrolysis of a metal alkoxide and the second consisting of a polycondensation reaction. *In situ* polymerization is an easy to handle and rapid method. The modified SiO_2 nanoparticles are dispersed in monomer or monomer solution and then polymerized. The most important factors that affect the properties of composites are the dispersion and the adhesion at the polymer and filler interfaces. Polymer–silica nanocomposites prepared by *in situ* polymerization have been reported in the literature.[84–88]

So far, most polymer nanocomposites contain only one type of nanofiller. Recent studies revealed that combination of clay and SiO_2 has a more enhanced effect on the polymer matrix.[89,90] In this chapter, we discuss the structure and properties of clay- and silica-based polymer nanocomposites prepared by *in situ* emulsion polymerization, especially polyacrylonitrile (PAN)–clay–silica ternary nanocomposites. The chapter consists of three parts: (1) synthesis and structure of polymer–clay–silica nanocomposites; (2) thermal properties of polymer–clay–silica nanocomposites; and (3) mechanical properties of polymer–clay–silica nanocomposites.

11.2 Synthesis and Structure of Polymer–Clay–Silica Nanocomposites

In situ emulsion polymerization in aqueous solution is an effective and successful method for the synthesis of polymer–clay nanocomposites. The structure of the nanocomposites formed is usually characterized by X-ray diffraction (XRD) and transmission electron microscopy (TEM). XRD offers a convenient method to determine the position, shape and intensity of the nanofillers.[91–98] The XRD patterns shows a basal (001) reflection when the spacing between the clay layers is small. However, X-rays cannot detect the (001) reflection when the layers are exfoliated in the composite. An increase in spacing between layers is correlated with an increase in the degree of exfoliation. In contrast, TEM observation gives a direct view of the internal structure, spatial distribution of the various phases and defect structures. By combining these methods, the structure of polymer nanocomposites can be determined.

Tang and co-workers prepared poly(ethyl acrylate) (PEA)–bentonite nanocomposites by *in situ* emulsion polymerization.[99] Bentonite was first dispersed in water and then ethyl acrylate as monomer, potassium persulfate as initiator and sodium dodecyl sulfate as surfactant were added to the bentonite solution. Under agitation, the mixture was reacted. Finally, the resulting emulsion was cast into polytetrafluoroethylene (PTFE) molds and dried in vacuum. They used XRD and TEM testing to examine the structure of the nanocomposites and the results are presented in Figures 11.2 and 11.3. Figure 11.2 shows a series of XRD patterns for PEA–clay hybrids with various compositions. With respect to clay, the diffraction peaks corresponding to the pristine silicate disappear in the PEA–clay hybrids, and a set of new peaks appear corresponding to the basal spacing of PEA–clay hybrids (from $d_{001} = 3.91$ 4.96 nm). This indicates that the clay is intercalated in the hybrids. However, it is difficult to draw definite conclusions about the structure of the hybrids from XRD results exhibiting diffraction patterns because the relatively featureless diffraction patterns of the exfoliated structure may be covered by the diffraction peak of the intercalated structure. Therefore, TEM is necessary for determining the nature of the hybrids and it also provides additional information that can aid the interpretation of the XRD results. A typical TEM image of a PEA–clay nanocomposite is given in Figure 11.3. The dark lines are the cross-sections of the silicate layers. Some single silicate layers and ordered intercalated assembled layers of clay are well dispersed in the PEA matrix. On the basis of the TEM and XRD results, the direct-cast PEA–clay sample is apparently a nanocomposite with the intercalated and exfoliated structures of clay.

Huang and Brittain reported the synthesis of poly(methyl methacrylate) (PMMA)–layered silicate by *in situ* emulsion polymerization.[100] The MMA monomer was first polymerized by emulsion polymerization. The surfactant was *n*-decyltrimethylammonium chloride or [2-(methacryloyloxy)ethyl]-trimethylammonium chloride and the initiator was 2,2′-azobis(isobutylamidine hydrochloride) (AIBA). Then, an aqueous dispersion of layered silicate (not organically modified) was added to the PMMA emulsion. Because the latex

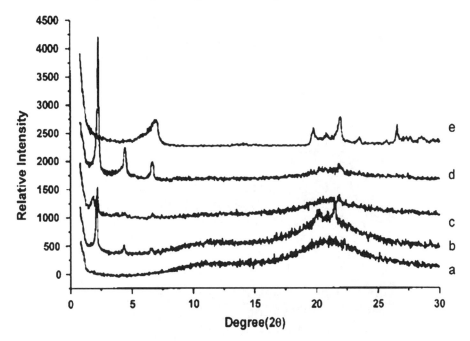

Figure 11.2 XRD patterns of PEA–clay nanocomposites with various PEA–clay compositions (w/w): 100:0 (a), 98:2 (b), 95:5 (c), 90:10 (d) and 0:100 (e). Reproduced with permission from reference 99.

particles have cationic surface charges and the silicate layers have anionic charges, the electrostatic forces promote an interaction between the silicate and polymer particles. Finally, after centrifugation, washing with distilled water and drying in vacuum, the PMMA–clay nanocomposite was obtained. XRD and atomic force microscopy (AFM) analyses showed that exfoliated PMMA–clay nanocomposites were produced. The prepared exfoliated structure is very stable, and even after melt processing the exfoliated structure can be preserved.

Wang's group demonstrated a new technique, ultrasonically initiated *in situ* emulsion polymerization, to prepare exfoliated polystyrene (PS)–Na montmorillonite (MMT) nanocomposites.[101] Under ultrasonic irradiation, the clay particles are pulverized and dispersed on the nanoscale and the polymerization reaction is initiated without any additional chemical initiator. The structure of the composite was characterized by combining TEM and XRD with Fourier transform infrared (FTIR) spectroscopy. The hydrophobic PS was confirmed to have been intercalated into the galleries of the hydrophilic MMT. It was found that properly reducing the emulsifier concentration is beneficial to widen the *d*-spacing between the clay layers. Ultrasonically initiated *in situ* emulsion polymerization is expected to achieve the effective breakup of the clay agglomerates and intercalation of polymer into the silicate layers to give useful polymer–clay nanocomposites.

Figure 11.3 Representative TEM image of exfoliated clay intercalated in a PEA matrix with 95:5 (w/w) PEA–clay. Reproduced with permission from reference 99.

Polymer–silica nanocomposites can also be prepared *via* the *in situ* emulsion polymerization method. In order to achieve effective encapsulation, surface modification of the SiO_2 particles is necessary. In fact, the nanocomposites formed are polymer-coated SiO_2 particles. For example, poly(styrene–methyl methacrylate)–SiO_2 and polystyrene–SiO_2 nanocomposites were synthesized *via* emulsion polymerization in aqueous solution.[102] Figure 11.4 shows a typical TEM image of polymer–silica nanocomposite particles. The morphology of the particles shows a core–shell structure, with the polymer as shell and silica as core. Figure 11.4 shows that the composites have only a single core. By altering the polymer to silica ratio, nanoparticles bearing two or more than two cores can be produced.

So far, most studies have concerned polymer nanocomposites containing only one type of nanofiller. Lin and co-workers reported the synthesis of polymer–clay–silica ternary nanocomposites by *in situ* emulsion polymerization

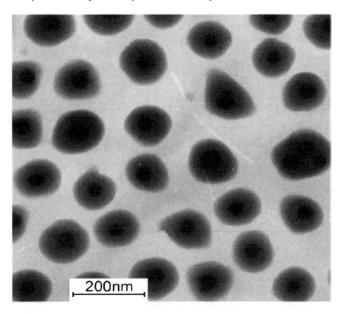

Figure 11.4 TEM image of PS–SiO$_2$ composite particles with core–shell structure. Reproduced with permission from reference 102.

in water.[89,90] The samples were prepared as follows. Na-MMT, SiO$_2$, emulsifiers 2-acrylamido-2-methyl-1-propanesulfonic acid (AMPS), polyoxyethylene-alkylphenol ether (OP-10), sodium pyrosulfite and acrylonitrile (AN) monomer were dispersed in deionized water to form an emulsion. The polymerization was then carried out under a nitrogen atmosphere at 55 °C. Finally, powder-like product was obtained by precipitating the reaction emulsion and drying.

They used the XRD testing to analyze the exfoliation of the MMT component in the PAN nanocomposites. Figure 11.5 shows the XRD diffraction patterns of Na-MMT and PAN nanocomposites containing Na-MMT and Na-MMT–SiO$_2$. Curve (a) shows a typical XRD pattern of pristine Na-MMT, which exhibits a diffraction peak of the (001) plane at 5.7° 2θ and its basal spacing is 1.55 nm. Curve (b) is the result obtained for PAN–Na-MMT. No diffraction peak is visible, suggesting that the Na-MMT layers are exfoliated. A similar result was obtained for the PAN–Na-MMT–SiO$_2$ composite, which indicates that the Na-MMT layers are exfoliated in this system.

The variation of the Na-MMT basal spacing with polymerization time was also studied. Figure 11.6 shows typical XRD results for PAN–Na-MMT–SiO$_2$ composites with 2.0 wt% Na-MMT and 2.0 wt% SiO$_2$. It can be seen that once the polymerization is initiated, the diffraction peak shifts to smaller angle and becomes broader. The disappearance of the diffraction peak indicates that the layers of Na-MMT are fully exfoliated.

TEM study offers direct observation of the morphology of PAN nanocomposites. Figure 11.7 shows typical TEM images of PAN–Na-MMT–SiO$_2$

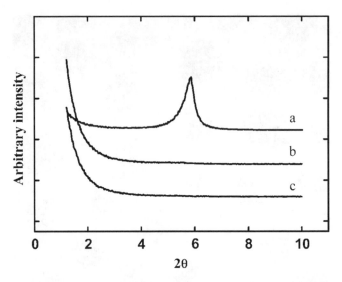

Figure 11.5 XRD patterns of pristine Na-MMT (a), PAN–Na-MMT with 4.0 wt%
Na-MMT (b) and PAN–Na-MMT–SiO$_2$ with 2.0 wt% Na-MMT and
2.0 wt% SiO$_2$ (c). Reproduced with permission from reference 89.

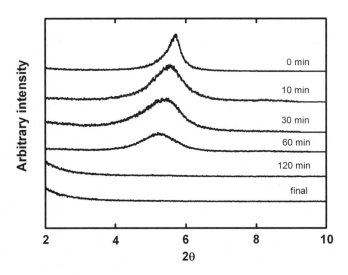

Figure 11.6 XRD patterns of PAN–Na-MMT–SiO$_2$ with 2.0 wt% Na-MMT and 2.0
wt% SiO$_2$, sampled at fixed time intervals during polymerization.
Reproduced with permission from reference 90.

composites with different nanofiller contents. The dark slices represent Na-
MMT layers and the spherical dark particles are SiO$_2$. The PAN matrix
appears as a light region. The Na-MMT layers together with the SiO$_2$ particles
show a good dispersion in the PAN matrix. Further, the Na-MMT layers are

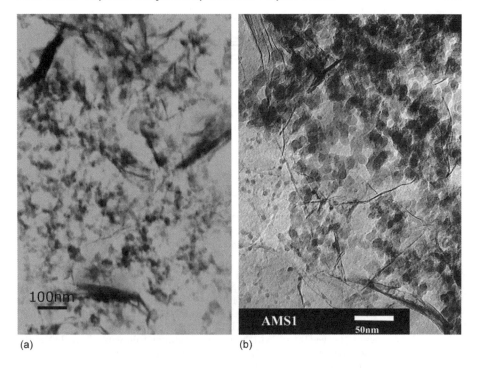

(a)   (b)

Figure 11.7 TEM images of PAN–Na-MMT–SiO$_2$ with 1.0 wt% Na-MMT and 1.0 wt% SiO$_2$ (a) and 2.0 wt% Na-MMT and 2.0 wt% SiO$_2$ (b). Reproduced with permission from reference 90.

exfoliated in the PAN matrix, as can be seen from these images. Such TEM observations are in line with the XRD results.

11.3 Thermal Properties of Polymer–Clay–Silica Nanocomposites

A major aim to prepare the polymer–nanofiller composites is to enhance the thermal stability of the materials. The thermal properties of polymer nanocomposites are usually evaluated by differential scanning calorimetry (DSC) and thermogravimetric (TG) analyses. From DSC and TG testing, the glass transition temperature (T_g) and decomposition temperature can be determined, from which the thermal properties of polymer nanocomposites can be evaluated.

Chung and co-workers studied the thermal properties of PS–clay nanocomposites that had been synthesized by *in situ* emulsion polymerization.[103] The XRD and TEM results indicated that clay is exfoliated in the polymer matrix. They used TG to analyze the thermal properties of the PS–clay and PAN–clay nanocomposites. As shown in Figure 11.8, the degradation onset

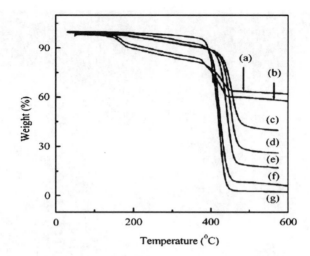

Figure 11.8 TG curves for the A0.3ST series, A0.3ST3% (a), A0.3ST1% (b), A0.3STD10% (c), A0.3STD5% (d), A0.3STD3% (e), A0.3STD1% (f) and A0.3STD0% (g), under nitrogen. In the sample code, A represents surfactant AMPS, S styrene, T pristine Na-MMT and D dodecylbenzenesulfonic acid sodium salt (DBS-Na) added. The number following A means the weight of AMPS and the number following T or TD is the weight percentage of clay relative to monomer. Reproduced with permission from reference 103.

temperature of the nanocomposites is higher than that of pure PS and shifts towards higher temperature as the amount of clay increases. In addition, the PS–clay nanocomposite with 10 wt% of clay shows a 40 °C increase in decomposition temperature at 20% weight loss.

Shim and co-workers used an ultrasonically assisted *in situ* emulsion polymerization to prepare electrically conducting copolymer poly(aniline-*co*-*p*-phenylenediamine) [poly(Ani-*co*-pPD)] and silica (SiO_2) nanocomposites.[104] It was found that the aggregation of SiO_2 nanoparticles could be reduced under ultrasonic irradiation. The thermal stability of the neat copolymer and nanocomposite with 5% SiO_2 was analyzed by TG and DSC, as shown in Figure 11.9. It was found that the thermal stability of the nanocomposite is increased. TG analysis revealed that the residual mass left at 800 °C increases from 33 to 38%, which is attributed to the interaction between SiO_2 and copolymer chains. DSC testing showed that the incorporation of SiO_2 results in an increase in the decomposition temperature of up to 5 °C relative to the neat copolymer. This behavior is as expected and is believed to be associated with the strong heterogeneous nucleation effect of SiO_2 in the nanocomposite system.

However, the synergistic effect of clay and silica on the thermal stability of polymer–clay–silica has been less studied. Pioneering work was conducted by Lin's group.[89,90] They investigated the thermal stability of PAN–clay, PAN–silica and PAN–clay–silica nanocomposites and also pure PAN. Figure 11.10 shows the TG graphs for matrix PAN and its nanocomposites. Thermal

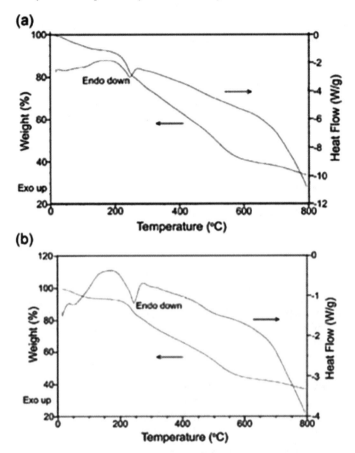

Figure 11.9 TGA and DSC curves for neat copolymer (a) and poly(Ani-co-pPD)–SiO₂ nanocomposite (5% SiO₂) (b) under nitrogen. Reproduced with permission from reference 104.

decomposition of the matrix PAN occurs at about 300–400 °C (curve a). Since the initial stage of the decomposition of PAN is mainly associated with the formation of ring compounds among the adjacent –CN groups, the presence of the nanofillers has less effect on the chemical reaction process. Therefore, no significant variation of the decomposition temperature with respect to the nanofiller content was observed. However, enhanced thermal stability was observed when temperature was further increased (curves b–f). All the PAN nanocomposites have enhanced thermal stability compared with the PAN matrix. Also, for nanocomposites with same nanofiller loadings, PAN–Na-MMT–SiO₂ showed more enhanced thermal stability than either PAN–Na-MMT or PAN–SiO₂.

Plots of decomposition temperature (at 40% weight loss) against nanofiller loading are shown in Figure 11.11. Curve a shows the result for PAN–SiO₂ nanocomposites and curves b and c show the results for the PAN–Na-MMT and

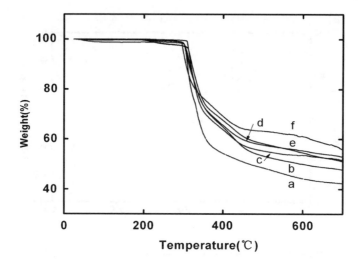

Figure 11.10 TG curves for PAN (a), PAN–SiO$_2$ with 4.0 wt% SiO$_2$ (b), PAN–Na-MMT with 4.0 wt% Na-MMT (c), PAN–Na-MMT–SiO$_2$ with 2.8 wt% Na-MMT and 1.2 wt% SiO$_2$ (d), PAN–Na-MMT–SiO$_2$ with 0.8 wt% Na-MMT and 3.2 wt% SiO$_2$ (e) and PAN–Na-MMT–SiO$_2$ with 2.0 wt% Na-MMT and 2.0 wt% SiO$_2$ (f). Reproduced with permission from reference 90.

PAN–Na-MMT–SiO$_2$ nanocomposites, respectively. For all these nanocomposite systems, the decomposition temperature increases as the nanofiller loading increases. Furthermore, the PAN–Na-MMT–SiO$_2$ ternary systems demonstrate a more evidently enhanced thermal stability than the binary systems. For instance, at 5.0 wt% nanofiller loading, the decomposition temperature of PAN–Na-MMT–SiO$_2$ (the weight ratio of Na-MMT and SiO$_2$ is 1:1) is 178 °C higher than that of PAN–Na-MMT and 241 °C higher than that of PAN–SiO$_2$.

The enhanced thermal stability of PAN–Na-MMT–SiO$_2$ nanocomposites is believed to be induced by the formation of a network structure, because the strong electronegative –CN groups in the PAN chain can interact with the SiO$_2$ particles and the exfoliated Na-MMT layers. In addition, the presence of Na-MMT can prevent the diffusion out of the volatile decomposition products because of its layer structure,[101] which makes a further contribution to the thermal stability. Consequently, the PAN–Na-MMT–SiO$_2$ nanocomposites exhibit superior thermal stability to the PAN–Na-MMT and PAN–SiO$_2$ systems.

11.4 Mechanical Properties of Polymer–Clay–Silica Nanocomposites

Improving the mechanical performance of the polymer is an important target of preparing the polymer–nanofiller composites. The storage modulus is one of

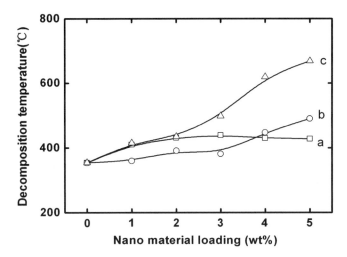

Figure 11.11 Decomposition temperature at 40% weight loss *versus* nanofiller loading for PAN–SiO$_2$ (a), PAN–Na-MMT (b) and PAN–Na-MMT–SiO$_2$ (c). For the PAN–Na-MMT–SiO$_2$ sample, the weight ratio of Na-MMT to SiO$_2$ was maintained at 1:1. Reproduced with permission from reference 90.

the most important of the mechanical properties of materials and many studies have focused on the enhancement of storage modulus by introducing nanofillers into the polymer matrix. It is widely observed that loading the nanofiller has a significant effect on the storage modulus of the nanocomposites. Here, we focus on the studies of storage modulus of polymer nanocomposites prepared by *in situ* emulsion polymerization.

Chung and co-workers prepared exfoliated PAN–clay nanocomposite and studied the storage modulus of the nanocomposite obtained.[105] Figure 11.12 shows a plot of storage modulus as a function of temperature at various clay loadings. The results show that the storage modulus of the nanocomposite increases with increase in the amount of silicate. For example, at 400 °C, the PAN–Na-MMT nanocomposites with 5 and 20% Na-MMT contents show 55 and 250% enhancements of storage modulus, respectively, over pure PAN. An increase in temperature induces a decrease in the storage modulus of the polymer and for samples containing higher contents of clay it decreases slightly. The exfoliated clay structure in the composites is believed to be the important factor inducing the large increase in the storage modulus of PAN–clay nanocomposites. In addition, pristine silicate and the polarity of PAN also affect the enhancement because the strongly negative –CN groups in PAN can interact with the exfoliated layers of hydrophilic Na-MMT.

Polymer–silica nanocomposites also show an enhancement of storage modulus compared with the corresponding pure polymer.[67] However, studies on the enhancement effect of storage modulus for the polymer–clay–silica ternary nanocomposites are lacking. Lin and co-workers studied the dependence of the

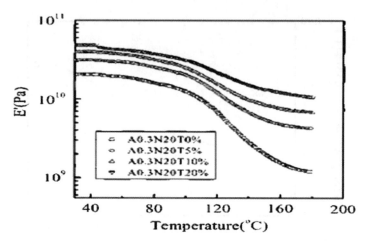

Figure 11.12   Dependence of storage modulus of A0.3N20T0% (a), A0.3N20T5% (b), A0.3N20T10% (c) and A0.3N20T20% (d) on temperature. In the sample code, A represents surfactant AMPS, N acrylonitrile (AN) and T pristine Na-MMT. The number following A means the weight of AMPS and the number following T is the weight percentage of clay relative to monomer. Reproduced with permission from reference 105.

storage modulus of PAN–clay–silica nanocomposites on temperature and nanofiller loading.[89,90] Figures 11.13 and 11.14 show the results obtained from the nanocomposites with total nanofiller contents of 2.0 and 4.0 wt%, respectively.[89] The E' values of all the PAN nanocomposites are higher than that of pure PAN over the entire range of temperature in both cases. Also, the reinforcement effect of Na-MMT–SiO$_2$ is appreciably better than that of either Na-MMT or SiO$_2$ alone (see Figures 11.13 and 11.14).

The dependence of the storage modulus of the composites on nanofiller content is shown in Figure 11.15, where the values of the storage modulus E' of the PAN–silica, PAN–clay and PAN–clay–silica nanocomposites are plotted against the total nanofiller loading (Na-MMT + SiO$_2$) at 50 °C.[90] For the PAN–clay–silica ternary nanocomposites, the loadings of SiO$_2$ and Na-MMT are equal. As shown in Figure 11.15, with increase in nanofiller loading, the storage modulus first increases to a maximum value, followed by a decrease. At a fixed nanofiller content, the PAN–clay–silica ternary nanocomposites exhibit the most elevated storage modulus among the three samples. Moreover, the reinforcement effect of the Na-MMT–SiO$_2$ is dependent on the Na-MMT to SiO$_2$ ratio. The inset of Figure 11.15 shows a plot of E' against Na-MMT to SiO$_2$ ratio for two systems with total nanofiller loadings of 2.0 and 4.0 wt% at 50 °C. It reveals that the E' value first increases (with an Na-MMT to SiO$_2$ ratio of about 5:5), then decreases with the change of the Na-MMT to SiO$_2$ ratio from 10:0 to 0:10.

The temperature dependence of tanδ (the ratio of the loss modulus to the storage modulus) for the PAN nanocomposites with various compositions was also examined,[90] and the results are presented in Figure 11.16. A relaxation

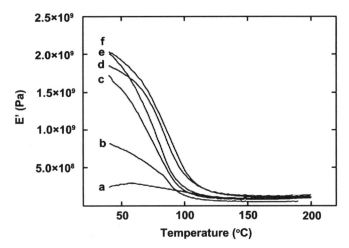

Figure 11.13 Dependence of storage modulus E' on temperature for PAN matrix (a), PAN–SiO$_2$ with 2.0 wt% SiO$_2$ (b), PAN–Na-MMT with 2.0 wt% Na-MMT (c), PAN–Na-MMT–SiO$_2$ with 1.4 wt% Na-MMT and 0.6 wt% SiO$_2$ (d), PAN–Na-MMT–SiO$_2$ with 0.6 wt% Na-MMT and 1.4 wt% SiO$_2$ (e) and PAN–Na-MMT–SiO$_2$ with 1.0 wt% Na-MMT and 1.0 wt% SiO$_2$ (f). Reproduced with permission from reference 90.

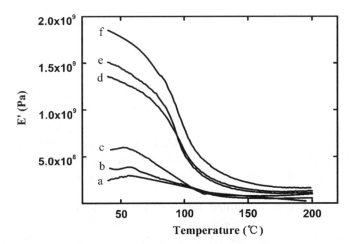

Figure 11.14 Dependence of storage modulus E' on temperature for PAN matrix (a), PAN–Na-MMT with 2.0 wt% Na-MMT (b), PAN–SiO$_2$ with 4.0 wt% SiO$_2$ (c), PAN–Na-MMT–SiO$_2$ with 2.8 wt% Na-MMT and 1.2 wt% SiO$_2$ (d), PAN–Na-MMT–SiO$_2$ with 1.6 wt% Na-MMT and 2.4 wt% SiO$_2$ (e) and PAN–Na-MMT–SiO$_2$ with 2.0 wt% Na-MMT and 2.0 wt% SiO$_2$ (f). Reproduced with permission from reference 89.

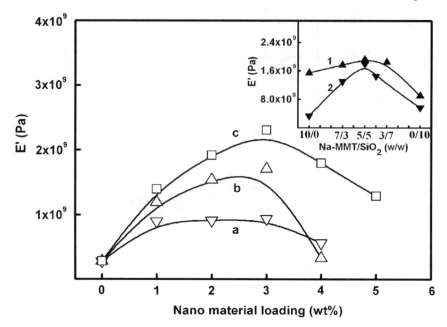

Figure 11.15   Dependence of storage modulus E' (50 °C) on the content of nanofillers
for PAN–SiO$_2$ (a), PAN–Na-MMT (b) and PAN–Na-MMT–SiO$_2$ (c).
For the PAN–Na-MMT–SiO$_2$, the Na-MMT to SiO$_2$ weight ratio was
maintained at 1:1. The inset shows a plot of E' (50 °C) against Na-MMT
to SiO$_2$ ratio for two systems with total nanofiller loadings of 2.0 wt%
(1) and 4.0 wt% (2). Reproduced with permission from reference 90.

peak of the PAN matrix is observed at 102.0 °C, which is the α transition
associated with backbone segmental motion at the glass transition temperature
(T_g). The T_α of PAN nanocomposites tends to be higher when the nanofillers
are charged (see Figure 11.16, curves b–d). Moreover, the PAN–clay–silica
nanocomposites tend to have a higher T_α than those of PAN–Na-MMT and
PAN–SiO$_2$ at same nanofiller loading. These results suggest that the segmental
motion of the polymer chains is restrained in the nanocomposites. Also, the
restraint could be the consequence of the interactions between the PAN chains
and nanofillers.

Lin and co-workers explained the superior mechanical properties of the
PAN–Na-MMT–SiO$_2$ nanocomposites as follows: The Na-MMT was exfoliated in the PAN nanocomposites (see Figure 11.7), and the enhancement of
the storage modulus results from the delamination of the silicates in the PAN
matrix and the strong interactions between the polymer chains and the Na-
MMT layers. However, the reinforcement could be anisotropic because of the
layer shape of the exfoliated Na-MMT layers. Cracks along the direction of the
Na-MMT layers may not be resisted. However, in PAN–clay–silica nanocomposites, the well-dispersed SiO$_2$ particles could bridge the cracks that are
not stopped by the Na-MMT layers. Therefore, the coexistence of the

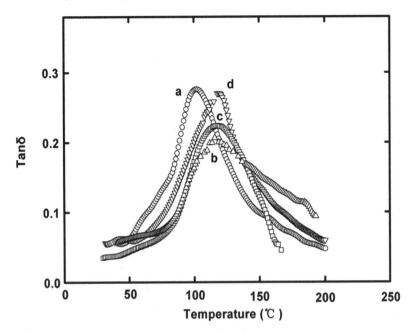

Figure 11.16 Temperature dependence of tanδ for PAN matrix (a), PAN–Na-MMT with 4.0 wt% Na-MMT (b), PAN–SiO$_2$ with 4.0 wt% SiO$_2$ (c), and PAN–Na-MMT–SiO$_2$ with 2.0 wt% Na-MMT and 2.0 wt% SiO$_2$ (d). Reproduced with permission from reference 90.

exfoliated Na-MMT layers and the nano-SiO$_2$ produces a synergistic effect on reinforcing the PAN matrix.

11.5 Conclusion

In situ emulsion polymerization is an effective and simple method to prepare polymer–clay nanocomposites. The clay can be easily exfoliated in the polymer matrix. The enhancement of the properties of the polymer matrix is significant. The thermal and mechanical properties of PAN–clay–silica ternary nanocomposites are significantly enhanced compared with PAN–clay and PAN–silica binary systems, which suggests that the synergistic effect of two kinds of nanofillers can make a further contribution to the enhancement of the thermal and mechanical properties of the polymer matrix.

References

1. A. C. Balazs, T. Emrick and T. P. Russell, *Science*, 2006, **314**, 1107.
2. Y.-W. Mai and Z.-Z. Yu (eds), *Polymer Nanocomposites*, Woodhead Publishing, Cambridge, 2006.

3. P. M. Ajayan, L. S. Schadler and P. V. Braun (eds), *Nanocomposite Science and Technology*, Wiley-VCH, Weinheim, 2003.
4. J. E. Mark, *Acc. Chem. Res.*, 2006, **39**, 881.
5. M. Benaglia, A. Puglisi and F. Cozzi, *Chem. Rev.*, 2003, **103**, 3401.
6. C. Park, Z. Ounaies, K. A. Watson, R. E. Crooks, J. Smith Jr, S. E. Lowther, J. W. Connell, E. J. Siochi, J. S. Harrison and T. L. St. Clair, *Chem. Phys. Lett.*, 2002, **364**, 303.
7. M. Avella, M. E. Errico and E. Martuscelli, *Nano Lett.*, 2001, **1**, 213.
8. J. W. Gilman, *Appl. Clay Sci.*, 1999, **15**, 31.
9. O. Breuer and U. Sundararaj, *Polym. Compos.*, 2004, **25**, 630.
10. R. Blake, Y. K. Gunko, J. Coleman, M. Cadek, A. Fonseca, J. B. Nagy and W. J. Blau, *J. Am. Chem. Soc.*, 2004, **126**, 10226.
11. Y. Fukushima, A. Okada, M. Kawasumi, T. Kurauchi and O. Kamigaito, *Clay Miner.*, 1998, **23**, 27.
12. S. Zulfiqar, Z. Ahmad and M. I. Sarwar, *Polym. Adv. Technol.*, 2008, **19**, 1720.
13. S. S. Ray and M. Okamoto, *Prog. Polym. Sci.*, 2003, **28**, 1539.
14. S. Pavlidou and C. D. Papaspyrides, *Prog. Polym. Sci.*, 2008, **33**, 1119.
15. H. Akat, M. A. Tasdelen, F. D. Prez and Y. Yagci, *Eur. Polym. J.*, 2008, **44**, 1949.
16. N. Sheng, M. C. Boyce, D. M. Parks, G. C. Rutledge, J. I. Abes and R. E. Cohen, *Polymer*, 2007, **45**, 487.
17. C. E. Powell and G. W. Beall, *Curr. Opin. Solid State. Mater. Sci.*, 2006, **10**, 73.
18. J.-J. Luo and I. M. Daniel, *Compos. Sci. Technol.*, 2003, **63**, 1607.
19. T. Lan and T. J. Pinnavaia, *Chem. Mater.*, 1994, **6**, 2216.
20. D. R. Robello, N. Yamaguchi, T. Blanton and C. Barnes, *J. Am. Chem. Soc.*, 2004, **126**, 8118.
21. W. J. Bae, K. H. Kim, W. H. Jo and Y. H. Park, *Macromolecules*, 2004, **37**, 9850.
22. Y. Q. Rao and J. M. Pochan, *Macromolecules*, 2007, **40**, 290.
23. E. P. GianneEs, *Adv. Mater.*, 1996, **8**, 29.
24. B. Chen, J. R. G. Evans, H. C. Greenwell, P. Boulet, P. V. Coveney, A. A. Bowdenf and A. Whiting, *Chem. Soc. Rev.*, 2008, **37**, 568.
25. R. Sengupta, S. Chakraborty, S. Bandyopadhyay, S. Dasgupta, R. Mukhopadhyay, K. Auddy and A. S. Deuri, *Polym. Eng. Sci.*, 2007, **47**, 1956.
26. M. Zanetti, S. Lomakina and G. Camino, *Macromol. Mater. Eng.*, 2000, **279**, 1.
27. B. Chen and J. R. G. Evans, *Carbohydr. Polym.*, 2005, **61**, 455.
28. P. Aranda and E. Ruiz-Hitzky, *Chem. Mater.*, 1992, **4**, 1395.
29. E. Ruiz-Hitzky and P. Aranda, *Adv. Mater.*, 1990, **2**, 545.
30. R. A. Vaia, H. Ishii and E. P. Giannelis, *Chem. Mater.*, 1993, **5**, 1694.
31. W. Krawiec, L. G. Scanlon, J. P. Fellner, R. A. Vaia, S. Vasudevan and E. P. Giannelis, *J. Power Sources*, 1995, **54**, 310.
32. Q. Zhang, Q. Zax, L. Jiang and Y. Lei, *Polym. Int.*, 2000, **49**, 1561.

33. N. Furuichi, Y. Kurokawa, K. Fujita, A. Oya, H. Yasuda and M. Kiso, *J. Mater. Sci.*, 1996, **31**, 4307.
34. Z. Shen, G. P. Simon and Y.-B. Cheng, *Polymer*, 2002, **43**, 4251.
35. J. W. Cho and D. R. Paul, *Polymer*, 2001, **42**, 1083.
36. P. Arada and E. Ruiz-Hitzky, *Adv. Mater.*, 1990, **2**, 545.
37. H. R. Fischer, L. H. Gielgens and T. P. M. Koster, *Acta Polym.*, 1999, **50**, 122.
38. J. J. Tunney and C. Detellier, *Chem. Mater.*, 1996, **8**, 927.
39. A. Usuki, Y. Kojima, M. Kawasumi, A. Okada, Y. Fukushima, T. Kurauchi and O. Kamigaito, *J. Mater. Res.*, 1993, **8**, 1185.
40. A. Okada and A. Usuki, *Macromol. Mater. Eng.*, 2006, **291**, 1449.
41. B. Chen, A. A. Bowden, H. C. Greenwell, P. Boulet, P. V. Coveney, A. Whiting and J. R. G. Evans, *J. Polym. Sci., Part B: Polym. Phys.*, 2005, **43**, 1785.
42. T. Lan, P. D. Kaviratna and T. J. Pinnavaia, *J. Phys. Chem. Solids*, 1996, **57**, 1005.
43. L. Biasci, M. Aglietto, G. Ruggeri and F. Ciardelli, *Polymer*, 1994, **35**, 3296.
44. P. B. Messersmith and E. P. Giannelis, *J. Polym. Sci., Part A: Polym. Chem.*, 1995, **33**, 1047.
45. L. Biasci, M. Aglietto, G. Ruggeri and A. D. Alessio, *Polym. Adv. Technol.*, 1995, **6**, 662.
46. P. B. Messersmith and E. P. Giannelis, *Chem. Mater.*, 1993, **5**, 1064.
47. K. Yano, A. Usuki, A. Okada, T. Kurauchi and O. Kamigaito, *J. Polym. Sci., Part A: Polym. Chem.*, 1993, **31**, 2493.
48. Y. Kojima, A. Usuki, M. Kawasumi, A. Okada, Y. Fukushima, T. Kurauchi and O. Kamigaito, *J. Mater. Res.*, 1993, **8**, 1185.
49. C.-H. Wang, Y.-T. Shieh and S. Nutt, *J. Appl. Polym. Sci.*, 2009, **114**, 1025.
50. M. Lai and J.-K. Kim, *Polymer*, 2005, **46**, 4722.
51. Y. Di, S. Iannace, E. Dimaio and L. Nicolais, *J. Polym. Sci., Part B: Polym. Phys.*, 2003, **41**, 670.
52. A. Rehab, A. Akelah, T. Agag and N. Shalaby, *Polym. Adv. Technol.*, 2007, **18**, 463.
53. H.-L. Tyan, Y.-C. Liu and K.-H. Wei, *Chem. Mater.*, 1999, **11**, 1942.
54. P. Govindaiah, S. R. Mallikarjuna and C. Ramesh, *Macromolecules*, 2006, **39**, 7199.
55. B. N. Jang, D. Wang and C. A. Wilkie, *Macromolecules*, 2005, **38**, 6533.
56. J.-M. Yeh, S.-J. Liou, M.-C. Lai, Y.-W. Chang, C.-Y. Huang, C.-P. Chen, J.-H. Jaw, T.-Y. Tsai and Y.-H. Yu, *J. Appl. Polym. Sci.*, 2004, **94**, 1936.
57. F.-A. Zhang, L. Chen and J.-Q. Ma, *Polym. Adv. Technol.*, 2009, **20**, 589.
58. M. Pan, X. Shi, X. Li, H. Hu and L. Zhang, *J. Appl. Polym. Sci.*, 2004, **94**, 277.
59. J. D. Sudha, V. L. Reena and C. Pavithran, *J. Polym. Sci., Part B: Polym. Phys.*, 2007, **45**, 2664.
60. D. C. Lee and L. W. Jang, *J. Appl. Polym. Sci.*, 1996, **61**, 1117.

61. M. W. Noh and D. C. Lee, *Polym. Bull.*, 1999, **42**, 619.
62. J. W. Kim, F. Liu, H. J. Choi, S. H. Hong and J. Joo, *Polymer*, 2003, **44**, 289.
63. M. Xu, Y. S. Choi, Y. K. Kim, K. H. Wang and I. J. Chung, *Polymer*, 2003, **44**, 6387.
64. Y. K. Kim, Y. S. Choi, K. H. Wang and I. J. Chung, *Chem. Mater.*, 2002, **14**, 4990.
65. H.-Q. Xie, Y.-M. Ma and J.-S. Guo, *Polymer*, 1998, **40**, 261.
66. D. Wang, J. Zhu, Q. Yao and C. A. Wilkie, *Chem. Mater.*, 2002, **14**, 3837.
67. H. Zou, S. Wu and J. Shen, *Chem. Rev.*, 2008, **108**, 3893.
68. T. Yu, J. Lin, J. Xu and W. Ding, *J. Polym. Sci., Part B: Polym. Phys.*, 2005, **43**, 3127.
69. C. L. Wu, M. Q. Zhang, M. Z. Rong and K. Friedrich, *Compos. Sci. Technol.*, 2005, **65**, 635.
70. M. Z. Rong, M. Q. Zhang, Y. X. Zheng, H. M. Zeng, R. Walter and K. Friedrich, *Polymer*, 2001, **42**, 167.
71. D. N. Bikiaris, G. Z. Papageorgiou, E. Pavlidou, N. Vouroutzis, P. Palatzoglou and G. P. Karayannidis, *J. Appl. Polym. Sci.*, 2006, **100**, 2684.
72. T. C. Merkel, B. D. Freeman, R. J. Spontak, Z. He, I. Pinnau, P. Meakin and A. J. Hill, *Science*, 2002, **296**, 519.
73. S. D. Kelman, S. Matteucci, C. W. Bielawski and B. D. Freeman, *Polymer*, 2007, **48**, 6881.
74. P. Hajji, L. David, J. F. Gerard, J. P. Pascault and G. Vigier, *J. Polym. Sci., Part B: Polym. Phys.*, 1999, **37**, 3172.
75. M. Fujiwara, K. Kojima, Y. Tanaka and R. Nomura, *J. Mater. Chem.*, 2004, **14**, 1195.
76. P. A. Charpentier, W. Z. Xu and X. S. Li, *Green Chem.*, 2007, **9**, 768.
77. T. Ogoshi, H. Itoh, K. M. Kim and Y. Chujo, *Macromolecules*, 2002, **35**, 334.
78. A. A. Kumar, K. Adachi and Y. Chujo, *J. Polym. Sci., Part A: Polym. Chem.*, 2004, **42**, 785.
79. F. Yang, Y. C. Ou and Z. Z. Yu, *J. Appl. Polym. Sci.*, 1998, **69**, 355.
80. Y. Li, J. Yu and Z. X. Guo, *J. Appl. Polym. Sci.*, 2002, **84**, 827.
81. W. T. Liu, X. Y. Tian, P. Cui, Y. Li, K. Zheng and Y. Yang, *J. Appl. Polym. Sci.*, 2004, **91**, 1229.
82. T. Kashiwagi, A. B. Morgan, J. M. Antonucci, M. R. VanLandingham, R. H. Harris, W. H. Awad and J. R. Shields, *J. Appl. Polym. Sci.*, 2003, **89**, 2072.
83. Y. Y. Yu, C. Y. Chen and W. C. Chen, *Polymer*, 2003, **44**, 593.
84. D.-M. Qi, Y.-Z. Bao, Z.-M. Huang and Z.-X. Weng, *J. Appl. Polym. Sci.*, 2006, **99**, 3425.
85. S. Reculusa, C. Poncet-Legrand, S. Ravaine, C. Mingotaud, E. Duguet and E. Bourgeat-Lami, *Chem. Mater.*, 2002, **14**, 2354.

86. R. Palkovits, H. Althues, A. Rumplecker, B. Tesche, A. Dreier, U. Holle, G. Fink, C. H. Cheng, D. F. Shantz and S. Kaskel, *Langmuir*, 2005, **21**, 6048.
87. V. Monteil, J. Stumbaum, R. Thomann and S. Mecking, *Macromolecules*, 2006, **39**, 2056.
88. L. Yao, T. Yang and S. Cheng, *J. Appl. Polym. Sci.*, 2010, **115**, 3500.
89. T. Yu, J. Lin, J. Xu, T. Chen and S. Lin, *Polymer*, 2005, **46**, 5695.
90. T. Yu, J. Lin, J. Xu, T. Chen, S. Lin and X. Tian, *Compos. Sci. Technol.*, 2007, **67**, 3219.
91. S. W. Zhang, S. X. Zhou, Y. M. Weng and L. M. Wu, *Langmuir*, 2005, **21**, 2124.
92. A. R. Mahdavian, M. Ashjari and A. B. Makoo, *Eur. Polym. J.*, 2007, **43**, 336.
93. Y. Zhong, Z. Zhu and S.-Q. Wang, *Polymer,* 2005, **46**, 3006.
94. Z. Sedlakova, J. Plestil, J. Baldrian, M. Slouf and P. Holub, *Polym. Bull.*, 2009, **63**, 365.
95. X. Kornmann, H. Lindberg and L. A. Berglund, *Polymer*, 2001, **42**, 4493.
96. Y. Xi, Z. Ding, H. He and R. L. Frost, *J. Colloid Interface Sci.*, 2004, **277**, 116.
97. J. W. Gilman, C. L. Jackson, A. B. Morgan and R. Harris Jr, *Chem. Mater.*, 2000, **12**, 1866.
98. X. Fu and S. Qutubuddin, *Polymer*, 2001, **42**, 807.
99. X. Tong, H. Zhao, T. Tang, Z. Feng and B. Huang, *J. Polym. Sci., Part A: Polym. Chem.*, 2002, **40**, 1706.
100. X. Huang and W. J. Brittain, *Macromolecules*, 2001, **34**, 3255.
101. C. Wang, Q. Wang and X. Chen, *Macromol. Mater. Eng.*, 2005, **290**, 920.
102. X. Ding, J. Zhao, Y. Liu, H. Zhang and Z. Wang, *Mater. Lett.*, 2004, **58**, 3126.
103. Y. K. Kim, Y. S. Choi, K. H. Wang and I. J. Chung, *Chem. Mater.*, 2002, **14**, 4990.
104. Y. Haldorai, P. Q. Long, S. K. Noh, W. S. Lyoo and J.-J. Shim, *Polym. Adv. Technol.*, 2010, DOI: 10.1002/pat.1577.
105. Y. S. Choi, K. H. Wang, M. Xu and I. J. Chung, *Chem. Mater.*, 2002, **14**, 2936.

CHAPTER 12

Polymer–Clay Nanocomposites Prepared in Miniemulsion Using the RAFT Process

EDDSON ZENGENI,[a] AUSTIN SAMAKANDE[b] AND
PATRICE C. HARTMANN[a, b]

[a] University of Stellenbosch, 7602 Matieland, South Africa; [b] Mondi
Packaging South Africa, Research and Development Centre, 7599
Stellenbosch, South Africa

12.1 Introduction

This chapter presents a review of polymer–clay nanocomposites (PCNs) prepared in miniemulsion using the reversible addition–fragmentation chain transfer (RAFT) process.

One of the most interesting research areas in nanotechnology is the inclusion of nanoparticles in polymers in order to enhance their physical properties, *e.g.* thermal stability, barrier properties and mechanical properties. In the early 1990s, the Toyota research group showed that the hydrated cations within the clay layers can be replaced by alkylammonium compounds.[1] A reactive clay modifier leads to strong interfacial adhesion between the clay and the polymer and subsequently to exceptional mechanical properties during load bearing. The use of tailor-made transfer agents in free-radical polymerization reactions allows one to achieve control of the polymerization process.[2] This results in polymers with low polydispersity indices (PDIs) and predictable molar masses. The discovery of controlled polymerization techniques, and in particular RAFT agents, was a milestone achievement.[3] RAFT polymerization now allows the preparation of polymer architectures that were never before envisaged to be possible.

RSC Nanoscience & Nanotechnology No. 16
Polymer Nanocomposites by Emulsion and Suspension Polymerization
Edited by Vikas Mittal
© Royal Society of Chemistry 2011
Published by the Royal Society of Chemistry, www.rsc.org

Thus, a combination of RAFT technology and clay nanotechnology for the synthesis of PCNs by RAFT-mediated polymerization can allow the preparation of tailor-made materials with specific properties for niche applications.

On the other hand, miniemulsion polymerization offers many advantages over other polymerization methods: it is environmentally friendly, the system has a high heat transfer capacity, which prevents local heat build-up, high conversions and high molar masses are achieved, high solids content can be attained, high rates of polymerization are achieved, due to compartmentalization, the viscosity remains low, inorganic particles can be incorporated into the system, it is compatible with highly hydrophobic monomers and it is easy to predict the kinetic parameters.[4,5] The many advantages of miniemulsion polymerization over other polymerization methods make it attractive to combine it with controlled polymerization techniques.[6] Hence it is not surprising that the RAFT technique has been applied to miniemulsions.

Due to the compatibility of the miniemulsion process with hydrophobic species, it can be used for the synthesis of PCNs, using modified clays, which will result in the clay particles being encapsulated inside the polymer particles.[7]

Clay particles are layered materials that occur naturally and synthetically, in good purity. The surface of each clay platelet is negatively charged and hydrated alkali or alkaline earth metals counterbalance the negative charges. These counter charge-balancing cations reside between the clay platelets, in a region commonly referred to as the intergallery region. The distance between adjacent clay platelets is known as the interlayer distance or the *d*-spacing. The weak van der Waals forces that hold the clay platelets together allow adjacent clay platelets to move away from each other to accommodate low molecular weight chemical species in between the clay galleries. This results in variations in the *d*-spacing, depending on the size of the species that is present in between the clay layers. The presence of inorganic hydrated cations in between the clay platelets makes the clays, in their natural state, compatible with only hydrophilic and/or polar species. However, in order to promote the compatibility of clay with hydrophobic species (polymers), the hydrated cations can also be replaced by organic cations, thus making the clay hydrophobic and hence compatible with hydrophobic species. The organic cations that can replace the hydrated alkali or alkaline earth metals include conventional cationic surfactants and/or functional cationic surfactants.[8–10]

12.2 Background to Controlled/Living Polymerization Using the RAFT Process

The RAFT process is one of the fastest growing and most robust controlled radical polymerization processes.[11,12] This is due to its versatility, such as tolerance to impurities and the numerous types of monomers that can be polymerized in a controlled manner. Furthermore, the experimental setup required for RAFT-controlled radical polymerization is relatively simple. The RAFT technique owes its success to a family of organic molecules containing the

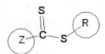

Scheme 12.1   General structure of a RAFT agent.

thio-carbonyl-thio group, known as RAFT agents.[13–15] The general structure of a RAFT agent is shown in Scheme 12.1.

The control of polymerization in RAFT is significantly influenced by the nature of both the leaving group (R) and the stabilizing group (Z), which makes their choice, to match the selected monomer, very important.[16,17] The Z group acts as an activator for the thio-carbonyl-thio group to react with radicals. It also acts as a stabilizing group for the transition-state radical that is formed between the thiocarbonyl group and a free radical. On the other hand, the R group should be a good leaving group by undergoing a homolytic scission: once it leaves the main RAFT agent as a free radical, it should be able to reinitiate polymerization by reacting with monomer units.

In the RAFT process, addition of a transfer agent (RAFT agent) in larger amounts relative to the initiator results in control of the polymerization process. The RAFT agent transfers the radical from a growing chain to itself to form the dormant form, which results in equilibrium between growing and dormant chains, with the equilibrium far to the dormant state. There are thus few active chains (radicals) and hence control is achieved. In RAFT polymerization, the product of chain transfer is also a chain transfer agent with similar activity to the precursor transfer agent. The generally accepted RAFT mechanism is outlined in Scheme 12.2.[17] As is the case with other controlled polymerization techniques, the RAFT process also undergoes termination reactions, either by radical coupling or disproportionation, in addition to other unwanted side reactions. The thio-carbonyl-thio group can be hydrolyzed by basic species such as hydroxide ions and primary and secondary amines.[2,18–21] The RAFT process also suffers from inhibition and retardation. Radical termination and side reactions have been reported to cause broadening of the molecular mass distribution (MMD).

The theoretical polymer molar masses in RAFT-mediated polymerizations are calculated using the following equation:[22]

$$M_n = \frac{[M]_0 M W_M x}{[RAFT]_0} + M W_{RAFT} \qquad (12.1)$$

where $[M]_0$ is the initial monomer concentration, MW_M the molecular mass of monomer, x the conversion, MW_{RAFT} the molecular mass of RAFT and $[RAFT]_0$ the initial concentration of RAFT. These theoretical molar masses are then compared with those obtained from size-exclusion chromatography (SEC), from where the level of control of the polymerization is determined.

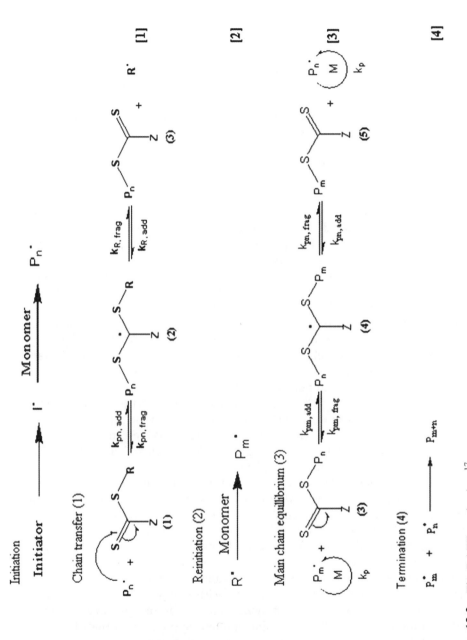

Scheme 12.2 The RAFT mechanism.[17]

12.3   Controlled Polymerization Techniques and Their Applicability to PCNs

To date, most of the research carried out into the polymer architecture in PCNs has been focused mainly on the use of uncontrolled radical polymerization.[9,23–26] The few articles available on the use of controlled living radical polymerization in PCNs focus on nitroxide-mediated polymerization (NMP).[27,28] Clay in its native state is hydrophilic and hence incompatible with hydrophobic monomers. Consequently, a prerequisite in preparing most PCNs is to modify the clay surface in order to make it compatible with the monomer–polymer system involved. This is generally done by ion exchange, using positively charged organic species such as quaternary ammonium surfactants.[9] In NMP, the control of polymerization is based on a nitrogen-containing species that can easily be quaternized and anchored on clay surfaces, resulting in modified clay, which can subsequently be used in a controlled free radical polymerization reaction. This leads to nanocomposites with both the polymer and clay architectures being controlled. This idea was pioneered by Weimer *et al.*,[26] who synthesized polystyrene–clay nanocomposites (PS–CNs) with controlled molar masses by first anchoring a tetramethylpiperidine oxyl (TEMPO) derivative on to clay layers prior to proceeding with the free radical polymerization of styrene. Di and Sogah[27] prepared nanocomposites based on this idea, with minor changes in the nitroxide mediating species. They also successfully synthesized block copolymers by the same method.

Other types of controlled living polymerization methods, such as ionic polymerization,[29] atom transfer radical polymerization (ATRP),[30,31] borane chemistry[32] and RAFT techniques,[33–37] have not been widely used in the preparation of PCNs. There are several reasons for this, particularly the techniques' susceptibility to impurities, *e.g.* the propagating species in ionic polymerization is destroyed by even small amounts of water (small amounts of water are always present in clays).[35] ATRP suffers from the presence of redox species,[36] the presence of which in clays is well known, as these might lead to the undesired reduction of the ionic copper species to zerovalent copper. This technique also suffers from a limited number of monomers that can be successfully synthesized under controlled conditions, and also the presence of toxic copper in the final product.[17] Although very little is known about controlling polymerization using borane chemistry, Yang *et al.*[32] successfully used this technique to synthesize PS–CNs with controlled molar masses.

Use of the RAFT technique has evolved from the use of a free RAFT agent (*i.e.* the RAFT agent is not attached to the clay layers) to the use of RAFT agents anchored on the clay platelets. Moad and co-workers[36,38,39] used RAFT-mediated polar polymers that are miscible with polypropylene (PP) and then melt blended both polymers (*i.e.* polymers and PP) in the presence of sodium montmorillonite to yield PP–CNs. Salem and Shipp[35] showed that the use of free RAFT agents to the prepare polystyrene–, poly(methyl methacrylate)– and poly(butyl acrylate)–clay nanocomposites by *in situ* intercalative polymerization in bulk led to products with well-controlled molecular weights and narrow polydispersity index values of less than 1.5 for all polymer samples studied. The resultant PCNs showed varying morphological structures from intercalated to exfoliated structures depending on the monomer system used. The use of anchored RAFT agents has also been reported by Zhang *et al.*[40] They anchored a cationic RAFT agent, 10-carboxylic acid-10-dithiobenzoate-decyltrimethyl-ammonium bromide (CDDA), on to clay and then used the modified clay for the controlled solution-based *in situ* intercalative polymerization of styrene and obtained exfoliated PCNs.

Ding *et al.*[33] anchored a negatively charged RAFT agent, 4-cyanopentanoic acid dithiobenzoate (CAD), on to a layered double hydroxide (LDH) (*i.e.* a positively charged, layered, clay-like material) and then used the modified LDH in the *in situ* intercalative polymerization of styrene to obtain LDH nanocomposites. The nanocomposites were either exfoliated or intercalated based on the LDH loading. It was shown that the RAFT agent was able control the molecular weight and PDI values of less than 1.5 were obtained.

Di and Sogah[34] used a dithiocarbamate-based modified clay for the synthesis of PS–CNs. However, dithiocarbamates are well known for their poor ability to control polymerization, except for the polymerization of specific monomer systems. The poor control is due to the generated dithiocarbamyl radicals undergoing several side reactions.[17] Positively charged RAFT agents can be anchored on to clay layers, thus allowing the synthesis of controlled RAFT-mediated PCNs, with the polymer growth taking place from the clay surfaces, as has also been reported by Di and Sogah[34] and Samakande *et al.*[41,42] Samakande *et al.* synthesized positively charged dithiocarbonate and trithiocarbonate RAFT agents,[19] which they anchored on to the clay platelets before using miniemulsion polymerization to control both the clay morphology and the polymer architecture.[41,42]

12.4 Preparation and Characterization of RAFT-mediated PCNs in Miniemulsion

Although PCNs have been extensively studied since the early 1990s, the use of RAFT in miniemulsion polymerization to prepare PCNs with controlled morphology and architecture is a relatively new approach. Samakande *et al.*[41,42] prepared the RAFT agents *N,N*-dimethyl-*N*-(4-{[[(phenylcarbonothionyl)thio]methyl}benzyl)ethanammonium bromide (PCDBAB) and *N*-4-[({[(dodecylthio)carbonothioyl]thio}methyl)benzyl]-*N,N* dimethylethanammonium bromide (DCTBAB) to modify montmorillonite (MMT) clay and the RAFT–MMT was used to prepare PS–RAFT–MMT and PS-*co*-BA–RAFT–MMT nanocomposites in miniemulsion polymerization using sodium dodecyl sulfate (SDS) as surfactant, hexadecane as cosurfactant and azobisisobutyronitrile (AIBN) as initiator. Various clay compositions, in the range 1–5% MMT, were used in the preparation of the nanocomposites. The resultant PS–PCDBAB–MMT and PS–DCTBAB–MMT, PS-*co*-BA–PCDBAB–MMT

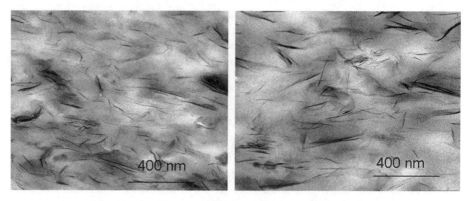

Figure 12.1   TEM images of PS–PCDBAB–3.6% MMT clay nanocomposite (left) and PS–DCTBAB–3.6% MMT clay nanocomposite (right). Reproduced from reference 41 with permission from Wiley.

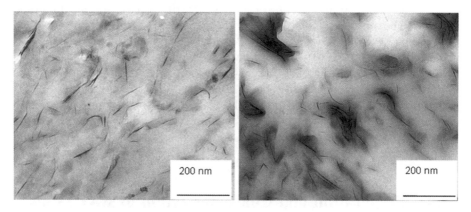

Figure 12.2   TEM images of PS-*co*-BA–DCTBAB–1% MMT (left) and PS-*co*-BA–PCDBAB–1% MMT (right). Reproduced from reference 42 with permission from Elsevier.

and PS-*co*-BA–DCTBAB–MMT nanocomposites exhibited intercalated morphology as confirmed by transmission electron microscopy (TEM) of microtomed films (Figures 12.1[41] and 12.2[42]) and small-angle X-ray scattering (SAXS) results (Figures 12.3[41] and 12.4[42]).

The SAXS pattern of the RAFT-mediated PCNs exhibited broad peaks at low clay loadings, which is an indication of partial exfoliation. This was in agreement with the TEM results. At higher clay loadings (*i.e.* 5% clay loading) more defined peaks, although still broad, which are characteristic of inter-calated morphology, were observed.

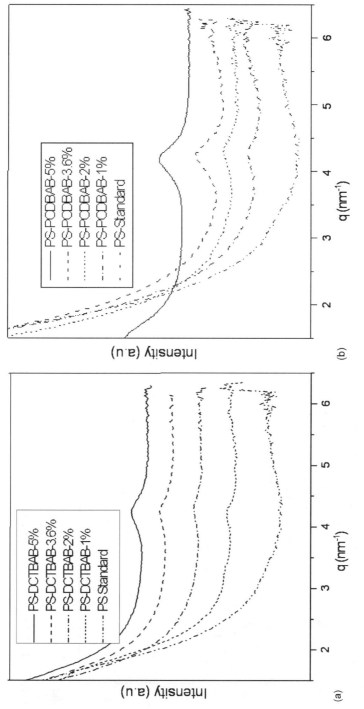

Figure 12.3 SAXS patterns of PS–CNs. Reproduced from reference 41 with permission from Wiley.

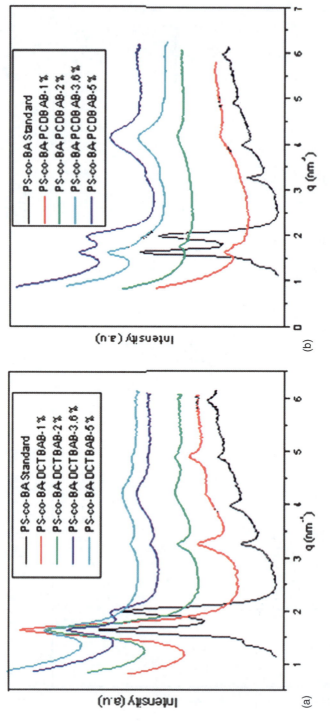

Figure 12.4 SAXS patterns of PS-*co*-BA–PCNs. Reproduced from reference 42 with permission from Elsevier.

12.5 Physical Properties of RAFT-mediated Polymer–Clay Nanocomposites

12.5.1 Thermomechanical Properties

The dynamic mechanical properties of dried PS–PCDBAB–MMT and PS–DCTBAB–MMT samples showed that all the nanocomposites had enhanced storage moduli in the glassy state relative to the neat polymer standard (see Figure 12.5[41]).

The increase in storage modulus is caused by the high aspect ratio of the dispersed clay platelets and the interaction between polymer chains and clay layers. This results in a decrease in the polymer segments' mobility near the polymer/clay interface.[43] However, within the two series of nanocomposites, the storage modulus was seen to decrease with increase in clay loading, except for the DCTBAB–2% clay loading, which had a slightly higher storage modulus in the glassy state than for the DCTBAB–1% clay loading.

These results differ from those reported for PCNs prepared by uncontrolled radical polymerizations, which yield high molar mass polymers and hence there is an increase in storage modulus with increase in clay loading.[44–47] The decrease in storage modulus with increase in clay content was attributed to (i) a decrease in the molar mass and (ii) a change in the nanocomposite morphology from semi-exfoliated to intercalated, as the clay loading increases,[48] for the PCNs prepared using RAFT agents in miniemulsion.[42] Figure 12.5 also shows that at 3.6 and 5% clay loadings the transition of the storage modulus from the glassy state to the rubbery state is relatively broad. This is attributed to the presence of nanoclay within the polymer matrix.[49]

The T_g values (taken from the peak of loss modulus) of PS–PCDBAB–MMT and PS–DCTBAB–MMT nanocomposites increased with increase in clay loading, regardless of the fact that the molar mass was decreasing (see Figure 12.6).[41]

A similar effect has been reported of other PCNs: Tyan *et al.*[44] reported an increase in T_g of a polyimide 3,3',4,4'-benzophenonetetracarboxylic acid dianhydride-4-4'-oxydianiline (BTDA-ODA)–clay nanocomposite with an increased clay content. Furthermore, Kim *et al.*[50] reported an increase in T_g of intercalated PS–organophilic MMT clay, prepared *via* the solvent casting method, with increasing clay content. Moraes *et al.*[7] reported that clay addition did not alter the T_g of their polystyrene-*co*-butyl acrylate nanocomposites prepared *via* miniemulsion. In contrast, Fu and Qutubuddin[47] reported a decrease in T_g with increase in clay loading for exfoliated PS–clay nanocomposites, which they attributed to a decrease in molar mass. Figure 12.7 shows the variation of tanδ with temperature of PS–DCTBAB and PS–PCDBAB clay nanocomposites prepared using the RAFT process, in miniemulsion.[41] The tanδ peak is associated with partial loosening of the polymer structure, so that small groups and chain segments can move. In the phase transition zone, tanδ measures imperfections in the elasticity of a material.

Peak broadening and a shift of the tanδ peaks of the nanocomposites to higher temperatures relative to neat polystyrene were observed. These shifts are

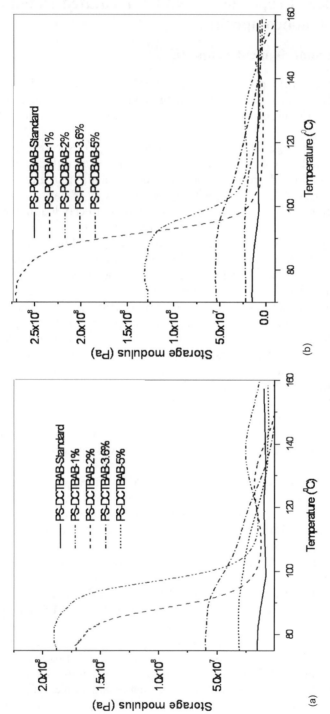

Figure 12.5 Storage modulus, as a function of temperature, of PS–DCTBAB–clay nanocomposites (left) and PS–PCDBAB–clay nano-composites (right), at 1, 2, 3.6 and 5% clay loadings. Reproduced from reference 41 with permission from Wiley.

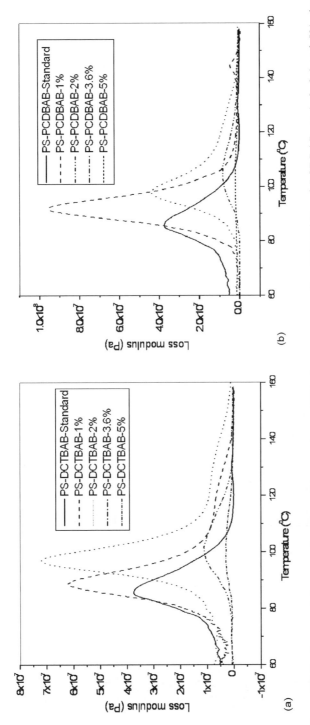

Figure 12.6 Loss modulus, as a function of temperature, of PS–clay nanocomposites, DCTBAB and PCDBAB, at 1, 2, 3.6 and 5% clay loadings. Reproduced from reference 41 with permission from Wiley.

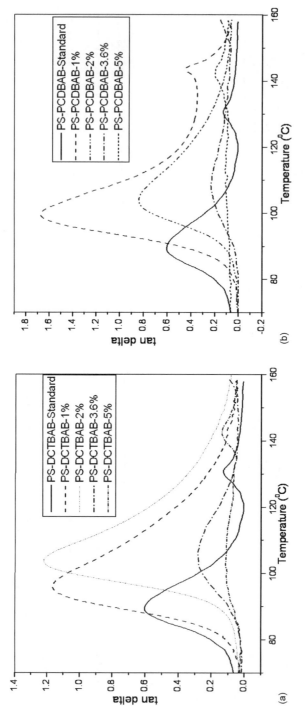

Figure 12.7 Tanδ as a function of temperature for PS–clay nanocomposites DCTBAB and PCDBAB, at 1, 2, 3.6 and 5% clay loadings. Reproduced from reference 41 with permission from Wiley.

a result of the restricted chain mobility brought about by the clay filler.[51,52] The peak broadening effects are more pronounced for low clay loadings, *i.e.* 1 and 2%. The level of peak broadening is directly attributed to the different levels of exfoliation and the higher PDI occurring at low clay loadings.

12.5.2 Thermal Stability

Figure 12.8 shows the TGA results for PS–DCTBAB–MMT and PS–PCDBAB–MMT nanocomposites[41] and Figure 12.9 shows those for PS-*co*-BA–PCDBAB–MMT and PS-*co*-BA–DCTBAB–MMT.[42] Only a slight improvement in the thermal stability of PCNs was observed above 50% degradation, relative to the neat polystyrene (see Figure 12.8). Jan *et al.*[53] also reported that epoxy–clay nanocomposites only showed enhanced thermal stability from 40 to 50% weight degradation. The thermal stability of PS–CNs was also found to increase slightly when the clay loading increased. This has been a characteristic feature of different PCNs, irrespective of their preparation route.[9,47,48,50,54–57] The formation of clay char, which acts as a mass transport barrier and insulator between the polymer and the superficial zone where the polymer decomposition takes place, is the cause of the improvements in the thermal stability of PCNs.[9,48,50,58] Concurrently, the restricted thermal motion of the polymer chains localized in the clay galleries can also bring about these thermal stability improvements.[9,59]

The labile thio-carbonyl-thio moiety is also believed to play a role in the thermal stability of the PCNs. PCNs made using DCTBAB were more stable, especially in the 200–300 °C region where degradation starts to occur, than PCNs made using PCDBAB. This temperature range is similar to that used by Postma *et al.*[60] for the removal of the thio-carbonyl-thio group from polymers by thermolysis.

On the other hand, PS-*co*-BA–PCDBAB–MMT and PS-*co*-BA–DCTBAB–MMT did not show any improvement in thermal stability.[43] This was attributed to the presence of low molar mass oligomers in the nanocomposites. The PDI values of the nanocomposites are high and these small oligomers could be accelerating further decomposition of the PCNs.[51,61–63]

12.5.3 Rheological Properties

Melt rheological properties of the PCNs are important in the consideration of their possible processing. Figures 12.10 and 12.11 show typical isothermal frequency sweeps for PCNs prepared by miniemulsion polymerization, using the RAFT process.[42]

The DCTBAB- and PCDBAB-mediated P(S-*co*-BA) polymer standards without clay showed typical Newtonian behavior.[64] The complex viscosity (η^*) increased linearly with decrease in the angular frequency up to about $10 \, \mathrm{rad \, s^{-1}}$, whereas at lower frequencies η^* tends to become independent of the angular frequency. This phenomenon, referred to as Newtonian behavior, is typical of

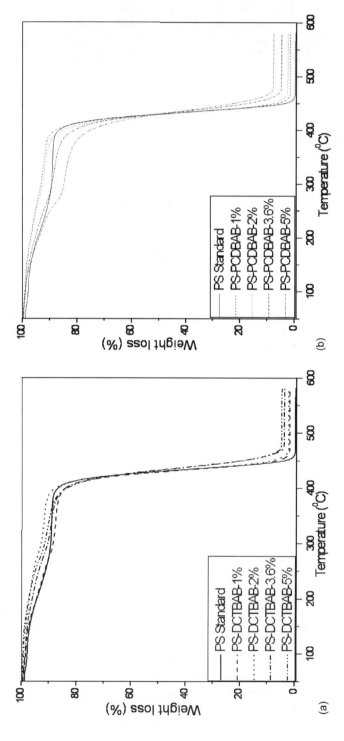

Figure 12.8 TGA thermograms of PS–DCTBAB–clay nanocomposites (left) and PS–PCDBAB–clay nanocomposites (right). Reproduced from reference 41 with permission from Wiley.

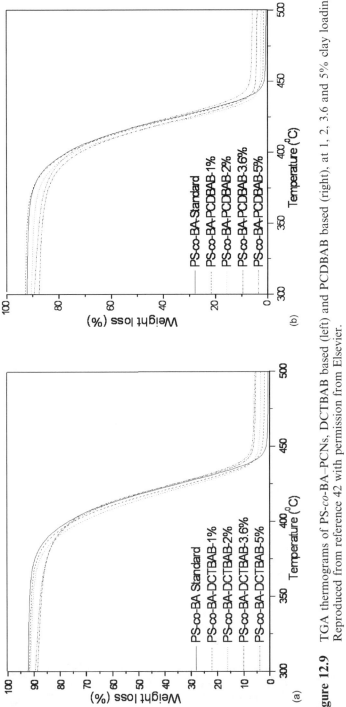

Figure 12.9 TGA thermograms of PS-*co*-BA–PCNs, DCTBAB based (left) and PCDBAB based (right), at 1, 2, 3.6 and 5% clay loadings. Reproduced from reference 42 with permission from Elsevier.

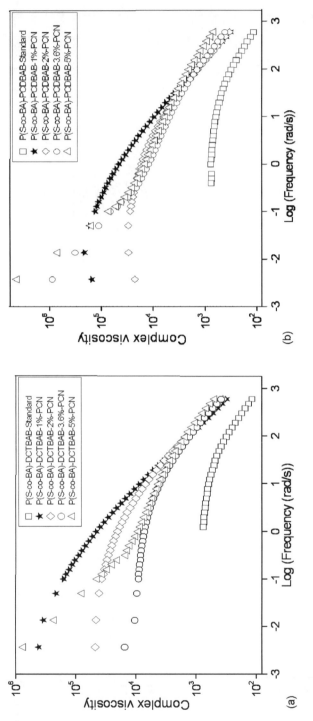

Figure 12.10 Complex viscosity as a function of angular frequency for P(S-*co*-BA)–DCTBAB-PCNs (left) and P(S-*co*-BA)–PCDBAB-PCNs (right), at 0% (□), 1% (★), 2% (◇), 3.6% (○) and 5% (△) clay loadings. Reproduced from reference 42 with permission from Elsevier.

unfilled polymers.[65,66] The value of the onset of the complex viscosity (high-frequency region) was found to increase monotonically with increase in clay loading. The complex viscosity as a function of frequency for 1 and 2% clay loadings was found to be dominated by the molar mass and showed little dependence on the clay loading and morphology. The molecular weight was found to decrease with increase in clay loading.

Here the increase in complex viscosity observed as the angular frequency decreased was more pronounced for the PCNs with 1% clay loading and which ultimately had higher molar mass than the PCNs with 2, 3.6 and 5% clay loadings. The viscosities of the samples with 1% clay loading became independent of angular frequency for values below 0.01 rad s^{-1}, whereas for samples with 2% clay loading this effect was observed for angular frequencies below 0.1 rad s^{-1}, which indicated a dependence of the viscosity on molar mass and PCN morphology. PCNs with clay content greater than 2% showed non-Newtonian behavior,[65,66] *i.e.* the viscosity increased with decrease in angular frequency.

The variation of the moduli G^{I} and G^{II} of the PS-*co*-BA with angular frequency followed a similar pattern to the complex viscosity: the onset (high-frequency region) of both storage and loss moduli increases monotonically with clay loading. This is typical of PCNs, irrespective of the preparation method, and is a result of the enhanced stiffness brought about by the incorporated clay. [50,64,67–70]

However, storage and loss moduli of the nanocomposites at 1 and 2% clay loadings were dominated by the molar mass of the polymer matrix as they did not show the solid–liquid character typical of PCNs (see Figure 12.11).[65,67,68,71] The fact that PCNs with 1 and 2% clay loadings showed G^{I} plots with plateau-like features whereas those with 3.6–5% clay loadings did not show such a plateau is attributed to the dependence of rheological properties on the molar mass of the matrix.[70,72]

In low clay loading PCNs, the rheological properties are dominated by the polymer matrix and PDI values rather than the clay, since a sharp decrease in both loss and storage moduli was observed with decreasing angular frequency, which is characteristic of molecular movement. At high clay loadings, pseudo-non-terminal, pseudo-solid–liquid-like behavior is observed,[7,64–68,71,73–75] meaning that the dependence of the storage and loss moduli on angular frequency diminishes and becomes more dependent on the clay loading.[68,74] This effect is a result of the increased frictional interactions of the clay particles, and also the strong polymer–clay interactions.[64,71] The pseudo-solid–liquid-like behavior results at clay loadings higher than 3% agree well with those of Krishnamoorti and Giannellis,[67] who also observed similar effects for their clay-attached (end-tethered) poly(ε-caprolactone)–clay nanocomposites, whereas Ren *et al.*[68] only observed this effect at 6.7% clay loading for non-clay-attached polystyrene–polyisoprene–clay nanocomposites. The differences in the clay levels where pseudo-solid–liquid-like behavior is observed can therefore be attributed to whether the polymers are attached to the clay layers or not.

Another important parameter related to the melt rheology of PCNs is the crossover frequencies (*i.e.* frequencies at which values of storage and loss

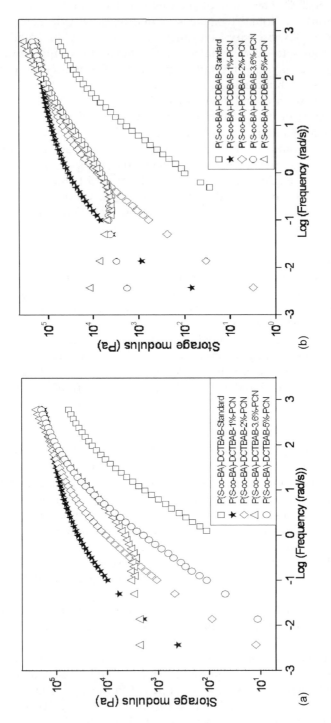

Figure 12.11 Variation of storage modulus (G^I) and loss modulus (G^{II}) with angular frequency (ω) for (left) P(S-*co*-BA)–DCTBAB–PCNs and (right) P(S-*co*-BA)–PCDBAB–PCNs, at 0% (□), 1% (★), 2% (◇), 3.6% (○) and 5% (△) clay loadings. Reproduced from reference 42 with permission from Elsevier.

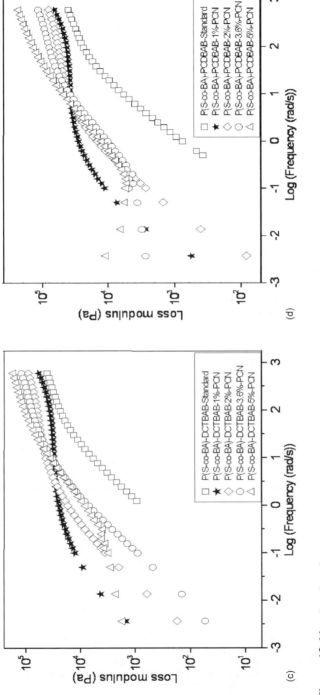

Figure 12.11 Continued.

moduli are equal). These angular frequency points give an indication of the relaxation times of the polymer chains in the PCNs. The number of these crossover points also gives an indication of the arrangement of the clay platelets relative to each other.[66] On the other hand, the terminal slope of the storage and loss moduli versus angular frequency gives an indication of the clay morphology and the polydispersity of the polymers. Monodisperse unfilled polymers obey the following relationships in the low angular frequency region: $G^I \propto \omega^2$ and $G^{II} \propto \omega^1$. In essence, the terminal gradient of the variation of storage modulus with angular frequency theoretically should be equal to 2 and the terminal slope of the loss modulus equal to 1, for a homogeneous and monodisperse homopolymer.[67–68,71] Deviations from these values (*i.e.* terminal gradients) may be due to polydispersity, the PCN morphology and/or clay loading.

G^I and G^{II} crossover frequencies and corresponding relaxation times are reported in Table 12.1.[42] The relaxation times of all PS-*co*-BA–PCNs were found to be greater than that of the neat P(S-*co*-BA) copolymer, although within the same series of PS-*co*-BA–PCNs, the relaxation time decreases as the clay loading increases. Although the relaxation time of polymer chains was expected to increase in the clay,[74] this decrease was attributed to molar mass having a more dominating effect than the clay. The presence of two crossover frequencies (the second being at low angular frequency) for the P(S-*co*-BA)–P–3.6% PCN, P(S-*co*-BA)–P–5% PCN and P(S-*co*-BA)–D–5% PCN show that for these nanocomposites there is a long-range relaxation process taking place, which is caused by the presence of clay.

This is to be expected, since an increase in clay loading results in highly confined polymer chains, hence long-range relaxation processes take place over a longer period of time. Moreover, two crossover angular frequencies are an

Table 12.1 Terminal gradients and the associated relaxation times of P(S-*co*-BA)–PCNs.[a] Reproduced from reference 42 with permission from Elsevier.

Polymer	M_n (g mol^{-1})	PDI	$\Delta G^I / \Delta \omega$	$\Delta G^{II} / \Delta \omega$	T_1 (s)	T_2 (s)
P(S-*co*-BA)–St	235 826	3.75	1.46	0.94	0.04	–
P(S-*co*-BA)–P-1%	114 215	2.09	1.93	1.21	12.57	–
P(S-*co*-BA)–P-2%	73 495	1.94	1.86	1.22	1.02	–
P(S-*co*-BA)–P-3.6%	42 196	1.68	0.41	0.10	0.26	39.27
P(S-*co*-BA)–P-5%	28 297	1.73	0.25	0.05	0.08	125.66
P(S-*co*-BA)–D-1%	191 824	2.42	1.38	0.82	25.13	–
P(S-*co*-BA)–D-2%	91 265	2.51	1.50	0.97	1.61	–
P(S-*co*-BA)–D-3.6%	49 075	2.22	0.73	0.93	0.26	–
P(S-*co*-BA)–D-5%	36 837	2.16	0.00	0.21	0.08	209.44

[a] M_n, molecular weight as determined by SEC; PDI, polydispersity index, M_w/M_n; $\Delta G^I / \Delta \omega$, terminal gradient, *i.e.* increase in storage modulus/increase in angular frequency (ω); $\Delta G^{II} / \Delta \omega$, terminal gradient, *i.e.* increase in loss modulus/increase in angular frequency (ω); T_1, relaxation time associated with first crossover; T_2, relaxation time associated with second crossover; T_1 and T_2 were calculated using the relation $T = 2\pi/\omega$.

indication of the clay platelets having reached the percolation threshold.[66] It is a phenomenon where clay platelets will be touching each other, *i.e.* a clay network. These results also indicate that the clay platelets were homogeneosly dispersed in the PCDBAB-based nanocomposites relative to the DCTBAB nanocomposites, as clay platelets formed a network at 3.6% clay loading for PCDBAB compared with 5% clay loading in the DCTBAB system. The terminal gradients of both the PCDBAB- and DCTBAB-based nanocomposites decreased with increase in clay loading, as reported by others.[64,73,75] The PDI values of the two series appeared to be important: the narrow PDI values of PCDBAB-based nanocomposites resulted in better adherence to the expected relationships between G^I and G^{II} as a function of ω.

References

1. Y. Kojima, A. Usuki, M. Kawasumi, A. Okada, T. Kurauchi and O. Kamigaito, *J. Polym. Sci., Part A: Polym. Chem.*, 1993, **31**, 983.
2. W. A. Braunecker and K. Matyjaszewski, *Prog. Polym. Sci.*, 2007, **32**, 93.
3. T. P. Le, G. Moad, R. Ezio and S. H. Thang, *PCT Int. Appl.*, WO 9801478 1998.
4. H. Matahwa, J. B. McLeary and R. D. Sanderson, *J. Polym. Sci., Part A: Polym. Chem.*, 2006, **44**, 427.
5. K. Landfester, *Annu. Rev. Mater. Res.*, 2006, **36**, 231.
6. A. Butte, G. Storti and M. Morbidelli, *Macromolecules*, 2001, **34**, 5885.
7. R. P. Moraes, A. M. Santos, P. C. Oliveira, F. C. T. Souza, M. Amaral, T. S. Valera and N. R. Demarquette, *Macromol. Symp.*, 2006, **245**, 106.
8. J. W. Jordan, *J. Phys. Colloid Chem.*, 1949, **53**, 294.
9. A. Samakande, P. C. Hartmann, V. Cloete and R. D. Sanderson, *Polymer*, 2007, **48**, 1490.
10. P. A. Wheeler, J. Z. Wang and L. J. Mathias, *Chem. Mater.*, 2006, **18**, 3937.
11. G. Moad, E. Rizzardo and S. H. Thang, *Aust. J. Chem.*, 2005, **58**, 379.
12. E. Rizzardo, J. Chiefari, R. Mayadunne, G. Moad and S. Thang, *Macromol. Symp.*, 2001, **174**, 209.
13. J. Chiefari, R. T. A. Mayadunne, G. M. Catherine L. Moad, E. Rizzardo, A. Postma, M. A. Skidmore and S. H. Thang, *Macromolecules*, 2003, **36**, 2273.
14. B. Y. K. Chong, J. Krstina, T. P. T. Le, G. Moad, A. Postma, E. Rizzardo and S. H. Thang, *Macromolecules*, 2003, **36**, 2256–2272.
15. S. H. Thang, B. Y. K. Chong, R. T. A. Mayadunne, G. Moad and E. Rizzardo, *Tetrahedron Lett.*, 1999, **40**, 2435.
16. S. Perrier and P. Takolpuckdee, *J. Polym. Sci., Part A: Polym. Chem.*, 2005, **43**, 5347.
17. A. Favier and M. T. Charreyre, *Macromol. Rapid Commun.*, 2006, **27**, 653.
18. J. Baussard, J. Habib-Jiwan, A. Laschewsky, M. Mertoglu and J. Storsberg, *Polymer*, 2004, **45**, 3615.

19. A. Samakande, R. D. Sanderson and P. C. Hartmann, *Synth. Commun.*, 2007, **37**, 3861.
20. G. Levesque, P. Arsene, V. Fanneau-Bellenger and T. Pham, *Biomacromolecules*, 2000, **1**, 400.
21. G. Levesque, P. Arsene, V. Fanneau-Bellenger and T. Pham, *Biomacromolecules*, 2000, **1**, 387.
22. A. Postma, T. P. Davis, G. X. Li, G. Moad and M. S. O'Shea, *Macromolecules*, 2006, **39**, 5307.
23. M. Okamoto, in *Encyclopedia of Nanoscience and Nanotechnology*, **Vol. 8**, ed. Hari Singh Nalwa, American Scientific Publishers, Valencia, CA, 2004, p. 791.
24. B. Ray, Y. Isobe, K. Morioka, S. Habaue, Y. Okamoto, M. Kamigaito and M. Sawamoto, *Macromolecules*, 2003, **36**, 543.
25. M. Rosorff, *Nano Surface Chemistry*, Marcel Dekker, New York, 2002.
26. M. W. Weimer, H. Chen, E. P. Giannelis and D. Y. Sogah, *J. Am. Chem. Soc.*, 1999, **121**, 1615.
27. J. B. Di and D. Y. Sogah, *Macromolecules*, 2006, **39**, 5052.
28. X. W. Fan, Q. Y. Zhou, C. J. Xia, W. Cristofoli, J. Mays and R. Advincula, *Langmuir*, 2002, **18**, 4511.
29. R. Advincula, Q. Y. Zhou, Y. Nakamura, S. Inaoka, M. K. Park and Y. F. Wang, *J. Mays. Abstr. Pap. Am. Chem. Soc.*, 2000, **219**, U498.
30. P. A. Wheeler, J. Z. Wang, J. Baker and L. J. Mathias, *Chem. Mater.*, 2005, **17**, 3012.
31. P. A. Wheeler, J. Z. Wang and L. J. Mathias, *Chem. Mater.*, 2006, **18**, 3937.
32. Y. Y. Yang, J. C. Lin, W. T. Yang and G. J. Jiang, *Polym. Prep.*, 2003, **44**, 855.
33. P. Ding, M. Zhang, J. Gai and B. Qu, *J. Mater. Chem.*, 2007, **17**, 1117.
34. J. B. Di and D. Y. Sogah, *Macromolecules*, 2006, **39**, 1020.
35. N. Salem and D. A. Shipp, *Polymer*, 2005, **46**, 8573.
36. G. Moad, G. Li, E. Rizzardo, S. H. Thang, R. Pfaendner and H. Wermter, *Polym. Prepr.*, 2005, **46**, 376.
37. P. Bera and S. K. Saha, *Polymer*, 1998, **39**, 1461.
38. G. Moad, K. Dean, L. Edmond, N. Kukaleva, G. X. Li, R. T. A. Mayadunne, R. Pfaendner, A. Schneider, G. Simon and H. Wermter, *Macromol. Symp.*, 2006, **233**, 170.
39. G. Moad, G. Li, R. Pfaendner, A. Postma, E. Rizzardo, S. Thang and H. Wermter, *ACS Symp. Ser.*, 2006, **944**, 514.
40. B. Q. Zhang, C. Y. Pan, C. Y. Hong, B. Luan and P. J. Shi, *Macromol. Rapid Commun.*, 2006, **27**, 97.
41. A. Samakande, R. D. Sanderson and P. C. Hartmann, *J. Polym. Sci., Part A: Polym. Chem.*, 2008, **46**, 7114.
42. A. Samakande, R. D. Sanderson and P. C. Hartmann, *Polymer*, 2009, **50**, 42.
43. J. Luo and I. M. Daniel, *Compos. Sci. Technol.*, 2003, **63**, 1607.
44. H. Tyan, K. Wei and T. Hsieh, *J. Polym. Sci., Part B: Polym. Phys.*, 2000, **38**, 2873.

45. M. Xu, Y. S. Choi, Y. K. Kim, K. H. Wang and W. J. Chung, *Polymer*, 2003, **44**, 6387.
46. W. Zhang, D. Z. Chen, Q. B. Zhao and Y. Fang, *Polymer*, 2003, **44**, 7953.
47. X. Fu and S. Qutubuddin, *Polymer*, 2001, **42**, 807.
48. W. A. Zhang, D. Z. Chen, H. Y. Xu, X. F. Shen and Y. E. Fang, *Eur. Polym. J.*, 2003, **39**, 2323.
49. E. A. Turi, *Thermal Characterization of Polymeric Materials*, **Vol. 1**, Academic Press, San Diego, 1997.
50. T. H. Kim, S. T. Lim, C. H. Lee, H. J. Choi and M. S. John, *J. Appl. Polym. Sci.*, 2003, **87**, 2106.
51. M. W. Noh and D. C. Lee, *Polym. Bull.*, 1999, **42**, 619.
52. Y. Yu, J. Yeh, S. Liou and Y. Chang, *Acta Mater.*, 2004, **52**, 475.
53. I. N. Jan, T. M. Lee, K. C. Chiou and J. J. Lin, *Ind. Eng. Chem. Res.*, 2005, **44**, 2086.
54. T. H. Kim, L. W. Jang, D. C. Lee, H. J. Choi and M. S. John, *Macromol. Rapid Commun.*, 2002, **23**, 191.
55. C. Tseng, J. Wu, H. Lee and F. Chang, *J. Appl. Polym. Sci.*, 2002, **85**, 1370.
56. G. Chigwada and C. A. Wilkie, *Polym. Degrad. Stabil.*, 2003, **80**, 551.
57. D. B. Zax, D. K. Santos, H. Hegemann, E. P. Giannelis and E. Manias, *J. Chem. Phys.*, 2000, **112**, 2945.
58. J. Wang, J. Du, J. Zhu and C. A. Wilkie, *Polym. Degrad. Stabil.*, 2002, **77**, 249.
59. A. Bluimstein, *J. Polym. Sci. Part A: Polym. Chem.*, 1965, **3**, 2665.
60. A. Postma, T. P. Davis, R. A. Evans, G. X. Li, G. Moad and M. S. O'Shea, *Macromolecules*, 2006, **39**, 5293.
61. A. Samakande, J. J. Juodaityte, R. D. Sanderson and P. C. Hartmann, *Macromol. Mater. Eng.*, 2008, **293**, 428.
62. Y. S. Choi and I. J. Chung, *Macromol. Res.*, 2003, **11**, 425.
63. A. Leszczynska, J. Njuguna, K. Pielichowski and J. R. Banerjee, *Thermochim. Acta*, 2007, **453**, 75.
64. K. Okada, T. Mitsunaga and Y. Nagase, *Korea–Aust. Rheol. J.*, 2003, **15**, 43.
65. T. D. Fornes, P. J. Yoon, H. Keskkula and D. R. Paul, *Polymer*, 2001, **42**, 9929.
66. J. Zhao, A. B. Morgan and J. D. Harris, *Polymer*, 2005, **46**, 8641.
67. R. Krishnamoorti and E. P. Giannelis, *Macromolecules*, 1997, **30**, 4097.
68. J. X. Ren, A. S. Silva and R. Krishnamoorti, *Macromolecules*, 2000, **33**, 3739.
69. M. J. Solomon, A. S. Almusallam, K. F. Seefeldt, A. Somwangthanaroj and P. Varadan, *Macromolecules*, 2001, **34**, 1864.
70. O. Meincke, B. Hoffmann, C. Dietrich and C. Friedrich, *Macromol. Chem. Phys.*, 2003, **204**, 823.
71. G. Galgali, C. Ramesh and A. Lele, *Macromolecules*, 2001, **34**, 852.

72. M. Xu, Y. S. Choi, Y. K. Kim, K. H. Wang and I. J. Chung, *Polymer*, 2003, **44**, 6387.
73. H. Y. Ma, L. F. Tong, Z. B. Xu and Z. P. Fang, *Polym. Degrad. Stabil.*, 2007, **92**, 1439.
74. S. T. Lim, C. H. Lee, H. J. Choi and M. S. John, *J. Polym. Sci., Part B: Polym. Phys.*, 2003, **41**, 2052.
75. T. T. Chastek, A. Stein and C. Macosko, *Polymer*, 2005, **46**, 4431.

CHAPTER 13

Polymer–Clay Nanocomposite Particles and Soap-free Latexes Stabilized by Clay Platelets: State of the Art and Recent Advances

ELODIE BOURGEAT-LAMI,[a] NIDA SHEIBAT-OTHMAN[b] AND AMILTON MARTINS DOS SANTOS[c]

[a] Laboratoire de Chimie, Catalyse, Polymères et Procédés (C2P2), LCPP Group, Université Lyon 1, CPE Lyon, CNRS, UMR 5265, 43 boulevard du 11 Novembre 1918, 69616 Villeurbanne, France; [b] Laboratoire d'Automatique et de Génie des Procédés (LAGEP), Université Lyon 1, CPE Lyon, CNRS, UMR 5007, 43 boulevard du 11 Novembre 1918, 69622 Villeurbanne, France; [c] Escola de Engenharia de Lorena – EEL, Universidade de São Paulo – USP, Estrada Municipal do Campinho s/n, 12602-810 Lorena/SP, Brazil

13.1 Introduction

The use of inorganic fillers to improve the properties of polymeric materials dates back to the earliest years of the polymer industry.[1] Initially used as extending agents to reduce the cost of polymer-based products, fillers were soon recognized to be an integral component in many applications involving polymers, particularly for reinforcement purposes.[2,3] Since the reinforcement efficiency of inorganic fillers is strongly related to their aspect ratio, anisotropic particles such as clay platelets, carbon nanotubes, gold nanorods and

RSC Nanoscience & Nanotechnology No. 16
Polymer Nanocomposites by Emulsion and Suspension Polymerization
Edited by Vikas Mittal
© Royal Society of Chemistry 2011
Published by the Royal Society of Chemistry, www.rsc.org

semiconductor nanowires have attracted particular interest over the last 10 years.[4] These nanocomposite materials can find applications in the fields of coatings and adhesives, for instance, where there is a continuing quest for new materials with enhanced mechanical performance, better hardness, excellent gas barrier properties and scratch resistance, to satisfy the evolving consumer needs.[5] Whereas a profusion of solvent-borne synthetic strategies have been reported in the recent literature, it appears that much less attention has been dedicated to waterborne processes such as emulsion, suspension and mini-emulsion polymerization.[14,15] However, these techniques offer the great advantage over solvent-borne methods of generating composite latex particles (hereafter referred to as colloidal nanocomposites) of controlled morphology at the nanoscale (*i.e.* within each individual particle). In addition, if the organic component is film forming, such particles can be further used to produce paints and coatings with ordered structures at larger scales.

Within the last 15 years, the field of colloidal nanocomposites has grown from being the subject of a few specialized papers to being the focus of a great deal of academic[6,7] and industrial research.[8,9] Several strategies have been reported for the preparation of nanocomposite particles, including hetero-coagulation, layer-by-layer assembly techniques and *in situ* polymerization. A detailed review of these routes was published by Bourgeat-Lami in 2002.[10] Of the various methods, emulsion polymerization, a free radical polymerization process widely employed industrially to manufacture a variety of products such as paints, adhesives, impact modifiers and so on, has been especially well documented.[11] Most recent examples of syntheses of composite latexes by emulsion polymerization are based on the polymerization of organic monomers in the presence of preformed inorganic colloidal particles used as seeds, which will be the approach described in this chapter.

Among the variety of inorganic solids (*e.g.* silica, iron oxides, titanium diox-ide, metals), clay minerals have recently attracted considerable attention. Indeed, polymer–clay nanocomposites (PCNs) have been the topic of extensive research worldwide since scientists at the Toyota Central Research laboratories reported 10 years ago that the incorporation of small amounts of montmorillonite (MMT) into nylon-6 resulted in a remarkable enhancement of the thermal and mechanical properties of the nanocomposite material.[12] Whereas there has been a tremendous number of studies on the synthesis, properties and applications of polymer–clay nanocomposites in the recent literature,[13] surprisingly only a small number of reports have dealt with emulsion polymerization.

The present review aims to cover and discuss recent developments in the synthesis of polymer–clay nanocomposites through emulsion polymerization. The first part summarizes the research status of *in situ* emulsion polymerization performed in the presence of pristine or organically modified clays (either MMT or Laponite). Then, the chapter highlights work on the synthesis of clay-armored latexes through soap-free polymerization. It reports the state of the art in the field and presents recent advances in the synthesis and characterization of high solids content latexes stabilized by Laponite clay platelets, with particular attention to mechanistic aspects and properties of the film materials issued from these latexes. Finally, the chapter considers the interest in studying process aspects and modeling in order to investigate the behavior of the system such as the nucleation mechanism and absorption of radicals into the polymer particles. These aspects are frequently considered in conventional polymerization processes but are usually omitted during the development of new materials, such as colloidal nanocomposites considered in this review.

13.2 Clay Structure

Composite materials based on layered silicates have been studied for a long time, probably because these natural minerals are readily available and can significantly improve the properties of the host polymer. Layered silicates commonly involved in nanocomposite synthesis belong to the structural family of the 2:1 phyllosilicates and have a sheet-like structure that consists of silica tetrahedra bonded to alumina or magnesia tetrahedra in a number of ways. They are usually from the clay group and more specifically from the smectite group. Among them, MMT and Laponite (a synthetic hectorite clay) are the clays by far the most frequently used. Their lamellar structure consists of two-dimensional layers with a central sheet of $M_{2-3}(O)_6$ octahedra (M being either a divalent or a trivalent cation), sandwiched between two external sheets of $Si(O,OH)_4$ tetrahedra (Figure 13.1).

The layer thickness is around 1 nm and the lateral dimensions of these layers may vary from a few tens nanometers to several micrometers depending on the particular silicate. Individual particles consist of a stack of a given number of face-to-face associated unit layers. Isomorphic substitution of Si(IV) by Al(III) in the tetrahedral sheet or of Al(III) by Mg(II) in the octahedral sheet generates

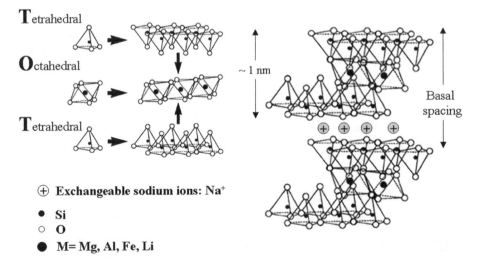

Figure 13.1 Schematic representation of formation and structure of smectite clay minerals.

negative charges. These negative charges are counterbalanced by hydrated alkali or alkaline earth metal cations (Na^+, Ca^{2+}) located on the exterior surfaces of the unit layer packages, and also in the interlayer. These cation layers are exchangeable and weakly bonded. Smectite clays are characterized by a moderate cation-exchange capacity (CEC) generally expressed in milli-equivalents per 100 grams (mequiv. per 100 g), typical CECs being in the range 60–120 mequiv. per 100 g. In addition to interlayer charges, under-coordinated metal ions (Mg^{2+}, Si^{4+}, Fe^{3+} or Al^{3+}), located on the broken edges of the crystals, can react with water molecules to form hydroxyl groups in order to complete their coordination sphere. The contribution of these edge sites to the CEC is approximately 20% and depends on the size and shape of the clay particles. The smaller the particle size, the more important is the edge con-tribution. Another attribute of smectite clays is their capacity to adsorb water and other polar molecules between the sheets, thus producing a significant expansion of the interlayer spacing. Only smectite clays have this particular property of increasing the interlamellar space. Other 2:1 clay minerals such as mica and vermiculite do not expand due to their excessively high layer charge, which results in strong irreversible electronic interactions between the sheets. For comprehensive and detailed information on clay colloidal chemistry and surface properties, the reader is referred to excellent reviews and text books.[14,15]

The reason for which clays have attracted such great interest in recent years is that they often allow a substantial improvement in mechanical strength, modulus and toughness while retaining optical transparency. These remarkable enhancements of material properties can be attributed to the high aspect ratio and the unique intercalation/exfoliation characteristics of clay minerals men-tioned above. Consequently, the addition of only a few weight percent of clay platelets that are properly dispersed throughout the polymer matrix can have a great impact on properties. The end-use properties of polymer–clay nano-composites are also greatly dependent on their nanostructure. Three main morphologies are usually reported: segregated, intercalated and exfoliated. The exfoliated morphology consists of individual silicate layers dispersed in the polymer matrix as a result of extensive polymer penetration and delamination of the silicate crystallites, whereas a finite expansion of the clay layers produces intercalated nanocomposites (Figure 13.2). In general, the greatest property enhancements are observed for exfoliated nanocomposites, which could be regarded as the 'ideal' morphology, although in practice many systems fall short of this idealized nanostructure.

13.3 Polymer–Clay Nanocomposites Produced by Conventional Emulsion Polymerization

13.3.1 Polymer–MMT Composite Latexes

Although both natural and synthetic aluminosilicates have been reported to be effective free radical initiators for the aqueous polymerization of vinylic

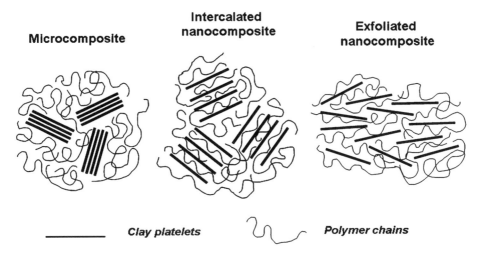

Microcomposite

Intercalated nanocomposite

Exfoliated nanocomposite

Clay platelets   *Polymer chains*

Figure 13.2 Different morphologies encountered in polymer-layered silicate nano-composites: (a) phase segregated, (b) intercalated and (c) exfoliated nanocomposites.

monomers for a long time,[16–18] their utilization in the synthesis of nano-composite materials was reported only recently. Compared with step-growth polymerizations, free radical polymerization and in particular heterophase polymerization present several advantages. In addition to providing the ability to conduct the reaction in water, a non-toxic medium, heterophase polymerization allows the easy removal of the resulting product from the reactor. In addition, this technique provides access to high molecular weight polymers with fast polymerization rates due to radical compartmentalization within growing particles. Since the polymer molecules are contained within the particles, the viscosity remains moderate and is not dependent on the molecular weight. Further, the final product can be used as is and generally does not need to be altered or processed. The methods used for the preparation of PCN materials through heterophase polymerization can be divided into three main categories depending on whether the clays have been organically modified or not. These methods are briefly reviewed below.

13.3.1.1 Polymer Latexes Synthesized in the Presence of Pristine MMT Platelets

Early Polymer–MMT Nanocomposite Syntheses. The first papers on poly-mer–MMT nanocomposite syntheses through emulsion polymerization were published in the 1990s by Lee and co-workers[19–21] and involved pristine clays. Intercalated nanocomposites based on bentonite (which consists of 70 wt% MMT) and poly(methyl methacrylate) (PMMA),[19] polystyrene (PS)[20] or styr-ene–acrylonitrile copolymers[21] were successfully obtained without any chemical modification of the clay using potassium persulfate (KPS) as initiator and sodium lauryl sulfate as anionic surfactant. Confinement of the polymer chains in the interlayer gallery space was evidenced by differential scanning calorimetry (DSC) and thermogravimetric analyses (TGA) and was suspected to originate from ion–dipole interactions between the organic polymer and the MMT surface. The products obtained were exposed to boiling toluene for 5 days in order to extract the non-bonded polymer, which was further analyzed in terms of polymer molecular weight (M_w). It appeared that the M_w of extracted polymer was comparable to that of the pure polymer. A non-negligible part of polymer was adsorbed on the clay surface after extraction, indicating significant polymer–clay interaction. Unfortunately, since the latex was coagulated before characterization, no information was given on particle morphology.

Following a very similar procedure involving KPS as initiator and sodium dodecyl sulfate (SDS) as surfactant, Tong *et al.* reported the *in situ* emulsion polymerization of ethyl acrylate in the presence of pristine bentonite.[22] Drying of the latex suspension produced films exhibiting enhanced mechanical and thermal properties and also reduced permeability to water vapor. Generally, and regardless of synthetic procedures, nanoscale dispersion of layered silicates in plastics produces glassy modulus enhancements of 1–2-fold and rubbery modulus increases of 5–20-fold. Improved thermal properties are also observed in many systems, including intercalated nanocomposites obtained by emulsion polymerization. Such improvements are mainly due to the nanometric dimensions of the platelets. TEM analysis of an ultramicrotomed section of the film indicated in the case described above, an intercalated (partially exfoliated) morphology as shown in Figure 13.3. Fairly recently, Pan *et al.* also showed enhanced mechanical properties for PVC–MMT nanocomposites elaborated in a similar way,[23] and Kim *et al.* observed a shift of the onset of the thermal degradation towards higher temperatures as a function of clay content in exfoliated PS–MMT nanocomposites.[24] Further, Bandyopadhyay *et al.* synthesized PMMA–MMT nanocomposites through emulsion polymerization with SDS as the emulsifier and showed enhanced thermal stability and an increase in T_g by 6 °C.[25]

High Solids Content Latexes. In addition to properties which most of the time are enhanced upon adding clay, another crucial issue of PCN materials prepared by emulsion polymerization is their solids content. Although this point is less frequently addressed, most polymer–clay composite latexes reported in literature have solids contents below 20 wt%. However, solids contents between 40 and 60 wt% and sometimes higher are required for industrial applications. Using a seeded semi-batch emulsion polymerization process and a procedure otherwise very similar to that described above for bentonite, Diaconu and co-workers recently reported the successful preparation of 45 wt% solids content P(BA–MMA)–MMT latexes containing 3 wt% of clay.[26] To maintain good colloidal stability and a low viscosity of the

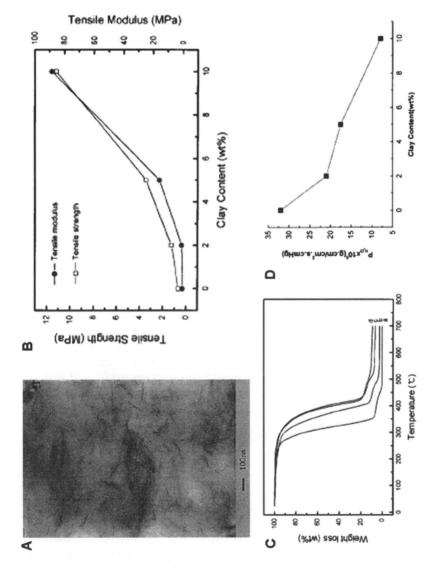

Figure 13.3 (A) Representative TEM image of exfoliated-intercalated clay in a poly(ethyl acrylate) matrix containing 5 wt% of clay. Dependence of (B) mechanical, (C) thermal and (D) permeation properties on clay loading. Adapted from reference 22 with permission of Wiley.

latex suspension, the clay was split between the seed and the feed. The resulting nanocomposite materials displayed an intercalated morphology and showed better mechanical, thermal and permeability properties than composites prepared by blending the pristine latex and Na-MMT or than pristine copolymer synthesized under the same conditions.

13.3.1.2 *In Situ Emulsion Polymerization using 'Reactive' MMT Platelets as Seeds*

The very simple methods depicted above show that, by virtue of their high swelling capabilities in water, pristine clays are valuable candidates for the elaboration of exfoliated or intercalated nanocomposites through *in situ* emulsion polymerization. In spite of that, a number of groups have performed organic modification of clay particles with the objective of increasing interfacial interactions and control particle morphology. In most of these studies, the organoclay is dispersed in water and the polymerization proceeds as in conventional emulsion polymerization by monomer diffusion from the droplets to the organophilic clay surface where polymer chain propagation takes place. However, in a few examples, the organoclay is dispersed in the monomer phase. This organoclay monomer suspension is next emulsified (sometimes with the aid of ultrasound to help dispersion and promote clay exfoliation), to form emulsion droplets that are subsequently polymerized.[27–32] These last processes appear closer to suspension or miniemulsion (depending on the nature of the initiator) than emulsion polymerization and will not be discussed further here. Organic modifiers are generally used to render clay surfaces hydrophobic and promote monomer penetration in the interlayer space. Organic modifiers carrying suitable reactive end groups can also react with the polymer matrix and further strengthen the interfacial polymer–clay interaction, as reported by many groups. Three major categories of clay modifications have been reported: cation exchange, silane grafting and physisorption of polar compounds (either polymers or small molecules). A non-exhaustive list of functional molecules carrying reactive groups used in PCN synthesis through emulsion polymerization is given in Table 13.1.

For instance, Qutubuddin *et al.* incorporated vinylbenzyldimethyldodecylammonium chloride (VDAC) in MMT by cation exchange.[33] The amount of VDAC was kept well below the CEC of the clay in order to preserve the clay's colloidal stability (an issue which is neglected surprisingly often in the literature). A partially exfoliated nanocomposite was obtained upon polymerization of styrene in the presence of the organoclay but the latex colloidal stability was only moderate as there was no surfactant to stabilize the resulting particles. Recently, Sedlakova *et al.* followed a similar approach using the cationic 2-(acryloyloxy)ethyltrimethylammonium chloride monomer.[34] Choi *et al.* also reported the preparation of PS–MMT nanocomposites through *in situ* emulsion polymerization using SDBS as surfactant and 2-acrylamido-2-methyl-1-propanesulfonic acid (AMPS, Table 13.1) as functional comonomer to

Table 13.1 Chemical structures of reactive organic modifiers used to functionalize smectite clays during the synthesis of PCN latexes through emulsion polymerization.

Name	Chemical formula	Abbreviation	Ref.
Ammonium methylstyrene chloride	$H_3\overset{+}{N}$—⟨benzene⟩—CH=CH$_2$ Cl$^-$	AMS	27
N,N-Dimethyl-*n*-hexadecyl(4-*vinylbenzyl*) ammonium *chloride*	$CH_3(CH_2)_{15}$—$\overset{+}{N}(CH_3)_2$—CH$_2$—⟨benzene⟩—CH=CH$_2$ Cl	VB16	28
N,N-Dimethyl-*n*-dodecyl (4-*vinylbenzyl*)ammonium *chloride*	$CH_3(CH_2)_{11}$—$\overset{+}{N}(CH_3)_2$—CH$_2$—⟨benzene⟩—CH=CH$_2$ Cl	VDAC	33
2-Acryloyloxyethyl-trimethylammonium chloride	CH_3—$\overset{+}{N}(CH_3)_2$—CH$_2$CH$_2$—O—C(=O)—CH=CH$_2$	ADQUAT	34
2-Acrylamido-2-methylpropanesulfonic acid	HO—S(=O)$_2$—C(CH$_3$)$_2$—NH—C(=O)—CH=CH$_2$	AMPS	35,36, 37
Sodium 11-methacryloyl-oxyundecan-1-yl sulfate	Na$^+$ $^-$O—S(=O)$_2$—O—(CH$_2$)$_{11}$—O—C(=O)—C(=CH$_2$)CH$_3$	MET	37
N-Isopropylacrylamide	(CH$_3$)$_2$CH—NH—C(=O)—CH=CH$_2$	NIPAM	37
Sodium 1-allyloxy-2-hydroxypropyl sulfonate	Na$^+$ $^-$O—S(=O)$_2$—CH$_2$—CH(OH)—CH$_2$—O—CH$_2$CH=CH$_2$	Cops	37
2-Methacryloyloxyethyl-trimethylammonium chloride	CH_3—$\overset{+}{N}(CH_3)_2$—CH$_2$CH$_2$—O—C(=O)—C(=CH$_2$)CH$_3$	MADQUAT	42
3-Methacryloyloxypropyl-trimethoxysilane	(MeO)$_3$Si—CH$_2$CH$_2$CH$_2$—O—C(=O)—C(=CH$_2$)CH$_3$	MPTMS	42
3-Methacryloyloxypropyl-dimethylethoxysilane	(Me)$_2$(OEt)Si—CH$_2$CH$_2$CH$_2$—O—C(=O)—C(=CH$_2$)CH$_3$	MPDES	42

promote adhesion at the polymer/clay interface.[35a] Exfoliated nanocomposites
with clay contents up to 20 wt% with respect to polymer were successfully
achieved by this method. It was speculated that the hydrogen ions of the
sulfonic acid protonated the amido group of the AMPS comonomer, which
could then undergo exchange reactions with the clay counter-ions. The method
was further extended to acrylonitrile,[35b] styrene–MMA[35c] and acrylonitrile–
butadiene–styrene[35d] (co)polymers. However no indication was given in these
papers of particle size and latex colloidal stability. It should be noted that
AMPS was also used by Li *et al.* in the synthesis of P(BA–AA)–MMT latexes
by semi-batch emulsion polymerization for pressure-sensitive adhesive applications.[36] Finally, it is worth mentioning the recent work of Greesh and co-
workers, who systematically investigated the effect of the nature of the organic
modifier on the microstructure and thermomechanical properties of P(S–BA)–
MMT composite films containing 10 wt% of clay.[37] The ability of the clay
modifier to interact efficiently with the clay surface, its compatibility with the
monomer and its reactivity were all found to be determinant parameters to
achieve fully exfoliated morphologies. AMPS was shown to fulfill all these
requirements.

13.3.1.3 In Situ Emulsion Polymerization in the Presence of 'Non-reactive' Organically Modified MMT Platelets

In addition to 'reactive' organic modifiers, a few groups have also reported the
use of 'non-reactive' cationic compounds (which are usually alkylammonium
or zwitterionic surfactants) to promote clay exfoliation.[38,39] For instance,
Meneghetti and Qutubuddin synthesized partially exfoliated PMMA–MMT
nanocomposites using SDS as regular surfactant and a zwitterionic octade-
cyldimethylbetaine as organic modifier.[40] It was shown that the zwitterionic
modifier provided better colloidal stability than a quaternary ammonium surfactant, which can probably be interpreted in terms of charge neutralization of
the SDS-stabilized latex particles by the cationic surfactant (although this was
not discussed in the paper). Li *et al.* also found that emulsion polymerization of
styrene conducted in the presence of MMT functionalized by a zwitterionic
surfactant produced exfoliated nanocomposites, as proved by TEM and by the
disappearance of the d_{001} peak in the X-ray diffraction (XRD) pattern.[41]

13.3.2 Polymer–Laponite Nanocomposite Latexes

All the examples of PCN latexes described in Section 13.3.1 involved MMT as
the clay. Although MMT is by far the most commonly used layered silicate in
PCN syntheses, Laponite has attracted increasing interest in the recent literature. Laponite is a synthetic clay similar in structure to MMT, except for the
nature of the interlayer cation, which is Mg^{2+} for Laponite and Al^{3+} for MMT.
Laponite also differs from MMT in the clay dimensions. Each elementary
Laponite disk has a thickness of around 1 nm and a diameter in the range

25–30 nm. Further, Laponite offers the great advantage over natural clays of being chemically pure and free from external contaminants. All these properties make Laponite a good candidate for the synthesis of polymer–clay nano-composites. Recent studies on the synthesis of Laponite-armored latexes and Laponite encapsulation by emulsion polymerization are reviewed below.

13.3.2.1   Laponite-armored Latexes

Negrete-Herrera and co-workers reported emulsion copolymerization of styrene and *n*-butyl acrylate in the presence of organically modified Laponite.[42-44] The clay particles were functionalized through ion exchange of a cationic initiator (AIBA) or a cationic monomer, 2-methacryloyloxyethyltrimethyl-ammonium chloride (MADQUAT), or through the reaction of the edge hydroxyls with suitable organosilane molecules carrying a copolymerizable methacrylate moiety and one or three hydrolyzable alkoxysilyl groups.[42] These silanes will be hereafter referred to as monofunctional and trifunctional silanes, respectively. The overall synthetic process is illustrated in Figure 13.4.

It was shown that grafting of the monofunctional silane had almost no effect on the physicochemical properties of Laponite, whereas grafting of the tri-functional silane resulted in a lower porosity, greater interlamellar distance and higher hydrophobicity.[43] Indeed, whereas the former could only form covalent bonds with the broken edges of the clay platelets, the latter was capable of both reaction with the clay edges and formation of complex polysiloxane oligomers covalently linking the clay sheets together. Fourier transform infrared (FTIR) spectroscopy showed that at low concentrations, the trifunctional silane formed a monolayer lying on the surface with the methacryloxy groups in close contact with the border of the clay plates. Further addition of coupling agent led to the formation of multilayer coverage and oriented the carbonyl group of the silane moiety away from the clay surface. The organoclays were further suspended in water and used as seeds in emulsion polymerization reactions. For organoclays functionalized by cation exchange, a small amount of tetra sodium pyrophosphate was introduced in the suspension to reverse the edge positive charges and promote clay–clay repulsions. The clays grafted with the trifunctional silane could not be suspended in water due to clay bridging by the polysiloxane oligomers. The monofunctional silane provided an elegant means to solve this problem. Emulsion polymerization was then performed in batch mode in the presence of the organoclay (10 wt% based on monomer) using SDS as surfactant and KPS or 4,4'-azobis(4-cyanopentanoic acid) (ACPA) as initiator, except when AIBA was used as organic modifier.[44] Regardless of the nature of the organic modifier, colloidally stable composite latexes with diameters in the range 50–150 nm were successfully obtained provided that the original clay suspension was sufficiently stable and unaggregated. All three methods produced clay-armored composite particles characterized by a polymer core surrounded by an outer shell of clay platelets, as evidenced by energy-filtered cryomicroscopy, which provided enhanced contrast and led to an improved understanding of the platelet coverage.[44c]

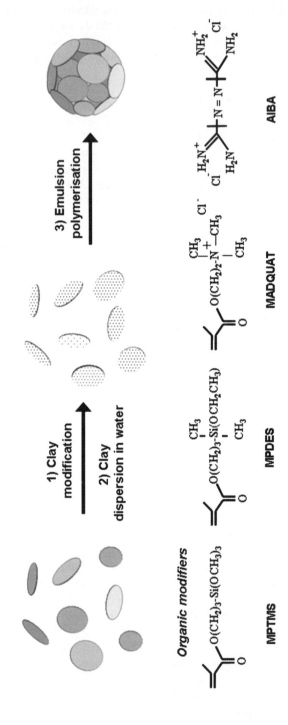

Figure 13.4 Schematic diagram illustrating the procedure used for the synthesis of Laponite-armored copolymer latexes through *in situ* emulsion polymerization.

To probe the properties of the materials, the composite latex suspensions were cast into films by pouring the suspension on to a Teflon plate and allowing the water to evaporate at room temperature until a crack-free and transparent film was formed. Analysis of the film microstructure by cryo-TEM revealed a honeycomb distribution of Laponite within the film, consistent with the armored morphology (Figure 13.5d and e).[44c] Moreover, the film retained optical transparency and exhibited remarkable mechanical properties characterized by a 50-fold increase in Young's modulus, an unusual necking behavior at low strain and a large elongation at break (Figure 13.6). The composite material also displayed a significant improvement in thermal behavior, with a substantial shift in T_g and an increase in the onset temperature of thermal degradation. These results demonstrate for the first time that strong reinforcement can be achieved even though the platelets are not fully exfoliated within the polymer matrix, provided that a filler network can be established, in a way that bears some similarities with the filler network that is present in carbon black-filled elastomers.[45,46] The formation of a well-defined filler network is likely to occur if strong interactions between the platelets can develop. Such interactions are in turn promoted if platelets located on neighboring latex particles come into contact or if interplatelet bonding is realized through polymer tie chains, again by analogy with the case of carbon black-reinforced elastomers.

Recently, Ruggerone *et al.* reported similar honeycomb morphologies for PS–Laponite films containing up to 50 wt% of clay.[47,48] Whereas only a relatively small reinforcement was found in the glassy state of the same order of magnitude as with pure PS, large stiffness increases were obtained above T_g, the highly filled nanocomposites presenting an improved storage modulus of more than two orders of magnitude with respect to the neat matrix (Figure 13.7). Classical micromechanical models (*e.g.* Halpin–Tsai or Mori–Tanaka) are able to account for the reinforcement in the glassy state, provided that both the effective Laponite volume fraction and aggregate aspect ratio are considered. In the rubbery state, these models were found to be completely inadequate and an effect of local matrix immobilization was invoked to fit the experimental results, based on material morphology and heat capacity measurements. Interestingly, while the stiffness increases in the rubbery state are related to overall Laponite content, in the glassy state the degree of Laponite exfoliation was also of certain importance. Effectively, the storage and tensile moduli of 5 wt% Laponite nanocomposites are greater than those for higher clay contents in the glassy state, and this was explained by a higher aspect ratio of the clay aggregates due to better dispersion.

The Laponite was also shown to affect the fracture properties of these materials, the tensile strength increasing up to about 5 wt% of clay and then strongly decreasing.[48,49] This was explained by a drastic change in matrix micro-deformation mechanisms when the Laponite content was increased beyond this amount, in agreement with the observation of reduced deformation ratios in the deformation zones by TEM and to the intrinsic weakness of the Laponite stacks and/or the PS/Laponite interface. The highly filled nanocomposites with Laponite contents comparable to the estimated threshold for

282

Chapter 13

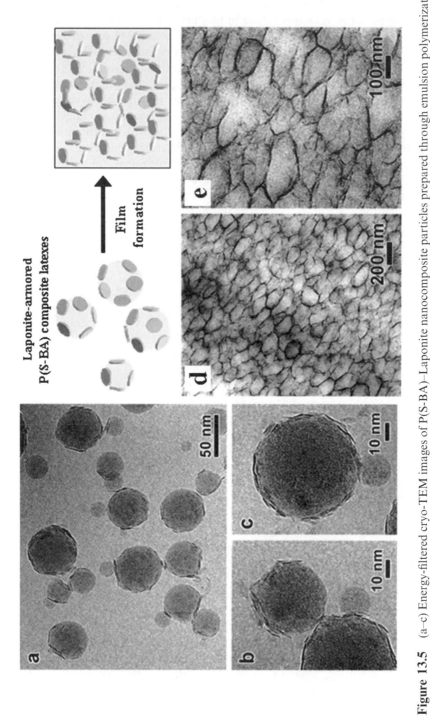

Figure 13.5 (a–c) Energy-filtered cryo-TEM images of P(S-BA)–Laponite nanocomposite particles prepared through emulsion polymerization using MADQUAT as modifying agent. The nanoparticles are seen embedded in a film of vitreous ice. (d, e) Electron micrographs at intermediate and high magnifications of an ultra-thin cryo-section of a film prepared from the composite latex suspension. The thin dark layer covering the surface of the polymer particles corresponds to 1 nm thick diffracting clay platelets that are oriented edge-on with respect to the direction of observation. Adapted from reference 44c with permission of Wiley-VCH.

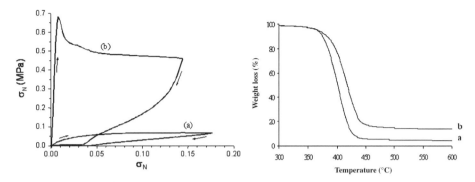

Figure 13.6 (Left) tensile mechanical behavior and (right) thermogravimetric analysis of the poly(styrene-*co*-butyl acrylate) copolymer matrix (a) and of the poly(styrene-*co*-butyl acrylate)–Laponite nanocomposite film (b). Reprinted from reference 44c with permission of Wiley-VCH.

percolation of contacts between the Laponite stacks showed extremely brittle behavior, associated with crack propagation along the interfaces between the latex particles, as observed by TEM and SEM (Figure 13.8).

13.3.2.2 Laponite Encapsulation

The afore-mentioned examples and a rapid survey of the literature on the preparation of polymer–clay nanocomposites reveal that the encapsulation of individual platelets by conventional emulsion polymerization is not straightforward. Although successful encapsulations have been reported for spherical particles such as titanium dioxide[50] and silica[51] using, for instance, organo-titanate or organosilane coupling agents to promote interaction of the growing polymer chains with the inorganic surface, it appears that similar approaches are unsuccessful for non-spherical particles such as clays. It seems that the anisotropy of the particle shape and their high aspect ratio associated with a high surface energy prevent them from efficient encapsulation. Recently, Voorn *et al.*[52] reported a strategy enabling Laponite or MMT platelets to be encapsulated. In this method, the clay is grafted with polymerizable organotitanate and organosilane molecules with some similarity to the work of Negrete-Herrera and co-workers reported above. However, instead of conducting the polymerization in batch, the authors performed starved-feed soap-free emulsion polymerization. Environmental scanning electron microscopy and cryo-TEM showed that the MMT-functionalized platelets were successfully encapsulated inside non-spherical dumbbell-like latex particles whereas Laponite was embedded inside spherical latex particles (Figure 13.9).

Miniemulsion polymerization, a variant of emulsion polymerization in which monomer droplets play a key role in particle nucleation, also appeared to be an efficient method for Laponite encapsulation in the recent literature.[53] For

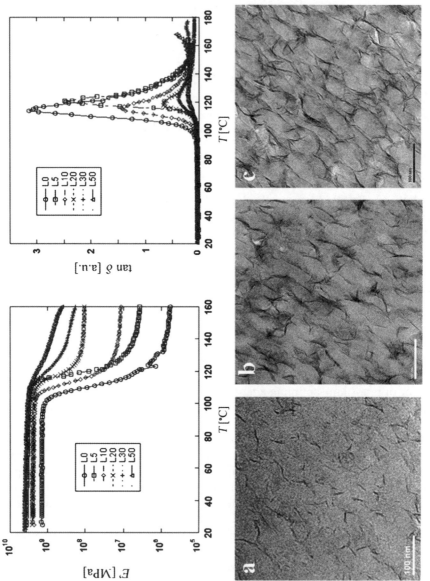

Figure 13.7 Top: E' and $\tan\delta$ at 1 Hz as a function of temperature for films prepared from PS latex particles coated with increasing amounts of Laponite. Bottom: TEM images of thin cross-sections of latex films containing (a) 10 wt%, (b) 20 wt% and (c) 50 wt% of Laponite respect to polymer. Adapted from references 47 and 49 with permission of Elsevier.

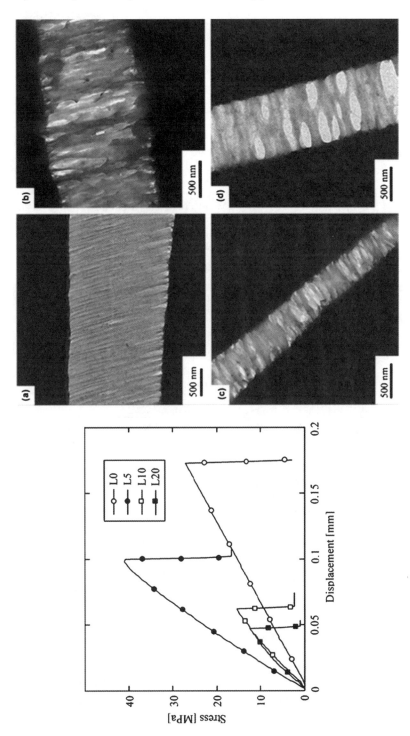

Figure 13.8 Left: stress–displacement curves of pressed films prepared from latexes containing 0, 5, 10 and 20 wt% Laponite. Right: TEM images of microdeformation in microtomed thin sections taken from the pressed films and strained at room temperature. Adapted from reference 49 with permission of Elsevier.

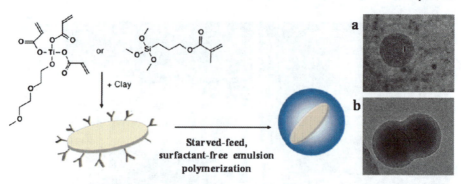

Figure 13.9 Example of clay encapsulation by starved-feed soap-free emulsion polymerization. The large MMT platelets are encapsulated inside dumbbell-like particles whereas small Laponite platelets are contained inside spherical latex particles. Adapted from reference 52 with permission of the American Chemical Society.

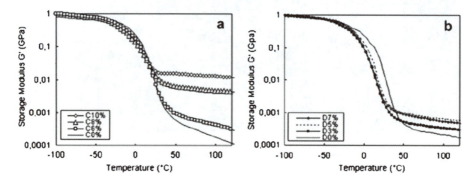

Figure 13.10 Storage modulus *versus* temperature obtained by dynamic mechanical analysis (DMA) of latex films obtained from (a) Laponite-armored latexes and (b) polymer-encapsulated Laponite. Adapted from reference 54 with permission of Elsevier.

instance, Faucheu *et al.* recently reported the synthesis of polymer–Laponite particles through miniemulsion polymerization.[54] The technique allowed the particle morphology to be precisely controlled, with the clay being either located on the surface of the latex spheres or embedded deeply inside the particles. The films obtained from the latex particles showed significantly different nanostructures and thus different mechanical behaviors and water uptakes. Whereas the clay-armored latexes led to the formation of a percolating clay network at high clay content, as described previously for emulsion polymerization, the Laponite-encapsulated latexes showed only limited clay–clay contacts due to confinement of the platelets inside the individual polymer particles. The clay-armored latexes thus led to very high mechanical reinforcement effects whereas lower reinforcements were observed for encapsulated samples of similar clay contents (Figure 13.10).

13.4 Soap-free Latexes Stabilized by Clay Platelets

13.4.1 Pickering Suspension and Miniemulsion Polymerizations: a Brief Overview

Over a century ago, Ramsden[55] and Pickering[56] observed that colloidal particles located at the oil/water interface were able to stabilize emulsions of both the oil-in-water and the water-in-oil type. These emulsions are referred to as either Pickering emulsions or solid-stabilized emulsions.[57] In line with these concepts, the polymer community has recently shown a surge of interest in the production of a vast range of hybrid colloidal materials using various nanoparticles as solid stabilizers for miniemulsion[58–63] or suspension[64–69] polymerization.

For instance Bon and co-workers reported the formation of polymer latexes armored with Laponite platelets by soap-free miniemulsion polymerization.[58,59] In this method, the clays adhere to the surface of the miniemulsion droplets and ultimately stabilize the hybrid particles by forming a protecting armor. It was shown that excess platelets were also present after polymerization, suggesting relatively low clay aggregation efficiency. The presence of 'free' platelets in the water phase is undesirable as this may significantly affect the long-term stability of the latex suspension. Moreover, the miniemulsion polymerization reaction was successful only when oil-soluble initiators were used and failed when the latter were replaced by water-soluble initiators such as KPS. Following these previous studies, Wang *et al.* recently reported the synthesis and mechanical properties of Laponite-armored latexes for pressure-sensitive adhesive applications. Latex films were produced by blending poly(lauryl acrylate) (PLA)–Laponite hybrid particles obtained by Pickering miniemulsion polymerization and a standard poly(butyl acrylate-*co*-acrylic acid) latex.[70] The clay-armored supracolloidal structure enabled a superior balance of viscoelastic properties to be achieved in a synergetic way. Indeed, the addition of small amounts (*e.g.* 2.7 wt%) of the PLA hybrid particles increased the tack adhesion energy considerably more than was found for the two individual components or for the sum of their individual contributions. In a related approach, Voorn *et al.*[60] showed that organically modified MMT platelets could efficiently stabilize inverse miniemulsions, producing MMT-armored polymer particles with a diameter in the range 700–980 nm (Figure 13.11). Finally, it is worth mentioning the recent work of Guillot *et al.*[63] on Pickering stabilization of monomer-containing lipid droplets by clays and their subsequent free radical polymerization.

13.4.2 Soap-free Emulsion Polymerization Stabilized by Inorganic Solids

In fact, the ability of inorganic solids to impart colloidal stabilization to polymer latexes is not new. Interest in these systems emerged in the early 1990s with the pioneering work of Armes and co-workers on the synthesis of silica–vinyl (co)polymer latexes stabilized by ultrafine silica particles.[71–74] However,

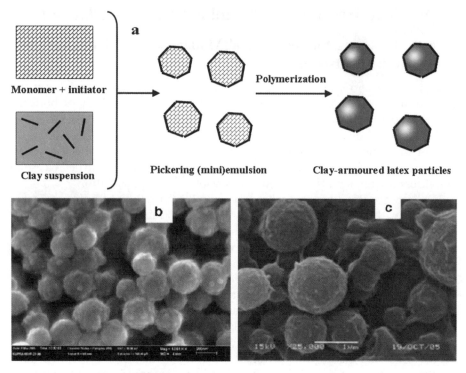

Figure 13.11   (a) Synthesis scheme for synthesis of Laponite-armored latexes *via*
Pickering polymerization. Redrawn after reference 58. (b) SEM image
of Laponite clay-armored polystyrene latex. Scale bar indicates 200 nm.
Reprinted with permission from reference 58. © 2005 American
Chemical Society. (c) MMT-armored polyacrylamide latex particles.
Scale bar indicates 1 ìm. Reprinted from reference 60 with permission of
the American Chemical Society.

in these early studies, the final particle size (hundreds of nanometers) was
significantly smaller than the diameter of the emulsion droplets (tens of
micrometers), indicating that the latter were not the main locus of polymerization as in Pickering polymerizations. Since then, a large variety of
inorganic solids have been reported to stabilize polymer latexes efficiently in the
absence of any added surfactant.[75–81] Indeed, soap-free emulsion polymerization offers some obvious advantages over conventional emulsion polymerization, such as ease of purification, due to the absence of surfactant and improved
mechanical and water-resistance properties of the resulting films. In addition,
emulsion polymerization presents the advantage over miniemulsion polymerization of being easily scalable as it does not require any specific homo-
genization device to generate the emulsion droplets. Among the large variety of
solid materials that can potentially stabilize polymer latexes, clay minerals are
particularly attractive due to their ability to accumulate at interfaces.[82,83] The
resulting polymer–clay nanocomposite latexes can find applications in the fields
of coatings and adhesives, for instance.

Despite the potential benefit of soap-free latexes in the coating industry, very few papers have reported the use of clay minerals as stabilizers in conventional emulsion polymerization.[84–87] In 2001, Choi and co-workers[84] reported the preparation of PMMA–MMT nanocomposites through *in situ* soap-free emulsion polymerization. However, at that time, the authors did not stress the role of the clay as stabilizer. Park and co-workers[85] synthesized PMMA–MMT nanocomposites for optical applications by soap-free emulsion polymerization, but again the article focused on properties and there was no mention of colloidal aspects. In the same period, Lin *et al.*[86] reported the formation of 300–600 nm diameter PMMA latex particles initiated by KPS and stabilized by MMT platelets. The authors argued that KPS intercalation inside the clay galleries was the driving force that enabled the polymerization to take place at the clay surface. Although the solids content was fairly low and the amount of MMT rather high, this article nevertheless remains very instructive. Using an original approach, Zhang *et al.* recently reported the synthesis of PMMA latex particles stabilized by MMT platelets tethered with poly(2-dimethylamino)-ethyl methacrylate (PDMAEMA) brushes (Figure 13.12).[87] The PDMAEMA polyelectrolyte brush was synthesized by atom transfer radical polymerization using a cationic initiator previously introduced in the clay galleries. The PDMAEMA-functionalized clay platelets were further used to stabilize the emulsion polymerization of MMA initiated by the remaining free radical initiator present on the clay surface.

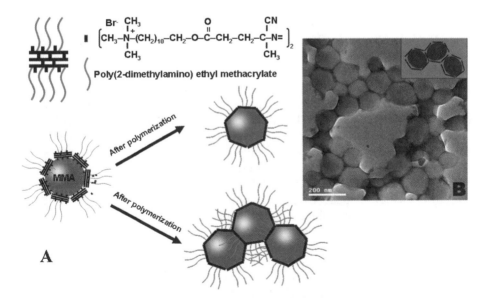

Figure 13.12 (A) Scheme illustrating the synthesis of clay-armored PMMA latexes through soap-free emulsion polymerization using PDMAEMA-tethered clay as stabilizer. (B) TEM image of the resulting PMMA–clay composite colloid. Reproduced from reference 87 with permission of Wiley.

13.4.3 Laponite-armored Latexes Produced by Soap-free Emulsion Polymerization

Our group recently developed a synthetic strategy leading to Laponite-armored polymer latexes *via* soap-free emulsion polymerization. Our first results in this field were reported in 2007.[88] In this preliminary work, we used a poly (ethylene glycol) methyl ether methacrylate (PEGMA) macromonomer with a molar mass of $1080\,g\,mol^{-1}$ to promote polymer–clay interactions and produce clay-armored particles. The interaction of clays with poly(ethylene oxide) (PEO) polymers is well known and has been extensively reported in the literature.[89–92] PEO can interact with clay surfaces through ion–dipole interactions between ethylene oxide units and clay ions in a manner similar to conventional PEO–salt complexes which form pseudo-crown ether type structures.[93] PEO molecules of different molar masses can be intercalated either from water solution[94] or by direct intercalation from the polymer melt.[95] Previous studies on PEO intercalation from water showed that low molar mass polymers adsorb on single clay particles and inhibit aggregation by steric hindrance,[96] whereas high molar mass polymers form bridges between particles and lead to the formation of large clusters or gels, particularly for high clay concentrations.[97,98] The interaction of PEGMA with Laponite was investigated by FTIR spectroscopy and XRD. These techniques, however, do not provide insights into the chemical reactivity of the intercalated macromonomer. PEGMA intercalation was evidenced by the increase in the d_{001} interlamellar spacing. Whereas for 5% PEGMA the variation of the basal spacing was moderate, the interlayer distance increased from 13 to $15.2\,\text{Å}$ upon adding 20 wt% PEGMA. A further proof of PEGMA–Laponite interaction was provided by FTIR spectroscopy. Infrared shifts were observed in the $1200–1500\,cm^{-1}$ region subsequent to PEGMA intercalation. Moreover, split bands around 1347 and $1360\,cm^{-1}$, attributed to CH_2 stretching vibrations, confirmed the formation of ion–dipole interactions between ethylene oxide units and clay ions.[99] In addition, the broad band around $3500\,cm^{-1}$ attributed to the stretching O–H vibration of water molecules decreased in intensity while sharp peaks corresponding to Mg–OH and Al–OH appeared, indicating that, as expected, the PEGMA macromonomer displaced water from the clay galleries. Finally, PEGMA adsorption on the Laponite surface was also clearly evident from zeta potential measurements. The absolute value of the zeta potential decreased with increase in the concentration of macromonomer due to screening of the clay surface charges upon PEGMA addition.

In the following, in-depth investigation of soap-free emulsion polymerization involving PEGMA and Laponite as stabilizer is reported as a case study. The goal of this section is to explore the underlying particle formation mechanism (nucleation, stabilization, coagulation and absorption of radicals) and to highlight existing differences with conventional emulsion polymerization.

13.4.3.1   Reaction Scheme

To assess the role of the macromonomer on nanocomposite particles formation, a series of experiments were carried out in the presence of PEGMA only, Laponite only, Laponite and PEGMA and Laponite and PEG oligomers (*i.e.* without any polymerizable end group). The clay content was fixed at 5 wt% based on styrene and only the initial composition of the suspension medium was changed. In order to promote clay dispersion and retard aggregation, a predetermined amount of tetrasodium pyrophosphate (10 wt% based on Laponite) was introduced in the reactor before polymerization. In salt-free water, Laponite platelets are stabilized by electrostatic repulsions between the negatively charged faces. However, if the interactions are partially screened by the presence of salt, the particles aggregate by the formation of electrostatic bonds between the positively charged rims and the negatively charged faces. By adsorbing on the positively charged rims, the tetravalent negatively charged pyrophosphate ions allow the rim charge to be effectively screened and the gelation rate to be decreased. As mentioned above, low molar mass PEO molecules have also been reported to retard Laponite aggregation.[97] Here, the adsorbed polymer chains form a steric barrier preventing the clay platelets from coming into contact. Figure 13.13 shows the evolutions of the monomer conversion with time and of the particle size with conversion for this series of experiments.

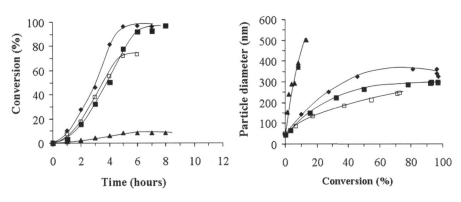

Figure 13.13   Conversion *versus* time and particle diameter *versus* conversion for soap-free emulsion polymerizations of styrene initiated by ACPA in the presence of PEGMA alone, Laponite alone, Laponite + PEGMA and Laponite + PEG oligomer. Clay content = 5 wt% based on styrene. PEGMA (1080 g mol⁻¹) or PEG (1000 g mol⁻¹) = 5×10⁻⁵ mol g⁻¹ clay. Total solids content = 17%.

As expected, the polymerization performed in the presence of PEGMA alone (*i.e.* without any added surfactant and without Laponite) was very slow and gave very large particles with limited conversions. In contrast, the polymerization performed in the sole presence of 5 wt% of Laponite (based on styrene) yielded small latex particles around 250 nm in diameter. However, these particles were unstable and coagulated after around 70% conversion. When a mixture of Laponite and PEGMA macromomomer or PEG oligomer of similar molar mass were used, stable latex particles, around 300 nm in diameter, could be successfully obtained, with a final conversion of 95%, the polymerization rate being lower for PEG than for PEGMA. The observed differences in polymerization rates suggest that the PEGMA macromonomer effectively participates in the polymerization and promotes the capture of oligoradicals, which accelerates the reaction. It could then be concluded that PEGMA and PEG play an important role in the system as they enable stabilization and high conversions to be achieved, but are not mandatory for nucleation.

Polymerizations were also performed in the presence of PEGMA of different molar masses. The polymerization performed in the presence of low molar mass PEGMA ($300 \, \text{g mol}^{-1}$) behaved in exactly the same way as the polymerization performed in the absence of PEGMA, suggesting that a minimum molar mass is necessary to stabilize the objects formed. The use of PEGMA with a molar mass of $2080 \, \text{g mol}^{-1}$ also gave stable latexes. However, the particles formed were larger than those synthesized in the presence of PEGMA of lower molar mass, likely because high molar mass macromonomers are capable of bridging neighboring latex particles and promoting aggregation.

Based on the above experiments, we suggest the following mechanism for soap-free polymerization of styrene in the presence of PEGMA. The polymerization first starts in the water phase as in conventional emulsion polymerization. The oligoradicals then undergo frequent collisions with the clay platelets and copolymerize with adsorbed PEGMA. This leads to unstable primary hybrid particles due to collapse of polymer chains on the clay surface. The primary particles subsequently aggregate to decrease the surface area of the clusters formed and promote colloidal stability by increasing surface charge repulsions. A fraction of the PEGMA chains is presumed to be buried inside the particles while another part is likely adsorbed on the external surface of the clay-armored latex particles and confer extra colloidal stability on them. This mechanism was supported by TEM observations, dynamic light scattering (DLS) measurements and zeta potential measurements.

13.4.3.2   Radical Entry into Particles

The assumption has been made up to this point that polymerization takes place at the vicinity of the clay surface, considering that the oligoradicals formed in the aqueous phase are captured by the methacrylate group of the adsorbed PEGMA macromonomer moving likewise the locus of polymerization from the aqueous phase to the clay surface. This is supported by the fact that the polymerization rate is lower for PEG than for PEGMA. However, it has been seen that bare Laponite or Laponite with adsorbed PEG oligomers could also successfully stabilize the resulting latex particles. It therefore seems clear that PEG or PEGMA is required to stabilize the growing particles in the long term, but are not necessary at the beginning of polymerization during the nucleation step. It was then suspected that the initiator could also play a determinant role in particle nucleation. Indeed, many reports suggest that clays can form an active complex with free radical initiators such as KPS and increase the polymerization rate.[100] However, to our knowledge, nothing similar has been reported for diazoic initiators such as ACPA. To investigate the effect of the nature of the initiator on kinetics and particle formation, a series of emulsion polymerization reactions was carried out with a fixed amount of Laponite in the presence of PEGMA ($1080 \, g \, mol^{-1}$) and four different types of initiator: ACPA, KPS, 2,2'-azobis[2-methyl-N-(2-hydroxyethyl)propionamide] (VA86) and a redox pair made of ascorbic acid and hydrogen peroxide (Table 13.2).

Figure 13.14 shows that the polymerization initiated by KPS led to significantly smaller latex particles and displayed higher polymerization rates than when ACPA was used as initiator. In contrast, the polymerization performed with VA86 gave large particles with a low reaction rate whereas the polymerization initiated with the redox pair was totally inhibited in the presence of the clay platelets.

The decrease in particle size when KPS is used as initiator can be attributed to a decrease in the ionic strength of the suspension medium. Indeed, as ACPA is not water soluble, it was dissolved in 1 M sodium hydroxide solution. According to the phase diagram of Laponite dispersions in the presence of tetrasodium pyrophosphate,[97] the Laponite platelets form aggregates under

Table 13.2 Chemical formulae of the free radical initiators used for the synthesis of Laponite-armored latexes through soap-free emulsion polymerization.

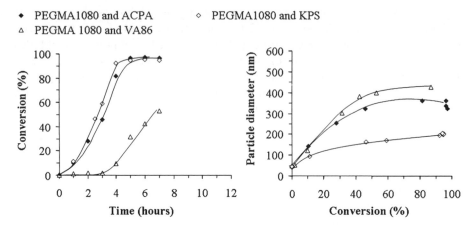

Figure 13.14 Conversion *versus* time and particle diameter *versus* conversion curves for 17% solids content (SC) latexes synthesized in the presence of 5 wt% Laponite, PEGMA (1080 g mol^{-1}) and different initiators.

these conditions. This was confirmed by DLS measurements, which showed an increase in the Laponite particle size upon addition of ACPA and NaOH. Consequently, the real number of platelets available to stabilize the growing latex particles was less important for ACPA than for KPS, which led to significant differences in particle sizes. Moreover, the fact that the four initiators displayed significantly different kinetic behaviors suggests that they may undergo different levels of interaction with the clay surface. Indeed, it has been reported previously that KPS can form intercalated structures with clay layers.[79,101] Although the exact mechanism of interaction is still not completely understood, it is generally accepted that KPS interacts with clays through ionic bonds.[102–104] As VA86 is a neutral molecule, it is therefore presumed to display less interaction with Laponite. As regards the redox pair, it is possible that only one component is adsorbed on the clay layers, which would account for the observed loss of reactivity. Although these observations clearly highlight the importance of the nature of the initiator in particle formation and polymerization kinetics, the precise underlying mechanism remains unclear and requires further investigation. For the sake of simplicity, we will consider in the following that both ACPA and KPS can adsorb on the Laponite surface and contribute to the formation of clay-armored latex particles in the presence or absence of PEGMA by initiating the polymerization near the clay surface.

Figure 13.15 shows the evolution of monomer conversion with time for increasing clay contents and a fixed solids content of 17 wt%. It can be seen that the polymerization rate increases with increasing clay content whereas the particle size decreases slightly from 205 to 167 nm.[88] The fact that the final particle size is almost independent of clay content suggests that the concentration of Laponite is much higher than the amount of platelets necessary to saturate the surface of the polymer particles. As will be discussed in detail in

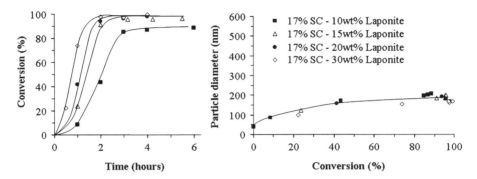

Figure 13.15 Conversion *versus* time and particle diameter versus conversion curves for 17% solids content (SC) soap-free polystyrene latexes stabilized by increasing amounts of Laponite expressed in wt% of monomer. Adapted from reference 88 with permission of Wiley-VCH.

Section 13.4.4.2, the increase in reaction rate is likely due to an increase in the radical absorption rate on increasing the Laponite concentration.

13.4.3.3 Film-forming Latexes Stabilized by Laponite–Clay Platelets

Despite the great interest in PCNs in the coating and adhesive industries, examples of the synthesis of film-forming polymer–clay latexes by soap-free emulsion polymerization are still sparse.[105] Based on the promising results obtained in the synthesis of Laponite-armored polystyrene latexes, we performed similar experiments with a low-T_g film-forming poly(styrene-*co*-*n*-butyl acrylate) composition (styrene:butyl acetate = 45:55 by weight). Unfortunately, polymerizations performed in the presence of PEGMA and KPS did not allow the formation of stable nanocomposite particles under the chosen conditions. The latex suspension systematically turned into a gel and increasing the clay content or decreasing the total amount of solid did not allow the latex colloidal stability to be improved. After many efforts, the nature of the macromonomer was finally found to be the key to success. Stable latexes were indeed obtained only when the PEGMA macromonomer was replaced by a methyl ether acrylate-terminated poly(ethylene glycol) macromonomer (hereafter referred to as PEGA). This result shows that there is an additional and important effect of the macromonomer in the synthesis of such latexes, *i.e.* its reactivity. In order to understand the role of the PEGA macromonomer in nanocomposite particle formation and latex stability, a series of experiments were then carried out, varying the PEGA and Laponite concentrations for different solids contents. The clay content was calculated based on the amount of monomers whereas the amount of PEGA was adjusted based on the clay content. Figure 13.16 shows the evolution of monomer conversion and particle size with time for this series of experiments.

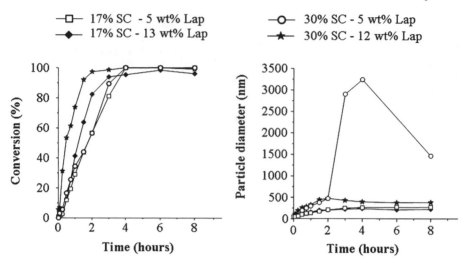

Figure 13.16   Conversion and particle diameter *versus* time curves for poly(styrene-*co*-butyl acrylate) latexes of increasing solids contents (SC) and containing different Laponite contents expressed in wt% of monomers.

It can be seen that high conversions could be achieved whatever the solids content and the clay loading. As for polystyrene, the larger the amount of Laponite, the higher was the polymerization rate.[105] This confirms the strong effect of clay on the polymerization kinetics, as mentioned above. The highest polymerization rate was obtained for 12 wt% of Laponite and 30% solids content, which correspond to the highest concentration of clay in the water phase ($52\,\mathrm{g\,l^{-1}}$). These experimental conditions led to a very stable latex with a relatively small particle size (383 nm) and high particle surface coverage (187%). Moreover, it was found that the final particle size was affected in this case by both the clay content and the solids content. The lowest particle size (224 nm) was obtained for 13 wt% Laponite and 17% solids content, which again correspond to more than 100% coverage of the latex particles by the clay platelets. Decreasing the clay content to 5 wt% resulted in slightly larger latex particles (269 nm in diameter) with a low polydispersity index (0.03), indicating that under such conditions of low solids content, 5 wt% of Laponite was sufficient to form small, colloidally stable latex particles. On the other hand, increasing the solids content to 30% while maintaining the same clay content (5 wt%) led to a decrease in the latex stability, as attested by a pronounced increase in particle diameter after around 50% monomer conversion. The particle size increased up to a maximum value of 3241 nm and then decreased to around 1466 nm after the reaction was completed. The increase in particle size was accompanied by an increase in viscosity. It is clear that the presence of a large amount of Laponite in the water phase plays a determinant role in the mechanism of particle nucleation, accelerating the polymerization reaction and leading to low particle sizes (*e.g.* a high number of particles). However, for the solids content targeted in this experiment, the sudden increase in particle size after around 50% conversion suggests that the amount of Laponite surrounding the newly nucleated particles was not sufficient to maintain their colloidal stability. The composite particles therefore coalesced to decrease the surface to volume ratio in order to achieve a minimum surface coverage necessary to achieve colloidal stability. The subsequent decrease in particle size indicates that the aggregation was reversible. A similar phenomenon was observed by Sheibat-Othman and Bourgeat-Lami during the synthesis of PS–SiO$_2$ and poly(styrene-*co*-methyl methacrylate)–SiO$_2$ composite latex particles *via* soap-free emulsion polymerization.[81] According to these authors, some aggregates were formed during polymerization and were subsequently disrupted due to the presence of SiO$_2$ particles that could prevent the polymer particles from sticking together. The above results support the assumptions of Sheibat-Othman and Bourgeat-Lami, in particular if one considers that in the present system, the polymer particles have a very low T_g. The observed decrease in particle size at the end of the reaction thus confirms that the adsorbed clay platelets are capable of preventing the latex particles from coalescing in an irreversible way.

Figure 13.17 shows a cryo-TEM image of a 17% solids content hybrid latex containing 13 wt% Laponite. The clay platelets can be clearly seen at the particle surface and only a minor amount of platelets was found outside the particles. Some latex particles have a non-spherical shape. These particles

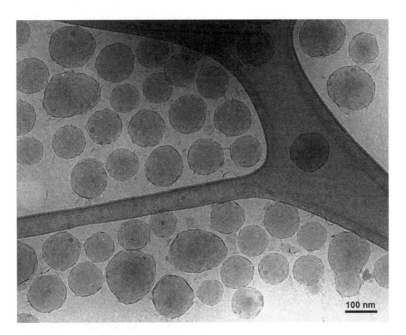

100 nm

Figure 13.17 Cryo-TEM image of a 17% solids content poly(styrene-*co*-butyl acrylate)–Laponite RD latex suspension containing 13 wt% Laponite RD. Image courtesy of J. L. Putaux/CERMAV-CNRS.

probably originate from the coalescence of unstable primary latexes. Therefore, the presence of embedded clay platelets cannot be completely ruled out in the present system, although most of the platelets are seen at the particle surface. In summary, the Laponite platelets form an outer stabilizing shell that should lead upon film formation to the formation of a percolating 3D clay network.

Mechanical Properties of the Latex Films.   Figure 13.18 shows a noteworthy effect of the addition of Laponite on the mechanical properties of the nano-composite film.[105] A remarkable increase in the storage modulus at the rub-bery plateau was observed with the addition of only 5 wt% Laponite. This improvement of the mechanical properties of the poly(styrene-*co*-butyl acry-late)–Laponite film was expected, considering the good dispersion of the clay platelets inside the polymeric matrix and the armored morphology of the hybrid particles. It is worth remembering here that similar results were repor-ted by Ruggerone and co-workers[47–49] and Faucheu *et al.*[54] for PS–Laponite films obtained from latexes prepared by emulsion and miniemulsion poly-merization, respectively (see Sections 13.3.2.1 and 13.3.2.2).

Finally, TGA showed that the incorporation of Laponite in the polymeric matrix did not improve the thermal stability of the nanocomposite film as usually reported in the literature. Indeed, the incorporation of clays into polymer matrices generally leads to enhanced thermal stability by acting as a superior insulator and mass transport barrier to the volatile products generated during decomposition and by assisting in the formation of char after thermal decomposition. For instance, Negrete-Herrera *et al.*[44c] found a significant effect of Laponite RD on the thermal stability of poly(styrene-*co*-butyl acrylate)–Laponite RD nanocomposite films produced through emulsion polymerization. The observed difference of thermal behaviors between this previous work and

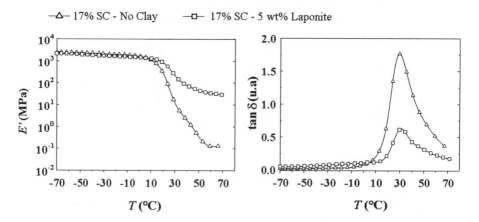

Figure 13.18 E' and tanδ at 1 Hz as a function of temperature for films prepared from 17% solids content poly(styrene-*co*-butyl acrylate) hybrid latexes without Laponite and with 5 wt% Laponite.

the present system may be due to difference in polymer–clay interfacial properties.

13.4.4 Modeling and Online Monitoring by Calorimetry of the Preparation of Polymer–Laponite Nanocomposite Particles

Reaction modeling and online monitoring can be of great value in the interpretation of the behavior of new systems and in the development of new materials such as colloidal nanocomposites considered in this chapter. However, they are often not considered because they usually require the use of large reactors, as in the case of calorimetry, for instance. Due to the high exothermic nature of polymerization processes, reaction calorimetry is, however, the most widely used technique to track processes online. In addition to applications in process optimization and control, reaction calorimetry can also be used for other purposes such as process safety and the estimation of kinetic parameters when it is combined with the material balance.[106–108] In the following section, some theoretical aspects of online calorimetry and process modeling are rehearsed. Calorimetric experiments were then performed to investigate the mechanism and kinetics of particle formation in the specific case study of soap-free emulsion polymerization involving Laponite clay platelets as solid stabilizer.

13.4.4.1 Theoretical Aspects

Reaction calorimetry allows the measurement of the amount of heat generated by the reaction based on temperature measurements in the different parts of the reactor and the heat balance of the reactor. The rate of heat production due to polymerization is then proportional to the reaction rate. This allows the estimation of the overall monomer conversion. Using this information, the material balance can be used to calculate the concentrations of monomer and radicals in the polymer particles.

Heat Balance – Monomer Conversion. The heat balance of a semi-continuous reactor is given by eqn (13.1). The reactor is surrounded by a jacket that provides heat exchange between the reactor and a coolant ($\dot{Q}_j$). The heat balance also involves the flow of heat caused by the addition of reagents ($\dot{Q}_{feed}$), the flow of heat dissipated by the stirrer ($\dot{Q}_{stiner}$) the heat produced by the reaction ($\dot{Q}_R$) and heat losses through the condenser or other non-jacketed parts of the reactor ($\dot{Q}_{loss}$).

$$\underbrace{MC_P\dot{T}}_{\dot{Q}_{cum}} = \dot{Q}_R + \underbrace{UA(T_j - T)}_{\dot{Q}_j} + \underbrace{F_{feed}C_{Pfeed}(T_{feed} - T)}_{Q_{feed}} - \dot{Q}_{loss} \qquad (13.1)$$

where T is the reactor temperature, F_{feed} is the input mass flow rate, T_{feed} is the input temperature, M is the total mass of species present in the reactor (comprising reactants, stirrer and baffles), C_P and C_{Pfeed} are the heat capacities of the reaction medium and the feed, respectively, U is the heat transfer coefficient and A is the heat transfer area between the reactor and the jacket (which can vary in a semi-continuous reactor).

The energy dissipated by the stirrer is often small compared with the other terms in the case of emulsion polymerization, because of the low viscosity and reduced agitation speeds, and can therefore often be neglected. Even if this is not entirely accurate, it is possible to include this quantity in the term for heat loss, generally used as a correction term for calorimetric calculations.

The heat produced by the reaction is related to the reaction rate (R_P) as follows:

$$\dot{Q}_R = (-\Delta H)R_P \tag{13.2}$$

where $-\Delta H$ is the reaction enthalpy. If all the terms in the heat balance are known and the temperatures are measured, the monomer conversion can be estimated from $\dot{Q}_R$ by the following relationship:

$$X = \frac{\int_0^t \dot{Q}_R}{(-\Delta H)N^T} \tag{13.3}$$

where N^T is the total number of moles of monomer introduced into the reactor and can be calculated by integrating the monomer flow rate added to the initial number of moles of monomer (N_0) according to:

$$N^T = N_0 + \int_0^t F\,dt \tag{13.4}$$

where F is the monomer molar flow rate.

Concentration of Monomer. The residual number of moles of monomer in the reactor can be calculated using the monomer conversion (X) as follows:

$$N = N^T(1 - X) \tag{13.5}$$

Calculation of the concentration of monomer in the polymer particles depends on the reaction interval. The polymer particles are nucleated in interval I and grow by consuming monomer during interval II. Particles are assumed to be saturated with monomer during this last interval and the excess of monomer is stored in the monomer droplets. In interval III, monomer droplets disappear and all the monomer is supposed to be in the polymer particles (this assumes that the monomer is not water soluble). It can be deduced that $[M^P]$ is constant during interval II and in interval III, all the residual monomer is in the polymer particles which gives the following equation:

$$[M^P] = \begin{cases} \dfrac{(1-\phi_P^P)\rho_m}{MW_m} & \text{if} \quad \dfrac{N}{\rho_m} - \dfrac{1-\phi_P^P}{\phi_P^P}\left(\dfrac{N^T-N}{\rho_P}\right) \geq 0 \\[2mm] \dfrac{N}{MW_m\left(\frac{N^T-N}{\rho_P}\frac{N}{\rho_m}\right)} & \text{else} \end{cases} \tag{13.6}$$

where ϕ_P^P is the fraction of polymer in the particles, ρ_m and ρ_p are the monomer and polymer densities and MW_m is the monomer molecular weight.

Concentration of Radicals. The reaction rate is related to the concentration of monomer and the number of moles of radicals in the polymer particles ($[M^P]$ and μ) by the following relationship:

$$R_P = k_P(T)\,\underbrace{\bar{n}N_P V/N_A}_{\mu}[M^P] \tag{13.7}$$

where $\bar{n}$ is the average number of radicals per particle, N_P the number of particles per liter, N_A Avogadro's number, V the reaction volume and k_P the propagation rate coefficient. With the measurement of the monomer conversion, the residual amount of monomer (N and $[M^P]$) can be calculated. The reaction rate can then be calculated using eqn (13.2). Subsequently, μ can be estimated. If the particle size is measured, an estimation of the particle number can be deduced, which allows the estimation of $\bar{n}$ with time.

13.4.4.2 Calorimetric- and Model-based Interpretations

The above-described soap-free emulsion polymerization of styrene involving Laponite and PEGMA was considered. Experiments were performed in a 1 l calorimeter. The heat and material balances were then applied to investigate the polymerization kinetics and estimate $\bar{n}$.

Comparison with Conventional Emulsion Polymerization. Let us consider two series of experiments with different KPS concentrations. One of these series is a conventional emulsion polymerization (involving a surfactant) and the second is a soap-free process involving Laponite and PEGMA.

In both cases, the initiator concentration affects radical concentration in the aqueous phase. Consequently, when the radical concentration in the aqueous phase increases, the probabilities of particle nucleation and radical entry into existing particles are also supposed to increase. Moreover, in the specific case of charged initiators, the initiator concentration also plays a role in particle stabilization. Indeed, in conventional emulsion polymerization of styrene, the

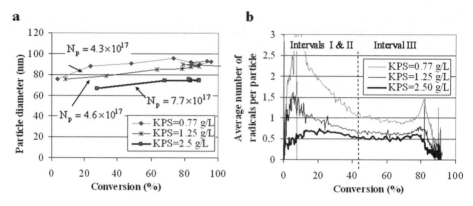

Figure 13.19   Effect of KPS concentration on the evolution of particle size (a) and on the average number of radicals per particle (b) for conventional emulsion polymerization of styrene stabilized by sodium dodecyl sulfate.

initiator concentration has a direct effect on the number of nucleated particles. Increasing the initiator concentration leads to nucleating more particles due to the higher entry rate of radicals into micelles. In turn, this leads to the formation of smaller particles at the end of the reaction for the same amount of polymer (Figure 13.19a). Therefore, the overall reaction rate increases with increasing initiator concentration. Also, since $\bar{n}$ is proportionally related to the particle size, it might decrease slightly on increasing the initiator concentration. Further, decreasing the initiator concentration might lead to prolongation of the nucleation period, which increases the broadness of the particle size distribution. Therefore, in conventional emulsion polymerization, the main initiator effect concerns the nucleation rate of particles.

A particular feature of conventional emulsion polymerization is the low value of $\bar{n}$ mainly for styrene (Figure 13.19b). In styrene polymerization, it is known that the system obeys 0–1 behavior, where the particles may only contain 1 radical at a time or 0 radicals (due to instantaneous termination of radicals). The equilibrium between the radical entry, exit and termination rates, leads to $\bar{n} = 0.5$ in interval II during which the particles are saturated with monomer. During interval III, the decrease in the concentration of monomer in the polymer particles leads to an increase in the particle viscosity and therefore to a decrease in the radical mobility and termination rate. This is known as the gel effect, during which the radical concentration in the polymer particles increases, which leads to an increase in the overall reaction rate. The gel effect is usually detected by an increase in the reaction temperature during interval III at high monomer conversions. The value of $\bar{n}$ usually increases during this interval and the polymer chains produced are usually longer than during interval II, although the monomer concentration is lower. These long chains hinder further diffusion of radicals into the particle, which inhibits the reaction unless more monomer is introduced. Therefore, some reactions may not reach 100% conversion.

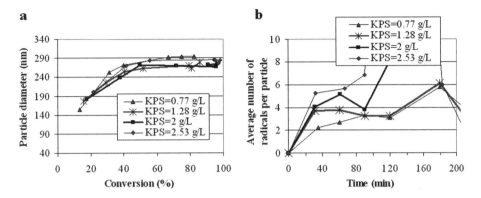

Figure 13.20   Effect of the KPS concentration on the evolution of particle size (a) and on the average number of radicals per particle (b) during soap-free emulsion polymerization of styrene stabilized by Laponite clay platelets. $N_P = 1.5 \times 10^{16} \, l^{-1}$.

In soap-free emulsion polymerization stabilized by Laponite platelets, the monomer conversion increases on increasing the concentration of KPS as in conventional emulsion polymerization (data not shown). However, due to fundamental differences in the nucleation mechanism between the two systems, increasing the initiator concentration (above a minimum amount) does not lead to an increase in the particle number (Figure 13.20a). Indeed, since the nucleation rate is no longer conditioned by entry of radicals into micelles, it is no longer related to the initiator concentration. In addition, in the present system, particle stabilization is no longer ensured by the initiator fragments as in conventional soap-free emulsion polymerization and is provided instead by the adsorbed Laponite platelets. Therefore, increasing the KPS concentration is not expected to increase stabilization and influence the nucleation rate.

Conversely, the presence of adsorbed inorganic colloids on the surface of the growing latex particles is expected to affect the absorption of radicals by creating a physical barrier to radical entry. For instance, Sheibat-Othman and Bourgeat-Lami[81] found that negatively charged silica particles located at the surface of polystyrene particles negatively affected the entry rate of radicals issued from the thermal decomposition of ammonium persulfate. In the current system, in contrast, the presence of Laponite platelets at the particle surface enhances the absorption of KPS radicals and probably decreases the desorption rate, which leads to a significant increase in $\bar{n}$ (Figure 13.20b). This consequently leads to unexpectedly high reaction rates, especially in view of the relatively small number of particles formed ($N_P = 1.5 \times 10^{16} \, l^{-1}$). It is worth mentioning here that the termination rate within the particles is not supposed to be affected by the nature of the surface. Hence the high values of $\bar{n}$ do not reveal a gel effect. Indeed, the polymer chains obtained had similar molar masses to those in conventional emulsion polymerization.

In conclusion, the above results show that the system no longer obeys the 0–1 kinetics model typical of styrene polymerization. The above-mentioned equilibrium between radical absorption, exit and termination is shifted towards absorption. Consequently, increasing the initiator concentration leads to an increase in the radical absorption rate and it can be concluded that in soap-free emulsion polymerization stabilized by Laponite in the presence of PEGMA, the main initiator effect concerns the rate of radical entry into particles.

Semi-Continuous Monomer Addition. It is known from the literature that the employment of a semi-continuous process allows better control of the nucleation period, the particle stability and the reaction rate. It also enhances the evacuation of heat from the reactor. As shown in the previous section, the particles are nucleated with a small size (about 140 nm), but their size rapidly increases in the batch process. This might be due to the presence of droplets [that can be calculated using eqn (13.6)] that destabilize the nucleated particles by absorbing a significant amount of platelets for their stabilization. According to this, the remaining amount of Laponite available for subsequent adsorption and stabilization of the polymer latex particles depends on the droplet size, which is in turn dependent on the stirring efficiency.[81] By introducing the monomer semi-continuously, the presence of monomer droplets can be avoided, which avoids irreversible destabilization of the polymer particles. Table 13.3 shows that much smaller particles could be obtained by changing only the addition mode of monomer while keeping the original formulation. It can be seen that introducing the monomer at a higher flow rate led to the accumulation of monomer and therefore to a diminution of the particle stability.

It is interesting to see also that the semi-continuous addition of monomer allowed control of the gel effect, as observed from the temperature measurements. The occurrence of the gel effect is to be avoided since it is considered as an uncontrolled step of the reaction. Indeed, this step is very rapid and therefore the heat produced by the reaction during this step is difficult to evacuate from an industrial reactor. Also, during this period, the polymer molecular weight usually increases considerably. Its influence on the quality of the final product then depends on the amount of polymer produced during this period.

Table 13.3 Effect of flow rate of monomer addition on particle diameter.

Flow rate of monomer $(g\ s^{-1})$	Initial particle diameter (nm) (about 5% solids content)	Final particle diameter (nm) (about 20% solids content)	Gel effect
Batch	180	280	Yes
0.2	140	250	Yes
0.05	140	190	No
0.022	140	190	No
0.017	140	190	No

Table 13.4 Effect of the initial monomer concentration on the particle diameter.

Initial amount of monomer (g)	*Initial particle diameter (nm) (about 5% solids content)*	*Final particle diameter (nm) (about 20% solids content)*
60	158	201
40	140	190
30	123	181
20	110	174

The semi-continuous process can be used to investigate the nucleation period by decreasing the initial amount of monomer, which reduces the probability of coagulation. In this system, the nucleation rate is controlled by the type and amount of monomer, Laponite, PEGMA and initiator. The effects of PEGMA, Laponite and initiator have been discussed above. Regarding the monomer concentration, it affects the particle size proportionally, as can be seen in Table 13.4. The lower the initial amount of monomer, the faster the droplets disappear, which ensures greater stability of the particles formed. Consequently, the number of particles decreases with increase in the initial amount of monomer.

In conclusion, the controlled introduction of monomer allows enhanced control of the particle size, polymer molecular weight and reaction temperature. The initial amount of monomer also affects the production of particles. A smaller amount of monomer ensures better stability of the nucleated particles.

13.5 Conclusions and Outlook

A great research effort is currently being conducted towards the development of products coming from sustainable technology owing to environmental concerns. This has led recently to increased interest in the synthesis of polymer–clay nanocomposites *via* waterborne processes such as emulsion and mini-emulsion polymerizations. Compared with bulk or solution polymerization, these processing routes have the advantage of generating hybrid particles of controlled morphologies at the nanoscale. By selecting appropriate synthetic procedures and performing adequate clay surface modifications, it is possible to form a variety of nanocomposite materials with exfoliated or intercalated morphologies. To date, most of the emphasis in the literature has been on the formation of polymer–clay latex particles with controlled morphologies and on the establishment of structure–property relationships. We have shown in this chapter that the properties of polymer–clay nanocomposite films are strongly dependent on particle nanostructure. For instance, nanocomposite films obtained by drying a suspension of clay-armored latex particles showed unconventional mechanical behavior and high mechanical resistance due to the formation of a clay percolating network within the film for high clay contents.

In contrast, randomly dispersed platelets originally confined (*e.g.* encapsulated) inside individual latex particles led to much lower reinforcements. The ability of clays to adsorb at interfaces can also be advantageously exploited to synthesize clay-armored latexes by soap-free emulsion polymerization. Latex particles stabilized by inorganic solids have attracted increasing attention in the recent literature as they offer many potential benefits in coating applications, such as fire retardancy, UV resistance and enhanced mechanical properties depending on the nature of the inorganic particles. However, surprisingly, there have been only a few reports describing the use of clay platelets as stabilizers in conventional emulsion polymerization. Indeed, until now, most research studies have focused on miniemulsion polymerization. In this chapter, some recent results from our group on the synthesis of Laponite-armored latexes by soap-free emulsion polymerization of styrene in the presence of a poly(ethylene glycol) methyl ether methacrylate (PEGMA) macromonomer have been briefly reviewed. The most important characteristics of this system were summarized, the emphasis being placed on the kinetics and mechanism of particle formation. In particular, it was shown that both PEGMA and PEG oligomers play a relevant role in the stabilization of latex particles, but are not mandatory for particle nucleation. The nature of the initiator was also found to be a determinant parameter. Ionic initiators such as KPS and ACPA are suspected to adsorb on Laponite and promote the formation of radicals in the vicinity of the clay surface. Online calorimetric measurements were performed to estimate the polymerization rate and provide insights into the mechanism of particle formation and stabilization. It was shown that the reaction rate was strongly affected by the presence of Laponite and increased with increase in KPS concentration, although the particle size remained constant. The system does not obey 0–1 behavior and is characterized by very high $\bar{n}$ values, likely due to higher rates of radical entry into particles. The latex stability could be increased and the particle size decreased by using a semi-continuous polymerization process in which a seed is formed with a minimum amount of monomer while the remaining part is introduced at a controlled feed rate. It was shown that the lower the amount of accumulating monomer, the smaller was the final particle size. The batch process was next successfully extrapolated to a soft styrene–*n*-butyl acrylate copolymer composition able to film form at room temperature. A remarkable effect of Laponite addition on the mechanical properties of the nanocomposite films was observed even for low clay concentrations. Such improvements in the mechanical properties were again ascribed to the formation of a 3D clay network within the film. Work is in progress to study the effect of clay incorporation on water uptake and gas permeation properties of these soap-free nanostructured films.

Acknowledgments

The authors are greatly indebted to Riccardo Ruggerone and Jean-Luc Putaux for their help during the preparation of this chapter.

References

1. H. F. Payne, *Organic Coatings Technology*, **Vol. II**, Wiley, New York, 1961, pp. 773–804.
2. K. Sill, S. Yoo and T. Emrick, in *Dekker Encyclopedia of Nanoscience and Nanotechnology*, ed. J. A. Schwarz, C. Contescu and K. Putyera, Marcel Dekker, New York, 2004, p. 2999.
3. B. J. Ash, A. Etan and L. S. Schadler, in *Dekker Encyclopedia of Nanoscience and Nanotechnology*, ed. J. A. Schwarz, C. Contescu and K. Putyera, Marcel Dekker, New York, 2004, p. 2917.
4. L. Chazeau, C. Gauthier, G. Vigier and J.-Y. Cavaillé, in *Handbook of Organic–Inorganic Hybrid Materials and Nanocomposites*, ed. H. S. Nalwa, **Vol. 2**, American Scientific Publishers, Valencia, CA, 2003.
5. G. Cao, *Nanostructures and Nanomaterials – Synthesis, Properties and Applications*, Imperial College Press, London, 2004.
6. E. Bourgeat-Lami, in *Encyclopedia of Nanoscience and Nanotechnology*, ed. H. S. Nalwa, American Scientific Publishers, Valencia, CA, 2004, p. 305.
7. E. Bourgeat-Lami and E. Duguet, in *Functional Coatings by Polymer Microencapsulation*, ed. S. K. Ghosh, Wiley-VCH, Weinheim, 2006, p. 85.
8. Z. Xue and H. Wiese, *US Pat.*, 7 094 830, 2006.
9. F. Tiarks, J. Leuninger, O. Wagner, E. Jahns and H. Wiese, *Surf. Coat. Int.*, 2007, **90**, 221.
10. E. Bourgeat-Lami, *J. Nanosci. Nanotechnol.*, 2002, **2**, 1.
11. E. Bourgeat-Lami and M. Lansalot, *Adv. Polym. Sci.*, 2010, DOI: 10.1007/12_2010_60.
12. Y. Kojima, A. Usuki, M. Kawasumi, A. Okada, Y. Fukushima and T. Kurauchi, *J. Mater. Res.*, 1993, **8**, 1185.
13. (a) M. Alexandre and P. Dubois, *Mater. Sci. Eng.*, 2000, **28**, 1; (b) S. S. Ray and M. Okamoto, *Prog. Polym. Sci.*, 2003, **28**, 1539; (c) A. Okada and A. Usuki, *Macromol. Mater. Eng.*, 2006, **291**, 1449; (d) B. Chen, J. R. G. Evans, H. C. Greewell, P. Boulet, P. V. Coveney, A. A. Bowden and A. Whiting, *Chem. Soc. Rev.*, 2008, **37**, 568; (e) V. Mittal, *Materias*, 2009, **2**, 992; (f) M. Paulis and J. R. Leiza, in *Advances in Polymer Nanocomposites Technology*, ed. V. Mittal, Nova Science Publishers, New York, 2009, Chapter 5.
14. P. F. Luckham and S. Rossi, *Adv. Colloid Interface Sci.*, 1999, **82**, 43.
15. (a) F. Wypych and K. G. Satyanarayana, *Clay Surfaces. Fundamentals and Applications*, Interface Science and Technology Series, Academic Press, London, 2004; (b) H. van Olphen, *An Introduction to Clay Colloid Chemistry*, Wiley-Interscience, New York, 1977.
16. (a) D. H. Solomon and M. J. Rosser, *J. Appl. Polym. Sci.*, 1965, **12**, 1261; (b) D. H. Solomon and B. C. Loft, *J. Appl. Polym. Sci.*, 1968, **12**, 1253.
17. S. Talapatra, S. C. Guhaniyogi and S. K. Chakravarti, *J. Macromol. Sci. Chem.*, 1985, **22**, 1611.

18. J. Bhattacharya, S. K. Chakravarti, S. Talapatra and S. K. Saha, *J. Polym. Sci., Part A: Polym. Chem.*, 1989, **27**, 3977.

19. (a) D. C. Lee and L. W. Jang, *J. Appl. Polym. Sci.*, 1996, **61**, 1117.

20. M. W. Noh and D. C. Lee, *Polym. Bull.*, 1999, **42**, 619.

21. (a) M. W. Noh, L. W. Jang and D. C. Lee, *J. Appl. Polym. Sci.*, 1999, **74**, 179; (b) M. W. Noh and D. C. Lee, *J. Appl. Polym. Sci.*, 1999, **74**, 2811.

22. X. Tong, H. Zhao, T. Tang, Z. Feng and B. Huang, *J. Polym. Sci., Part A: Polym. Chem.*, 2002, **40**, 1706.

23. M. Pan, X. Shi, X. Li, H. Hu and L. Zhang, *J. Appl. Polym. Sci.*, 2004, **94**, 277.

24. T. H. Kim, L. W. Jang, D. C. Lee, H. J. Choi and M. S. Jhon, *Macromol. Rapid Commun.*, 2002, **23**, 191.

25. S. Bandyopadhyay, E. Giannelis and A. Hsieh, *Polym. Mater. Sci. Eng.*, 2000, **82**, 208.

26. (a) G. Diaconu, J. M. Asua, M. Paulis and J. R. Leiza, *Macromol. Symp.*, 2007, **259**, 305; (b) G. Diaconu, M. Paulis and J. R. Leiza, *Polymer*, 2008, **49**, 2444.

27. M. Laus, M. Camerani, M. Lelli, K. Sparnacci and F. Sandrolini, *J. Mater. Sci.*, 1998, **33**, 2883.

28. D. Wang, J. Zhu, Q. Yao and C. A. Wilkie, *Chem. Mater.*, 2002, **14**, 3837.

29. G. Liu, L. Zhang, Z. Li and X. Qu, *J. Appl. Polym. Sci.*, 2005, **98**, 1010.

30. X. Feng, A. Zhong and D. Chen, *J. Appl. Polym. Sci.*, 2006, **101**, 3963.

31. W. T. Yang, T. H. Ko, S. C. Wang, P. I. Shih, M. J. Chang and G. J. Jiang, *Polym. Compos.*, 2008, **29**, 409.

32. H. Min, J. Wang, H. Hui and W. Jie, *J. Macromol. Sci. Part B: Physics*, 2006, **45**, 623.

33. S. Qutubuddin, X. Fu and Y. Tajuddin, *Polym. Bull.*, 2002, **48**, 143.

34. Z. Sedlakova, J. Plestil, J. Baldrian, M. Slouf and P. Holub, *Polym. Bull.*, 2009, **63**, 365.

35. (a) Y. S. Choi, K. H. Wang and M. Xu and. J. Chung, *Chem. Mater.*, 2002, **14**, 2936; (b) Y. K. Kim, Y. S. Choi, K. H. Wang and I. J. Chung, *Chem. Mater.*, 2002, **14**, 4990; (c) M. Xu, Y. S. Choi, Y. K. Kim, K. H. Wang and I. J. Chung, *Polymer*, 2003, **44**, 6387; (d) Y. S. Choi, M. Xu and I. J. Chung, *Polymer*, 2005, **46**, 531.

36. H. Li, Y. Yang and Y. Yu, *J. Adhes. Sci. Technol.*, 2004, **18**, 1759.

37. N. Greesh, P. C. Hartmann, V. Cloete and R. D. Sanderson, *J. Polym. Sci., Part A: Polym. Chem.*, 2008, **46**, 3619.

38. G. Chen, Y. Ma and Z. Qi, *Scr. Mater.*, 2001, **44**, 125.

39. C. S. Chou, E. E. Lafleur, D. P. Lorah, R. V. Slone and K. D. Neglia. *US Pat.*, 6 838 507, 2005.

40. P. Meneghetti and S. Qutubuddin, *Langmuir*, 2004, **20**, 3424.

41. H. Li, Y. Yu and Y. Yang, *Eur. Polym. J.*, 2005, **41**, 2016.

42. (a) N. Negrete-Herrera, J.-M. Letoffe, J.-L. Putaux, L. David and E. Bourgeat-Lami, *Langmuir*, 2004, **20**, 1564; (b) N. Negrete-Herrera, J.-L. Putaux and E. Bourgeat-Lami, *Prog. Solid State Chem.*, 2006, **34**, 121.

43. N. Negrete-Herrera, J.-M. Letoffe, J.-P. Reymond and E. Bourgeat-Lami, *J. Mater. Chem.*, 2005, **15**, 863.
44. (a) N. Negrete-Herrera, S. Persoz, J.-L. Putaux, L. David and E. Bourgeat-Lami, *J. Nanosci. Nanotechnol.*, 2006, **6**, 421; (b) N. Negrete-Herrera, J.-L. Putaux, L. David and E. Bourgeat-Lami, *Macromolecules*, 2006, **39**, 9177; (c) N. Negrete-Herrera, J.-L. Putaux, L. David, F. De Haas and E. Bourgeat-Lami, *Macromol. Rapid Commun.*, 2007, **28**, 1567.
45. A. Vidal and J. B. Donnet, *Prog. Colloid Polym. Sci.*, 1987, **75**, 201.
46. T. Witten, M. Rubinstein and R. H. Colby, *J. Phys. II*, 1994, **3**, 1845.
47. R. Ruggerone, C. J. G. Plummer, N. Negrete-Herrera, E. Bourgeat-Lami and J.-A. E. Manson, *Eur. Polym. J.*, 2009, **45**, 621.
48. R. Ruggerone, C. J. G. Plummer, N. Negrete-Herrera, E. Bourgeat-Lami and J.-A. E. Manson, *Solid. State Phen.*, 2009, **151**, 30.
49. R. Ruggerone, C. J. G. Plummer, N. Negrete-Herrera, E. Bourgeat-Lami and J.-A. E. Manson, *Eng. Fract. Mech.*, 2009, **76**, 2846.
50. (a) C. H. M. Caris, L. P. M. van Elven, A. M. van Herk and A. German, *Br. Polym. J.*, 1989, **21**, 133; (b) C. H. M. Hofman-Caris, *New J. Chem.*, 1994, **18**, 1087.
51. F. Corcos, E. Bourgeat-Lami, C. Novat and J. Lang, *Colloid Polym. Sci.*, 1999, **277**, 1142.
52. (a) D. J. Voorn, W. Ming and A. M. van Herk, *Macromolecules*, 2006, **39**, 4654; (b) D. J. Voorn, W. Ming and A. M. van Herk, *Macromol. Symp.*, 2006, **245**, 584.
53. V. Mellon Synthesis and characterization of waterborne polymer–Laponite nanocomposite latexes through miniemulsion polymerization, PhD thesis, Université Claude Bernard Lyon 1, 2009.
54. J. Faucheu, C. Gauthier, L. Chazeau, J.-Y. Cavaillé, V. Mellon and E. Bourgeat-Lami, *Polymer*, 2010, **51**, 6.
55. W. Ramsden, *Proc. R. Soc. London*, 1903, **72**, 156.
56. S. U. Pickering, *J. Chem. Soc. Trans.*, 1907, **91**, 2001.
57. B. P. Binks and T. S. Horozov, *Colloidal Particles at Liquid Interfaces*, Cambridge University Press, Cambridge, 2006.
58. S. Cauvin, P. J. Colver and S. A. F. Bon, *Macromolecules*, 2005, **38**, 7887.
59. S. A. F. Bon and P. J. Colver, *Langmuir*, 2007, **23**, 8316.
60. D. J. Voorn, W. Ming and A. M. van Herk, *Macromolecules*, 2006, **39**, 2137.
61. S. Lu, J. Ramos and J. Forcada, *Langmuir*, 2007, **23**, 12893.
62. F. F. Fang, J. H. Kim, H. J. Choi and C. A. Kim, *Colloid Polym. Sci.*, 2009, **287**, 745.
63. S. Guillot, F. Bergaya, C. de Azevedo, F. Warmont and J.-F. Tranchant, *J. Colloid Interface Sci.*, 2009, **333**, 563.
64. (a) W. P. Hohenstein, *Polym. Bull.*, 1945, **1**, 13; (b) W. P. Hohenstein and H. Mark, *J. Polym. Sci.*, 1946, **1**, 127.
65. (a) R. M. Wiley, *J. Colloid Sci.*, 1954, **9**, 427; (b) R. M. Wiley, *US Pat.*, 2 886 559, 1959.

66. (a) X. Huang and W. J. Brittain, *Polym. Prepr.*, 2000, **41**, 521; (b) X. Huang and W. J. Brittain, *Macromolecules*, 2001, **34**, 3255.
67. T. Hasell, J. Yang, W. Wang, J. Li, P. D. Brown, M. Poliakoff, E. Lester and M. Howdle, *J. Mater. Chem.*, 2007, **17**, 4382.
68. T. Chen, P. J. Colver and S. A. F. Bon, *Adv. Mater.*, 2007, **19**, 2286.
69. H. Liu, C. Wang, Q. Gao, J. Chen, X. Liu and Z. Tong, *Mater. Lett.*, 2009, **63**, 884.
70. T. Wang, P. J. Colver, S. A. F. Bon and J. F. Keddie, *Soft Matter.*, 2009, **5**, 3842.
71. (a) J. I. Amalvy, M. J. Percy and S. P. Armes, *Langmuir*, 2001, **17**, 4770; (b) M. J. Percy, J. I. Amalvy, D. P. Randall, S. P. Armes, S. J. Greaves and J. F. Watts, *Langmuir*, 2004, **20**, 2184.
72. C. Barthet, A. J. Hickey, D. B. Cairns and S. P. Armes, *Adv. Mater.*, 1999, **11**, 408.
73. J. A. Balmer, A. Schmid and S. P. Armes, *J. Mater. Chem.*, 2008, **18**, 5722 and references therein.
74. A. Schmid, P. Scherl, S. P. Armes, C. A. P. Leite and F. Galembeck, *Macromolecules*, 2009, **42**, 3721.
75. Y. Liu, X. Chen, R. Wang and J. H. Xin, *Mater. Lett.*, 2006, **60**, 3731.
76. J. H. Chen, C. Y. Cheng, W. Y. Chiu, C. F. Lee and N. Y. Liang, *Eur. Polym. J.*, 2008, **44**, 3271.
77. J. Zhang, K. Chen and H. Zhao, *J. Polym. Sci., Part A: Polym. Chem.*, 2008, **46**, 2632.
78. P. J. Colver, C. A. L. Colard and S. A. F. Bon, *J. Am. Chem. Soc.*, 2008, **130**, 16850.
79. K. F. Lin, S. C. Lin, A. T. Chien, C. C. Hsieh, M. H. Yen, C. H. Lee, C. S. Lin, W. Y. Chiu and Y. H. Lee, *J. Polym. Sci., Part A: Polym. Chem.*, 2006, **44**, 5572.
80. (a) X. Song, G. Yin, Y. Zhao, H. Wang and Q. Du, *J. Polym. Sci. Part A: Polym. Chem.*, 2009, **47**, 5728; (b) X. Song, Y. Zhao, H. Wang and Q. Du, *Langmuir*, 2009, **25**, 4443.
81. N. Sheibat-Othman and E. Bourgeat-Lami, *Langmuir*, 2009, **25**, 10121.
82. N. P. Ashby and B. P. Binks, *Phys. Chem. Chem. Phys.*, 2000, **2**, 5640.
83. N. Yan and J. H. Masliyah, *Colloids Surf. A*, 1995, **96**, 229.
84. (a) Y. S. Choi, M. H. Choi, K. H. Wang, S. O. Kim and Y. K. Kim, *Macromolecules*, 2001, **34**, 8978; (b) Y. S. Choi, K. H. Wang, M. Xu and I. J. Chung, *Chem. Mater.*, 2002, **14**, 2936; (c) Y. S. Choi, M. Xu and I. J. Chung, *Polymer*, 2005, **46**, 531.
85. E. H. Yeom, W. N. Kim, J. K. Kim, S. S. Lee and M. Park, *Mol. Cryst. Liq. Cryst.*, 2004, **425**, 85.
86. K. F. Lin, S. C. Lin, A. T. Chien, C. C. Hsieh, M. H. Yen, C. H. Lee, C. S. Lin, W. Y. Chiu and Y. H. Lee, *J. Polym. Sci., Part A: Polym. Chem.*, 2006, **44**, 5572.
87. J. Zhang, K. Chen and H. Zhao, *J. Polym. Sci., Part A: Polym. Chem.*, 2008, **46**, 2632.

88. E. Bourgeat-Lami, N. Negrete Herrera, J.-L. Putaux, A. Perro, S. Reculusa, S. Ravaine and E. Duguet, *Macromol. Symp.*, 2007, **248**, 213.
89. R. A. Vaia, S. Vasudevan, X. Krawiec, L. G. Scanlon and E. P. Giannelis, *Adv. Mater.*, 1995, **7**, 154.
90. P. Arranda, Y. Mosqueda, E. Perez-Cappe and E. Ruiz-Hitzky, *J. Polym. Sci., Part B: Polym. Phys.*, 2003, **41**, 3249.
91. Z. Shen, G. P. Simon and Y. Cheng, *Polymer*, 2002, **43**, 4251.
92. B. L. Papke, M. A. Ratner and D. F. Shriver, *J. Phys. Chem. Solids*, 1981, **42**, 493.
93. J. H. Wu and M. M. Lerner, *Chem. Mater.*, 1993, **5**, 835.
94. R. A. Vaia, B. B. Sauer, O. K. Tse and E. P. Giannelis, *J. Polym. Sci., Part B: Polym. Phys.*, 1997, **35**, 59.
95. W. Loyens, P. Jannasch and F. J. Maurer, *Polymer*, 2005, **46**, 915.
96. E. Loizou, P. Butler, L. Porcar, Y. Talmon, E. Kesselman and G. Schmidt, *Macromolecules*, 2005, **38**, 2047.
97. P. Mongondry, T. Nicolai and J.-F. Tassin, *J. Colloid Interface Sci.*, 2004, **275**, 191.
98. E. Loizou, P. Butler, L. Porcar and G. Schmidt, *Macromolecules*, 2006, **39**, 1614.
99. P. Aranda and E. Ruiz Hitzky, *Acta Polym.*, 1994, **45**, 59.
100. M. M. Al-Esaimi, *Macromolecules*, 2001, **34**, 367.
101. K. F. Lin, C. Y. Hsu, T. S. Huang, W. Y. Chiu, Y. H. Lee and T. H. Young, *J. Appl. Polym. Sci.*, 2005, **98**, 2042.
102. K. Haraguchi and T. Takehisa, *Adv. Mater.*, 2002, **14**, 1120.
103. K. Haraguchi, H.-J. Li, K. Matsuda, T. Takehisa and E. Elliott, *Macromolecules*, 2005, **38**, 3482.
104. K. Haraguchi, M. Ebato and T. Takehisa, *Adv. Mater.*, 2006, **18**, 2250.
105. E. Bourgeat-Lami, T. Roduguey Guimarães, A. M. Cenacchi Pereira, G. M. Alves, J. C. Moreira, J. L. Putaux, A. Martins dos Santos, *Macromol. Rapid Commun.*, DOI 10/1002/marc.201000305.
106. L. M. Gugliotta, M. Arotçarena, J. R. Leiza and J. M. Asua, *Polymer*, 1995, **36**, 2019.
107. N. Othman, G. Févotte and T. F. McKenna, *Polym. React. Eng.*, 2001, **9**, 271.
108. I. S. Sáenz de Buruaga, A. Echevarria, P. D. Armitage, J. C. de la Cal, J. R. Leiza and J. M. Asua, *Chem. Eng. Sci.*, 1996, **51**, 2781.

Subject Index

References to figures are given in *italic* type. References to tables are given in **bold** type.